Thermo-Hydro-Mechanical Coupling in Fractured Rock

Edited by
Hans-Joachim Kümpel

2003

Birkhäuser Verlag
Basel · Boston · Berlin

Reprint from Pure and Applied Geophysics
(PAGEOPH), Volume 160 (2003), No. 5-6

Editor(s):

Hans-Joachim Kümpel
Leibniz-Institute for Applied
Geosciences (GGA)
Stilleweg 2
D-30655 Hannover
Germany

e-mail: kuempel@gga-hannover.de

A CIP catalogue record for this book is available from the Library of Congress, Washington D.C., USA

Bibliographic information published by Die Deutsche Bibliothek
Die Deutsche Bibliothek lists this publication in the Deutsche Nationalbibliographie;
Detailed bibliographic data is available in the Internet at http://dnb.ddb.de

ISBN 3-7643-0253-4 Birkhäuser Verlag, Basel – Boston – Berlin

Member of the BertelsmannSpringer Publishing Group
Printed on acid-free paper produced from chlorine-free pulp

9 8 7 6 5 4 3 2 1

www.birkhauser.ch

Contents

Pure appl. geophys. 160 (2003) 809–812
0033 – 4553/03/060809–4

Pure and Applied Geophysics

Special issue 'Thermo-Hydro-Mechanical Coupling in Fractured Rock'

Introduction

The supply and protection of groundwater, the production of hydrocarbon reservoirs, land subsidence in coastal areas, exploitation of geothermal energy and the long-term disposal of critical wastes or of CO_2 at depth are all issues of obviously high socio-economic relevance. They all have in common to be closely related to fluid flow in porous and/or fractured rock. In many cases the conditions of fluid flow depend on the rheological behavior of rocks. Accordingly, mechanical coupling between the liquid phase and the rock matrix can generally not be neglected. The subject is receiving increasing interest from many researchers and reservoir engineers.

Comprehensive insight into coupled processes of rocks and fluids is often hampered due to the fact that rocks and rock formations are enormously complex. Each rock sample consists of a myriad of mineral particles, forming its matrix, and of fluid molecules residing in voids. Any two rock samples differ in their geochemical constituents, size, and shape of grains, structure of pore space, fracture networks, etc. Dealing with real rock material often means studying a system of a virtually infinite number of unknowns. How do geologists, petrophysicists, reservoir engineers handle such material?

A series of three Euroconferences on themes related to rock physics, rock mechanics and fluid flow in rocks, all supported by the European Union, took place in the years 1998, 1999, and 2000 (Boutéca and Guéguen, 1999; Couples, Main, and Meredith, 2000; Kümpel, 2001). For the last event, approximately eighty researchers met in November 2000 at Bad Honnef near Bonn, Germany, to exchange their latest knowledge and findings regarding thermo-hydro-mechanical coupling in fractured rock. Thematic sessions during this meeting were

- Experiments and models of induced fracturing
- Fracture networks and transport properties
- Coupling between thermal, mechanical, and hydraulic properties.

The participants agreed that new insights are generally based on both, experiments and numerical modeling, from the micro- to the macro- scale. For instance, the development of fractures around boreholes and the formation of compaction bands in sandstones as a function of porosity and grain size was reported to occur in a surprisingly variable manner. Laboratory tests have confirmed that deformation can be extremely localized, even for seemingly homogeneous rocks. Coupling between rock stress and fluid flow although affected by various petrophysical quantities has been shown to be strongly dependent on the geometry of fracture networks. Observations in a gallery have revealed that all permeable fractures in crystalline rock were critically stressed so that they can be thought as having formed under stress, whereas many non-permeable fractures were found to be critically stressed as well, testifying the transient character of the status quo. Also, seismicity that has been induced by massive hydraulic tests in many boreholes worldwide appears to indicate a low mechanical stability of water saturated crustal formations even down to depths of nine kilometers.

More general conclusions of the Bad Honnef Conference concern the suitability of standard methods and of widely used parameters in describing coupled phenomena in rocks. Models of a compact rock matrix with connected voids and those of bonded aggregates of individual grains are two opposing, though valid approaches to simulate transport of liquids through a porous medium. Application of stochastic models and effective media concepts are important means to bridge spatial gaps ranging from millimeters to kilometers, and the dynamics of processes on time scales from seconds to hundreds of years.

However, a fundamental effective parameter like permeability, which by definition is a measure of average transport of fluid volume through a hydraulically conductive medium per time increment, may not always be a useful quantity. Large-scale field experiments have shown that fluid flow may be highly channelized and that, depending on the specific site conditions, traditional upscaling can give erroneous results. Ultimately, better understanding of the various physical processes in fractured rock helps to improve predictions on how fluids can best be pumped out of or injected into subsurface reservoirs, in which way toxic fluids are absorbed or released by geologic formations, or to what extent rocks compact when pore fluids are extracted.

Twenty articles, most of which were presented at the Bad Honnef Conference, are contained in this special issue. Only a few of them deal with fully thermo-hydro-mechanical coupling. Rather, the individual investigations present contributions to processes where such coupling is relevant. The special issue in whole covers aspects of coupling from various points of view. Recent progresses made in understanding of coupled phenomena in rock physics/rock mechanics is well documented in this volume.

The first four articles describe findings from laboratory experiments that are related to different types of sandstones and granites. They include processes like the formation of compaction bands (contribution no. 1), fracture mechanics in the brittle deformation regime (2), the spatial distribution of damage zones in stressed rock samples (3), and the damage behavior of wet granites submitted to deviatoric stress

(4). The next three papers extend these views by taking a closer look on parameters that govern hydraulic diffusivity in sandstones and other types of rocks. Specific targets addressed are the influence of differential stress on permeability (5), imaging of the fracture geometry (6), and pressure induced variations in the pore geometry (7). Contributions no. 8 to 10 cover investigations of permeability-porosity relationships during rock evolution (8), of the formation, propagation, and roughness of fractures in a plexi-glass block (9), and pressure oscillation effects of two-phase flow under controlled conditions (10). The subsequent four articles focus on diverse modeling approaches. Issues considered are how the geometry and the mechanical behavior of fractures can be characterized by mathematical expressions (11), how the evolution of permeability in a microcracking rock can be expressed by an analytical model (12), deviations from the cubic law for a fracture of varying aperture (13), and the numerical simulation of scale effects in flow through fractures (14). Three further papers refer to *in situ* observations, being related to topics as the assessment of *in situ* permeability from the spatio-temporal distribution of an aftershock sequence (15), to the scale dependence of hydraulic pathways in crystalline rock (16), and to the significance of pore pressure – stress coupling in deep tunnels and galleries (17). The last three contributions deal with thermal aspects, namely, stress controlled pathways of fluid flow in Hot-Dry-Rock reservoirs (18), relationships between porosity and thermal conductivity for granite (19), and measurements of thermal conductivity through optical scanning (20).

Many problems in rock physics and rock mechanics are still unsolved. Future research should continue to focus on scaling rules – in both directions, i.e., on upscaling and downscaling. Particularly poor knowledge seems to exist on interactions between petrophysical and geochemical processes. What are the key parameters in such type of coupling, and how can they be obtained? Findings from experimental studies is and will remain an invaluable source for future progress in understanding hydro-thermo-mechanical coupling in fractured rock. Nonetheless, special attention must be paid to the question how site specific observations in the field are. Given the complexity of the target material, an ever broader collection of successful predictions and pitfalls forms our imagination of appropriate models, and what aspects will continue to sharpen our insight into the multi-scale phenomena we are encountering.

The papers collected in this volume provide some idea of the exciting meeting at Bad Honnef. I would like to thank all the authors for contributing their original articles to this special issue. The following reviewers are acknowledged for their assistance: S. Baraka-Lokmane, B. Bourbiaux, M. Boutéca, D.S. Daev, C. David, K. Evans, Y. Guéguen, B.C. Haimson, S. Heikamp, J. Heiland, T. Ito, E. Klein, F. Lanaro, B. Maillot, Y. Méheust, G. Nover, D. Pisarenko, Y.A. Popov, T. Popp, T. Reuschlé, E.H. Saenger, J. Schmittbuhl, A. Schubnel, S.A. Shapiro, G.D.H. Simpson, O. Stephansson, F. Surma, R.W. Zimmerman, G. Zimmermann.

References

Boutéca, M., and Guéguen, Y. (1999), *Rock Conference Looks at Pore Pressure, Scale Effects, and Deformation*, EOS (Transactions American Geophysical Union), April 20, 188.

Couples, G., Main, I., and Meredith, Ph. (2000), *Relationships between Damage and Localization: A Euroconference on Rock Mechanics and Rock Physics*, Newsletter Physical Properties of Earth Materials, October 2000, 3–8.

Kümpel, H.-J. (2001), *Conference Addresses Thermo-hydro-mechanical Coupling in Fractured Rock*, EOS (Transactions American Geophysical Union), May 29, 248.

Hans-Joachim Kümpel
Leibniz-Institute for Applied Geosciences (GGA)
Stilleweg 2
D-30655 Hannover
Germany
E-mail: kuempel@gga-hannover.de

Pure appl. geophys. 160 (2003) 813–831
0033 – 4553/03/060813–19

Pure and Applied Geophysics

Borehole Breakouts in Berea Sandstone Reveal a New Fracture Mechanism

BEZALEL C. HAIMSON[1]

Abstract—Vertical drilling experiments in high-porosity (22% and 25%) Berea sandstone subjected to critical true triaxial far-field stresses, in which σ_H (maximum horizontal stress) > σ_v (vertical stress) > σ_h (least horizontal stress), revealed a new and non-dilatant failure mechanism that results in thin and very long tabular borehole breakouts that have the appearance of fractures, and which counterintuitively develop orthogonally to σ_H. These breakouts are fundamentally different from those induced in crystalline rocks, as well as limestones and medium-porosity Berea sandstone. Breakouts in these rocks are typically dog-eared in shape, a result of dilatant multi-cracking tangential to the hole and subparallel to the maximum far-field horizontal stress σ_H, followed by progressive buckling and shearing of detached rock flakes created by the cracks. In the high-porosity sandstone a narrow layer of grains compacted normal to σ_H is observed just ahead of the breakout tip. This layer is nearly identical to "compaction bands" observed in the field. It is suggested that when a critical tangential stress concentration is reached along the σ_h spring line at the borehole wall, grain bonding breaks down and a compaction band is formed normal to σ_H. Debonded loose grains are expelled into the borehole, assisted by the circulating drilling fluid. As the breakout tip advances, the stress concentration ahead of it persists or may even increase, extending the compaction band, which in turn leads to breakout lengthening.

Key words: Rock mechanics, fractures, *in situ* stress, borehole breakouts, sandstone, compaction band.

Introduction

Stress-induced instability around boreholes drilled vertically into the earth's crust often results in 'breakouts' or zones of cross section elongation due to brittle fracture at the borehole wall. Extensive field evidence and laboratory experiments suggest that breakout orientation along the perimeter of such boreholes is typically aligned with the direction of the minimum horizontal *in situ* stress σ_h (BELL and GOUGH, 1979; SHAMIR and ZOBACK, 1992; HAIMSON and HERRICK, 1986; LEE and HAIMSON, 1993). The compressive stress concentration there, generated by the prevailing crustal stresses, is at its highest. Breakouts are at once both a cause for concern regarding the borehole structural integrity, and potentially a most

[1] Department of Materials Science and Engineering, and the Geological Engr. Program, University of Wisconsin, 1509 University Avenue, Madison, WI 53706-1595, U.S.A. E-mail: bhaimson@wisc.edu

useful *in situ* stress indicator (VERNIK and ZOBACK, 1992; BRUDY *et al.*, 1997; SONG and HAIMSON, 1997). The growing importance of borehole breakouts as a means of assessing *in situ* stress directions in the earth crust at locations where direct measurements are not feasible economically (too expensive) or physically (hostile wellbore conditions) has brought about intensified research related to breakout-forming failure mechanisms. The two major objectives of the borehole-stability laboratory study at the University of Wisconsin have been to identify mechanisms of fracture that bring about breakouts, and to examine their potential use as *in situ* stress-magnitude indicators. In this paper I primarily address the first objective, with particular reference to Berea sandstone.

The borehole failure mechanism that is common to several major rock types is exemplified by that displayed in laboratory tests of Westerly and Lac-du-Bonnet granites (unconfined compressive strengths, UCS, of about 200 MPa; porosities of about 1%), in which sufficiently high far-field stresses lead to the development of multitudes of subparallel extensile microcracks behind the borehole along the spring line of the least horizontal far-field stress σ_h (LEE and HAIMSON, 1993; SONG and HAIMSON, 1997). These dilatant microcracks, which are generally aligned with the direction of the maximum horizontal stress σ_H, create slender rock flakes that buckle owing to the high longitudinal stress. Broken flakes fall off in sequence starting with the one at the borehole wall. Slender cantilevers left by the broken first flake at both ends limit the span of the subsequent spall, and as this process continues flakes grow shorter until the breakout stabilizes in the form of a pointed dog ear (Fig. 1).

Drilling experiments in Indiana and Alabama limestones (unconfined compressive strength, UCS, of about 40 MPa; porosity: 17%), revealed similar dilatant mechanisms of breakout formation consisting of the development of extensile microcracks subparallel to σ_H behind the borehole wall along the direction of σ_h. As in the case of granite, rock flakes produced by these discontinuities, albeit thicker, spall off and create dog-eared breakouts (HAIMSON and HERRICK, 1986; HERRICK and HAIMSON, 1994). The process of breakout forming in much weaker limestones, such as the Cordova Cream (UCS: 18 MPa; porosity: 23%), although initiated by dilatant shear fracturing emanating from the borehole wall, also results in deep and pointed dog-eared breakouts similar to that in granite (HAIMSON and SONG, 1993).

However, recent laboratory drilling experiments in Berea sandstone have revealed disparate breakout characteristics and fundamentally different mechanisms of formation, depending on grain size, porosity, and cementation, and are the subject of this paper. One of the failure mechanisms has never been observed in rocks, and appears to be particularly significant in the efforts to understand the mechanical behavior of poorly consolidated sandstones, borehole stability, and the problem of sand production in oil-producing fields.

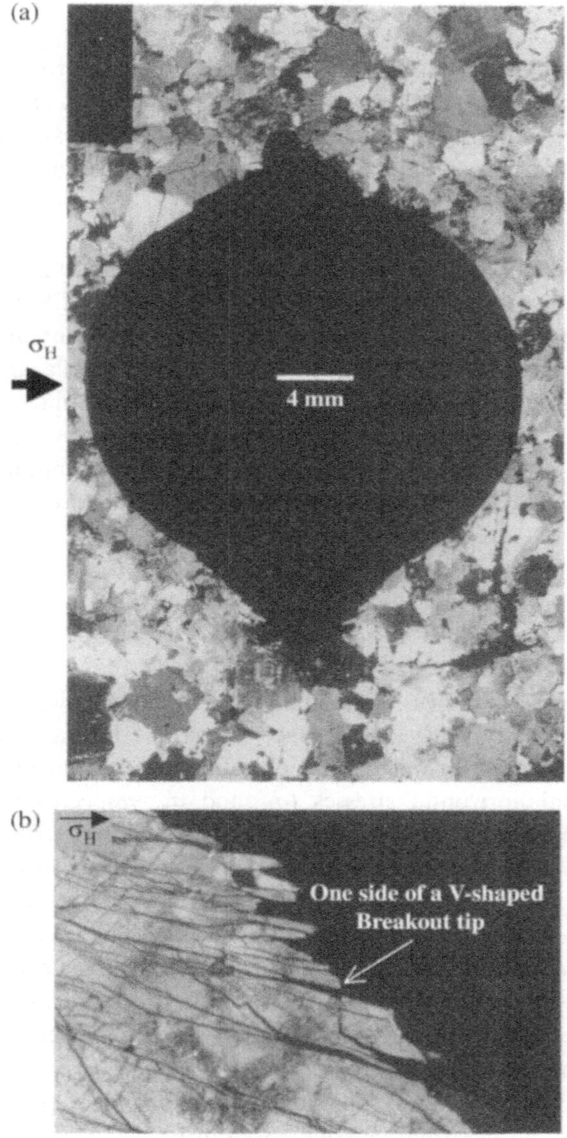

Figure 1
Typical dog-eared borehole breakout in granite. (a) Thin section micrograph (under optical microscope) of a borehole cross section in Lac du Bonnet granite drilled under high far-field stresses, showing two diametrically opposed breakouts aligned with the direction of the least horizontal far-field stress (normal to σ_H). (b) Detail of one side of a breakout tip revealing a multitude of dilatant subparallel extensile microcracks and remnants of rock flakes that buckled and fell into the borehole.

Berea Sandstone

Berea sandstone is early Mississippian, light gray to brown, flat-bedded to massive, and is quarried in the State of Ohio. In our laboratory experiments we have used three varieties of Berea, all from different sections of the same quarry, and each possessing a distinct porosity (17%, 22%, and 25%). Accordingly we named the three varieties BSs17, BSs22, and BSs25.

Thin sections were examined using a petrographic microscope. The sandstone is composed primarily of quartz grains, with significant amounts of feldspar. The BSs17 also contains some carbonate minerals. The grains in BSs22, and BSs25 are held together mostly by sutured contacts, in which adjacent grains have fused together along the contact areas due to compaction (Figs. 2a and b). In BSs17, however, cementation is provided mainly by iron rich clay minerals in addition to some sutured grain contacts (Fig. 2c). Grain size increases and sorting improves with porosity from 0.1 mm in BSs17, to 0.2 mm in BSs22, to 0.5 mm in BSs25. Table 1 presents some physical and mechanical properties of the three rocks as determined in our laboratory.

Experimental Setup and Procedure

Borehole drilling experiments were conducted in a newly designed University of Wisconsin high-capacity (3.8 MN in each direction) biaxial loading apparatus (Fig. 3a), which was used for the application of independent far-field maximum and minimum principal horizontal stresses (σ_H and σ_h, respectively) to rectangular prismatic rock specimens measuring 150 × 150 × 230 mm. Thin copper shims (0.25 mm) coated with a layer of Stearic acid were placed between the specimen faces and loading platens to reduce friction (LABUZ and BRIDELL, 1993). The biaxial cell was then positioned in a specially designed loading frame (Fig. 3b), which enabled both the application of the far-field vertical principal stress σ_v (1.3 MN load capacity) and the drilling of a vertical central hole in specimens that were already subjected to three far-field principal stresses. These were applied using independent servo-controlled systems.

The experimental procedure consisted of four major steps: loading, drilling, resting (controlled by acoustic emission monitoring), and recording of breakout characteristics. The three mutually perpendicular loads, giving rise to three unequal principal stresses in the specimen, were applied simultaneously. When the predetermined level of the least principal stress (typically σ_h) was reached, loading in that direction was stopped, but continued in the other two directions until the preset intermediate principal stress was attained. While holding both σ_h and σ_v constant, the largest principal stress σ_H was then increased until its planned level was reached. All three principal stresses were held constant for the remainder of the test. We used four

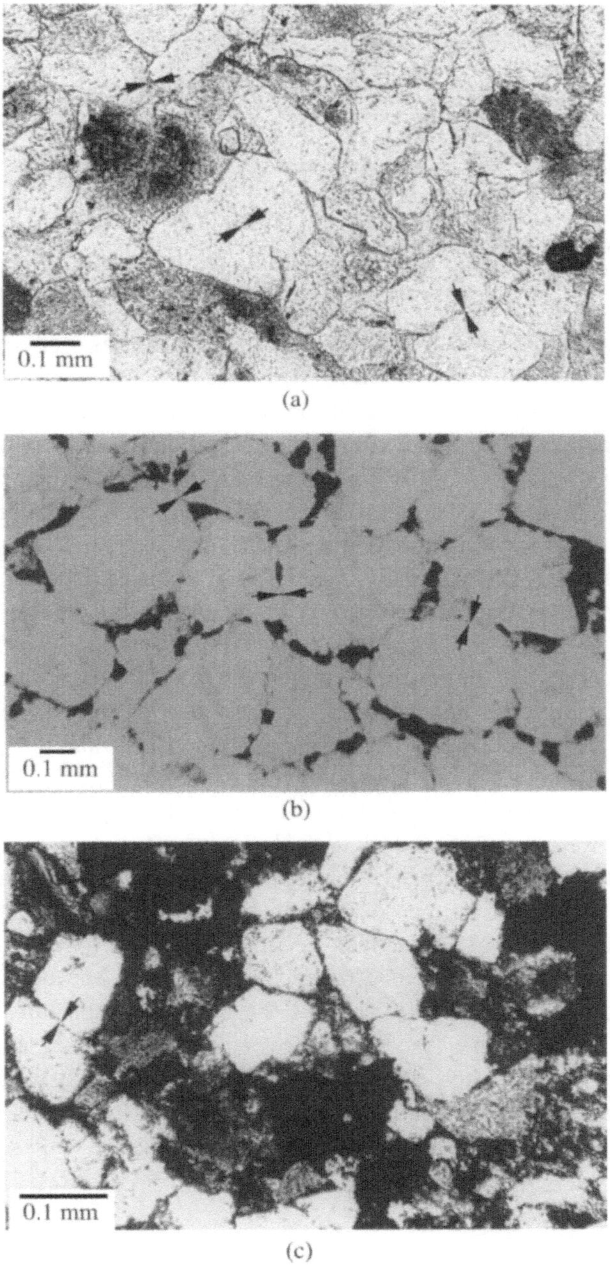

Figure 2
Micrographs of thin sections of three Berea sandstones under optical microscope. Arrows indicate examples of grain sutures. (a) BSs22 (22% porosity) and (b) BSs25 (25% porosity) reveal mainly sutured grains; (c) BSs17 (17% porosity) shows little suturing, with grain bonding dominated by iron-rich clay minerals.

Table 1

Some physical and mechanical properties of three Berea sandstones (as determined in our laboratory)

Property\Rock Type	BSs17	BSs22	BSs25
Average grain size (mm)	0.1	0.2	0.5
Main cementation	Iron rich clay	Sutured contacts	Sutured contacts
Porosity (%)	17	22	25
Permeability (mdarcy)	18	Not available	1800
Unconfined compressive strength (MPa)	53	42	22
Brazilian tensile strength (MPa)	4	2.2	1.7

levels of least horizontal stress: 20, 30, 40, and 50 MPa. Levels used for the maximum horizontal stress were 30–100 MPa in steps of ten.

Borehole drilling was achieved using an electrically powered drill-rig mounted atop the loading frame (Fig. 3b). Drilling of a central borehole along the vertical axis of the specimen was carried out using a water-cooled diamond-impregnated coring bit (Fig. 3b). The drilling bit used in most tests was 23 mm in diameter. A double-acting pneumatic cylinder supplied the thrust force on the coring bit. Drill-bit penetration was controlled manually using a potentiometer depth gage and a time-based drilling guide. A flow meter was employed to monitor the flow of the drilling fluid (tap water in our tests). The borehole was typically advanced 100–150 mm. The drill-bit was then extracted and drilling-fluid flow was stopped.

At the conclusion of drilling the specimen was allowed to rest while maintaining the initial state of far-field stress. A one-channel acoustic emission (AE) signal was monitored for any signs of instability following drilling (Fig. 3b). Acoustic emission gradually diminished within 30 to 45 minutes after the conclusion of drilling. The applied stresses were then relieved. A typical test record is shown in Figure 4.

Subsequently, the sample was removed from the biaxial cell and dried. The borehole was filled with epoxy to prevent the breakouts from being damaged during handling. Several 12 mm thick sections were cut to assess the breakout shape and measure dimensions at different depths within the borehole. Measurements of breakout length, angular span at the borehole wall, and breakout width away from the wall (applied specifically to breakouts in BSs22 and BSs25) were taken from both sides of each slice. Photographs of representative slices from each sample were used for comparative analysis. Thin sections were prepared from several slices and analyzed using a stereoscopic microscope with a polarized light source.

Breakout Characteristics and Mechanism of Formation in BSs17

As in all previously tested rocks, breakouts in BSs17 occurred at two diametrically opposed zones of the borehole wall along the σ_h spring line, where

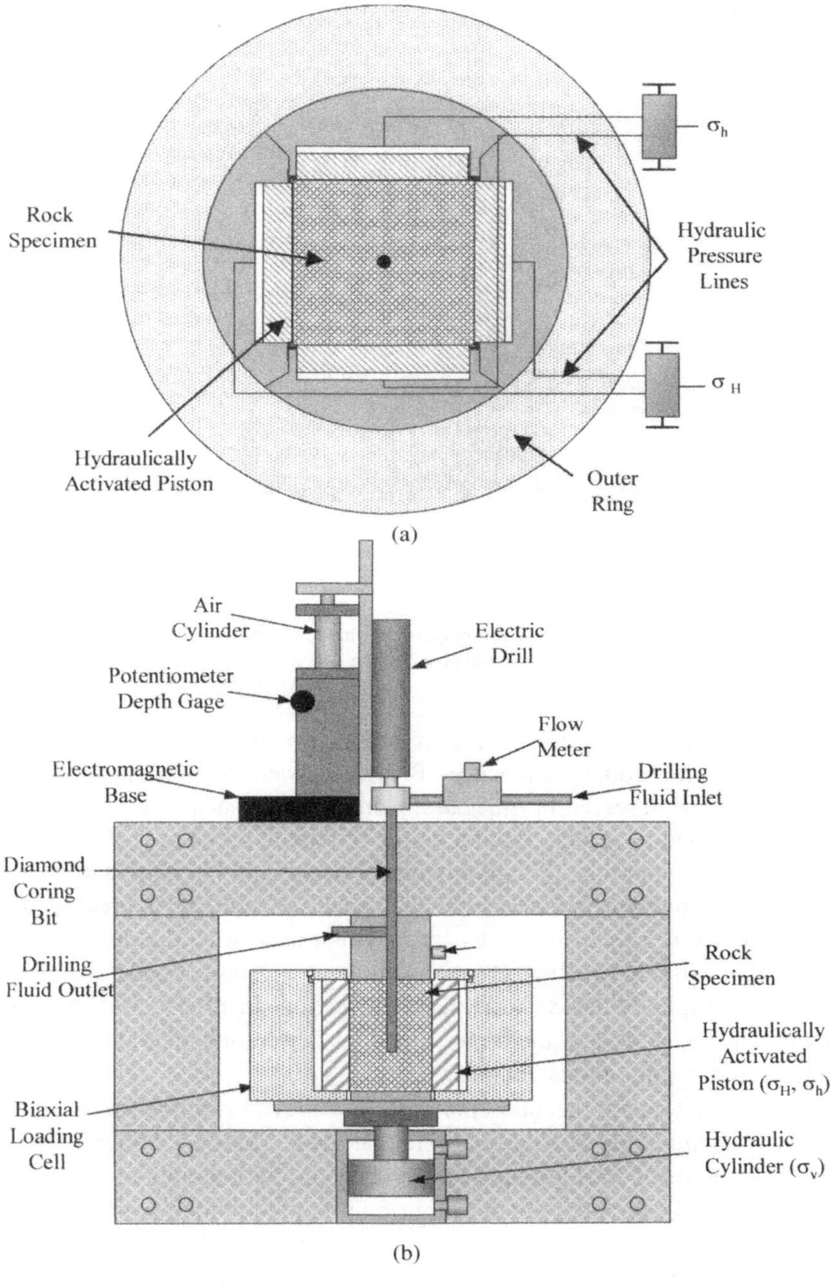

Figure 3

Schematics of (a) horizontal cross section of the University of Wisconsin biaxial cell, and (b) vertical cross section of an especially built loading frame, which enables drilling of a borehole while a prismatic rock specimen is subjected to three independent far-field principal stresses.

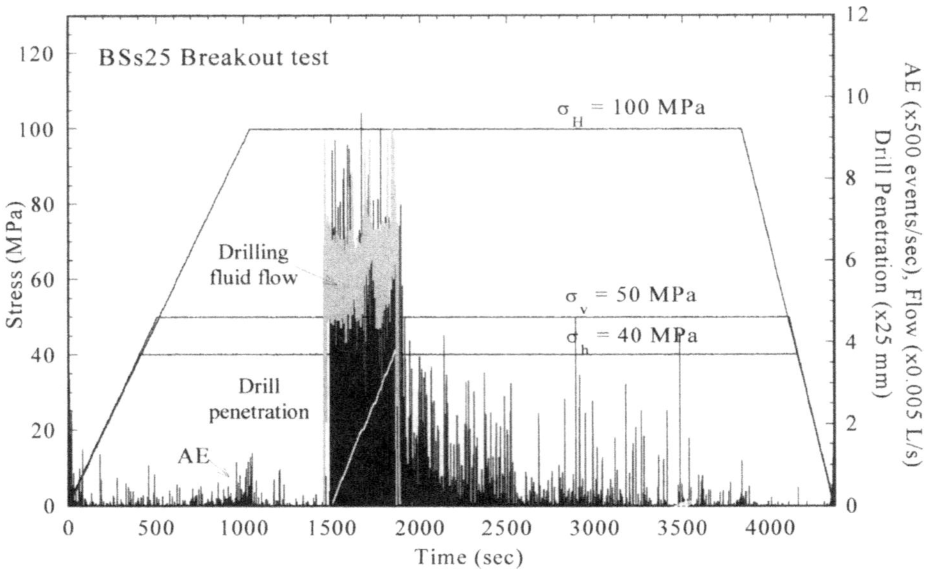

Figure 4
Typical test record, showing the sequence of applying the three far-field principal stresses, the drilling of the central borehole, and the post-drilling acoustic emission activity.

the highest tangential-stress concentration is achieved, and where it is reasonable to expect that failure should commence. Breakouts were *V*-shaped, or dog-eared, analogous to those observed in other sedimentary rocks such as several limestones (Haimson and Herrick, 1986; Herrick and Haimson, 1994) as well as in different granites (Lee and Haimson, 1993; Song and Haimson, 1997). Also, similarly to the findings of these previous studies, both the spans (or widths) of breakouts at the borehole wall, as well as their depths (or lengths), grew with the magnitude of σ_H when the other far-field principal stresses were kept constant. Figure 5 depicts a set of three photographs of drilled borehole cross sections in BSs17. The characteristic breakout *V*-shape, as well as the larger breakout depth and span at higher σ_H (for constant σ_h and σ_v) are unmistakable.

We also investigated the micromechanical behavior leading to breakouts in BSs17, by studying thin sections of borehole cross sections. Different stages of breakout development were observed depending on the magnitudes of the applied far-field stresses. During incipient breakout failure, predominantly intra- and transgranular microcracking occurred (Fig. 6). These microcracks were located immediately behind the borehole wall and along the σ_h spring line. The general direction of the cracks was subparallel to that of σ_H. The microcracks appeared open and, hence, dilatant. They were, in fact, similar to the extensile cracks observed in granite prior to breakout development. Under larger prevailing far-field stresses,

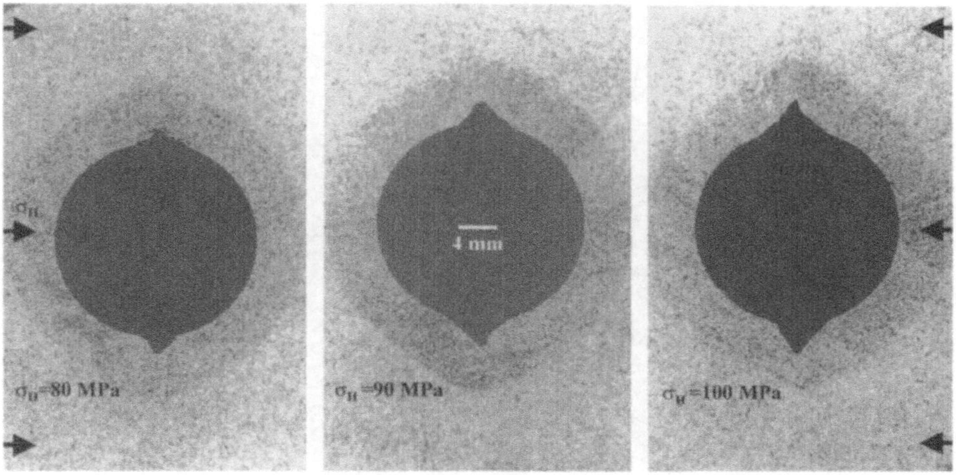

Figure 5
Photographs of three borehole cross sections in medium-porosity (17%) Berea sandstone (BSs17). They were all drilled under far field stresses $\sigma_h = 50$ MPa, $\sigma_v = 60$ MPa, and σ_H varying from 80 to 90 to 100 MPa. They all reveal dog-eared breakouts but both their width at the borehole wall and their length (or depth) increase with the magnitude of σ_H.

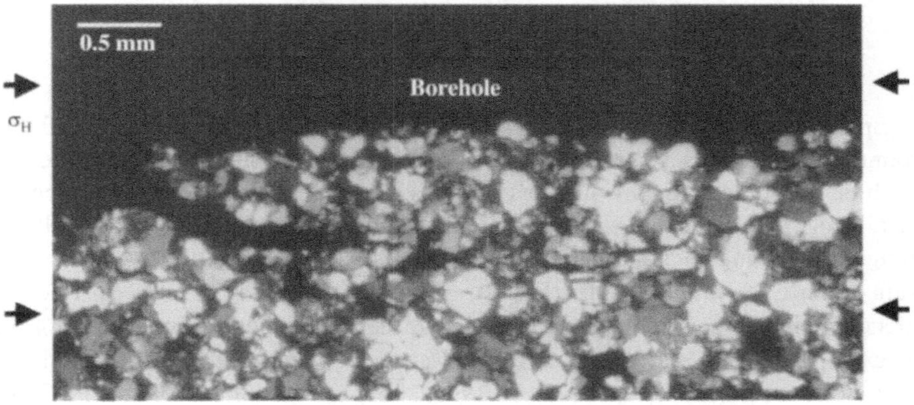

Figure 6
Micrograph of a borehole-wall segment (in cross section) in Berea sandstone of 17% porosity (BSs17) along the σ_h spring line. Specimen was subjected to far-field stresses that were subcritical in terms of producing a complete breakout. Visible incipient breakout failure is in the form of dilatant extensile intra- and trans-granular microcracking subparallel to the maximum horizontal stress σ_H.

complete dog-eared breakouts formed. A close up of a breakout tip (Fig. 7) shows subparallel dilatant microcracks that developed transgranularly and helped form the breakout in a similar way to that experienced in previously tested rocks.

Figure 7
Micrograph of a breakout cross section in Berea sandstone of 17% porosity (BSs17), exhibiting the commonly observed dog-eared shape. An extensile, mainly intra-granular crack just under the tip of the breakout and extending subparallel to the σ_H direction reveals the type of failure leading to the breakout.

Breakout Characteristics in BSs22 and BSs25

Breakouts in the higher-porosity BSs22 and BSs25 were dramatically different from those in the denser sandstone, and entirely unexpected since nothing of the kind had been encountered in any previous field observations or laboratory experiments in rock. Those breakouts did occur at the same locations around the borehole wall, i.e., along the σ_h spring line, but their shape is best described as long, narrow (about 5–10 grain diameters) tabular slots (Fig. 8). They resembled stress-induced fractures extending along a counterintuitive path, i.e., rather than being subparallel to σ_H (see for example JAEGER and COOK, 1979, p. 92), these fractures were orthogonal to it (HAIMSON and SONG, 1998; HAIMSON, 1999, 2001; KOVACICH and HAIMSON, 2000).

Figure 8 also indicates that while the widths of these breakouts remain practically constant even as stress conditions vary, the lengths clearly increase with the far-field

▶

Figure 8
Photographs of three borehole cross sections in each of two high-porosity Berea sandstones: (a) 22% porosity (BSs22), and (b) 25% porosity (BSs25). They were all drilled under far field stresses $\sigma_h = 50$ MPa, $\sigma_v = 60$ MPa, and σ_H varying from test to test as shown in Figure. Boreholes reveal long and thin tabular breakouts which have never been observed in any other rock types. The length of these fracture-like features suggests a direct dependence on the magnitude of σ_H.

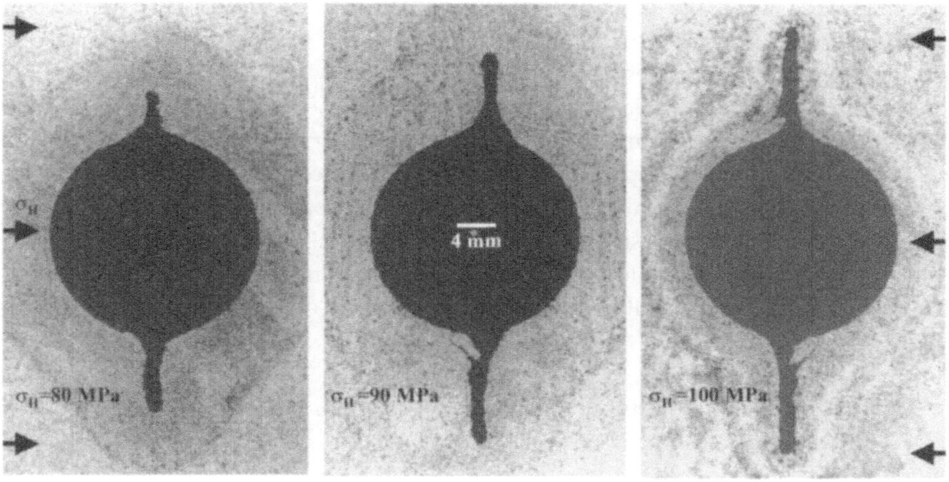
σH
4 mm
σH=80 MPa
σH=90 MPa
σH=100 MPa

(a)

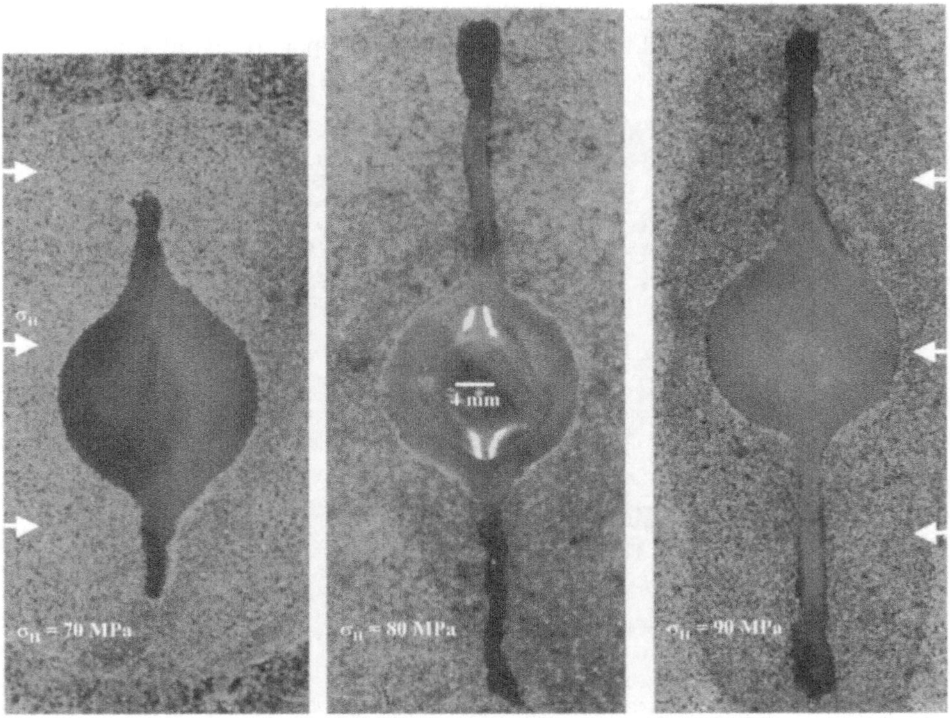
σH
4 mm
σH = 70 MPa
σH = 80 MPa
σH = 90 MPa

(b)

σ_H magnitude when the other principal stresses are unchanged. This indicates a potential use of this type of breakouts to estimate *in situ* stress magnitudes, provided breakout length can be assessed.

Breakouts in BSs25 are identical in shape to those in BSs22, except that they are considerably longer (reaching several borehole diameters in length) for similar far-field stress conditions (compare Figs. 8a with 8b). A comparison of Figure 5 with Figure 8 illustrates the disparity in breakout shape between the lower-porosity BSs17 and the sandstones of porosity higher than 20%. This shape discrepancy suggests a different mechanism of formation. There is, however, a similarity in terms of breakout location at the borehole wall and its length dependence on the magnitude of σ_H.

Possible Mechanisms of Compressive Failure in BSs22 and BSs25

Initial Model

Micromechanically, incipient breakout failure in BSs22 and BSs25 takes the form of grain debonding at the points of highest stress concentration (along the σ_h spring line) at the borehole wall. As shown in Figure 9, upon breakout initiation no

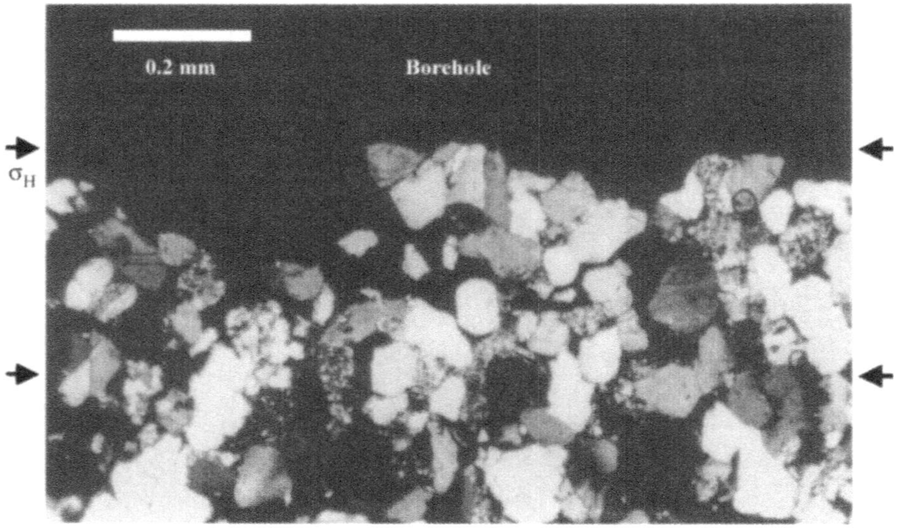

Figure 9
Micrograph of a borehole-wall segment (in cross section) in Berea sandstone of 22% porosity (BSs22) along the σ_h spring line. Specimen was subjected to far-field stresses that were subcritical in terms of producing a complete breakout. The shown incipient breakout failure takes the form of intergranular-matrix disintegration releasing individual grains at the points of highest stress concentration at the borehole wall. Practically no microcracking is visible. This apparent non-dilatant failure mechanism gives rise to the unexpected fracture-like breakouts.

microcracking subparallel to σ_H is visible, and the only failure observed is in the form of loose grains falling into the borehole.

The mechanism of failure leading to fracture-like breakouts is not straightforward. Our initial assumption was that it is related to the bonding between quartzitic grains. As mentioned above, in BSs22 and BSs25 sandstones grains are held together mainly by suturing (Figs. 2a and b). Under sufficient far-field stresses, sutured contacts could weaken at the points of highest stress concentration at the borehole wall, and some grains may be freed (Fig. 9). These grains are then ejected from the borehole wall with the help of the circulating drilling fluid. Some adjacent grains left hanging by the removal of the first grains are also debonded and flushed out by the fluid. As one layer of several grains along the hole wall is removed, the high stress concentration along the σ_h spring line is transferred to the next layer, bringing about more grain debonding and additional grain flushing. This process is repeated several times, resulting in the long facture-like breakout (Fig. 10). Breakout growth will

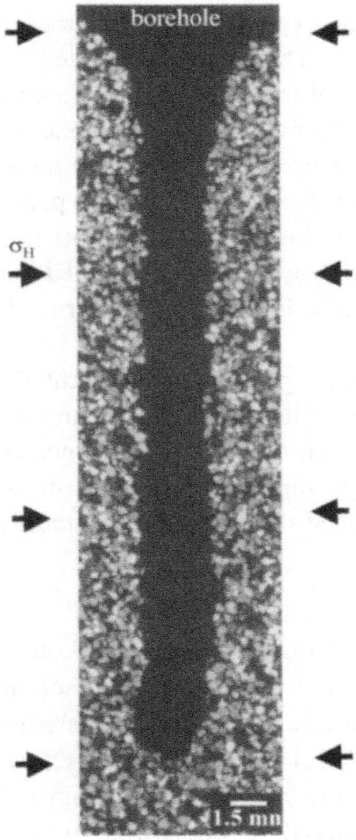

Figure 10
A mosaic of micrographs showing a thin section of a complete fracture-like breakout in Berea sandstone of 25% porosity (BSs25) extending along the σ_h spring line.

come to an end when diminishing drilling fluid circulation away from the borehole and/or the possible formation of a stable arch at the breakout tip prevent further grain loosening.

The importance of the circulating drilling fluid in the process of forming elongated breakouts became clear to us as a result of several tests in which we first drilled a hole in the center of the specimen, and only then applied the far-field stresses. In these tests the breakouts were typically short and not fracture-like. The reason for the narrow, 5–10 grain-size breakout widths, is that suturing is sufficiently strong to resist debonding as the stress concentration is reduced away from the σ_h spring line. Our interpretation of the micromechanics of failure in the high-porosity Berea sandstones has subsequently evolved to invoke the concept of compaction bands.

Compaction Bands

Contemporaneously with our borehole breakout studies in sandstone, MOLLEMA and ANTONELLINI (1996) discovered in the field the existence of a structure they called a "compaction band." Their paper presented evidence for the existence of tabular zones of localized deformation in the Navajo Sandstone of the Kaibab monocline, Utah, that accommodate pure compaction but zero offset or shear (Fig. 11). Compaction bands occur in sandstones possessing large grain size (0.3–0.8 mm) and high porosity (exceeding 20%), along planes that are orthogonal to the maximum compressive stress. ISSEN and RUDNICKI (1999) reported that the shear localization theory originated by RUDNICKI and RICE (1975) was found to admit solutions for compaction bands in the range of parameters representative of porous rock.

Within compaction bands, porosity is substantially reduced, thus creating a curtain of low permeability that could become an unwanted barrier to flow if it occurred in an aquifer or an oil reservoir. Also, since grains are crowded inside the compaction band by the high compressive stress acting across it, some grain splitting under Hertzian stresses applied through grain contacts is often observed.

Compaction-band Based Model

OLSSON (1999) suggested that the long breakouts we had obtained in high-porosity Berea sandstone could be "empty compaction bands." Consequently, we have conducted a thin section study of fracture-like breakout tips in BSs25 (HAIMSON, 1999, 2001). Several breakouts in BSs25 show the development of an apparent compaction band just ahead of the breakout tip. It is characterized by a thin band of grains orthogonal to σ_H, within which compaction has occurred, creating a zone of diminished porosity (Fig. 12). Some grain splitting is evident, a result of loosened grains crowding each other in the σ_H direction. On both sides of the apparent compaction band grains are visibly more separated as expected in a

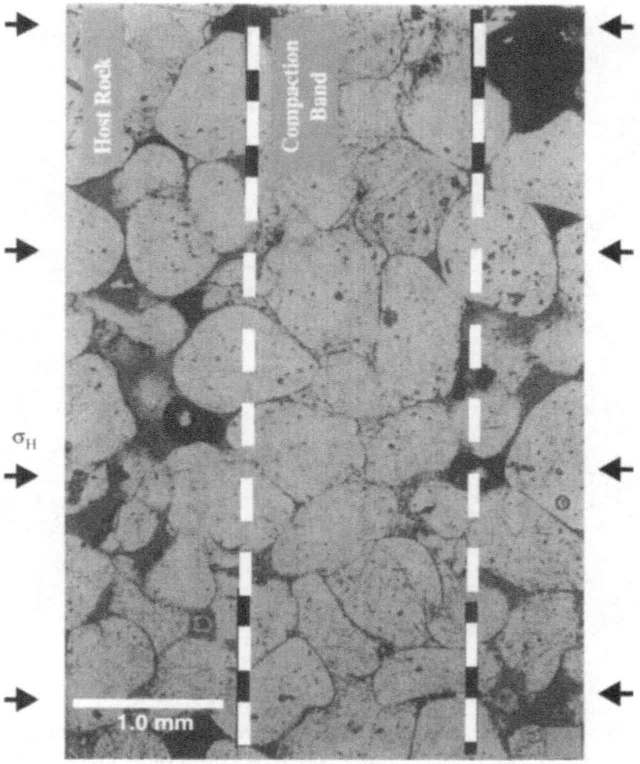

Figure 11
Micrograph of a compaction band observed in an outcrop of the Navajo sandstone (after MOLLEMA and ANTONELLINI, 1996). Note the clear loss of porosity within the compaction band (between the dashed lines).

high-porosity granular rock. The occurrence of compaction bands in BSs25 is practically identical to that described in field situations (MOLLEMA and ANTONELLINI, 1996) in terms of rock properties (porosity 25%, grain size 0.5 mm), band width (5–10 grain diameters), and counterintuitive orientation orthogonal to largest compressive stress (compare Figs. 11 and 12).

The discovery of the apparent compaction band formation ahead of the breakout tip (Fig. 12) has enabled us to amend our initial model. Our original inference that sufficiently high compressive stress concentration along the σ_h spring line in BSs25 causes loosening of sutured grains along and adjacent to that line is still valid. Grain debonding follows, enabling freed grains within that narrow zone to pack tightly in the σ_H direction. This creates a thin tabular belt of low porosity along the σ_h spring line that resembles the field-observed compaction band (Fig. 12). Some compacted grains at the free surface of the borehole wall are ejected if the line of loading of the high tangential compressive force to which they are subjected is

Figure 12
Micrograph of a thin section of an apparent compaction band in front of a fracture-like breakout tip in Berea sandstone of 25% porosity (BSs25). Grains between the two dashed lines delineating the compaction band appear substantially more crowded than those outside of the band.

favorably located. The expulsion of these grains is abetted by the radial expansion of adjacent grains behind them, which are also subjected to high tangential compression. Circulating drilling fluid assists in this process by flushing out any additional loose grains from the compaction band and into the borehole. As the breakout tip advances, the stress concentration ahead of it persists, extending the compaction band, which in turn leads to additional grain ejection and breakout lengthening (Fig. 10).

It is possible that in field situations this process may continue for substantial distances (at least several times the wellbore diameter), potentially leading to considerable sand production. Stabilization of a breakout may occur when fluid circulation is significantly reduced as the breakout tip reaches farther form the borehole wall, or when breakout tip grains become so well wedged that they create a stable arch, stopping any additional grain expulsion. The breakout maintains its constant narrow width because the adjacent grains, located away from the critical stress concentration zone, are still held together predominantly through suturing.

Discussion

An independent suggestion that long and narrow breakouts could develop in granular rocks was advanced by BESSINGER *et al.* (1997), who tested an artificial material made of sintered glass beads meant to simulate sandstone. They found that lightly sintered block samples subjected to uniaxial compressive loading yielded relatively long linear fractures perpendicular to loading direction. They interpreted the phenomenon as related to the bonding-strength to grain-strength ratio. Scanning electron microscopy showed that failure occurred along grain boundaries, after which the load was transferred to adjacent grains.

Supporting evidence that fracture-like breakouts occur in natural sandstones other than Berea, and that they are preceded by compaction, was provided by CUNDALL *et al.* (1999). They performed numerical simulations using the distinct element method to reproduce results obtained in tests of thick-wall hollow cylinders of high-porosity (>30%) Castlegate sandstone. In their simulation, progressive loading of the specimen outer surface caused breakouts to occur on the inner surface. Initially the failure appeared to be of a spalling type, but as loading increased the geometry evolved into a cusp-like form, eventually developing a long fracture-like object extending from the tip of the cusp. The inference from the numerical simulations is that cracks could form and propagate, counterintuitively, in a direction perpendicular to the maximum compressive stress, giving rise to the conjecture that the fracture tip formation mechanism is the result of material compaction in the form of a compaction band.

The process of fracture-like breakout formation is in effect the result of a compressive failure mode hitherto unrecognized. The resulting breakout fracture has the appearance of a mode I tensile crack, except that it is the result of high compressive (rather than tensile) stress acting across it (CUNDALL *et al.*, 1999). The failure mechanism observed in previously tested granites, limestones and sandstones is clearly dilatant, by which rock volume increases as extensile and/or shear microcracking occurs, culminating in a dog-eared shaped zone of material failure around the borehole. To the contrary, in BSs22 and BSs25 we observe an unexpected non-dilatant failure mechanism since breakout formation is not preceded by inter- and intra-granular microcracking, except for sporadic grain splitting (Fig. 12). In fact, the concept of breakout formation by emptying a compaction band gives rise to what could be termed an anti-dilatant failure mechanism, since the volume of the failed zone actually undergoes volumetric loss as a result of grain compaction.

Conclusions

Boreholes drilled in high-porosity Berea sandstone (22% and 25%) subjected to high far-field stresses create breakouts that are spectacularly different from those observed in granite, limestone, and the lower-porosity Berea sandstone (17%). The new breakouts are not dog-eared, as is normally observed in other rocks; rather they resemble narrow (5–10 grain size) tabular slots that develop orthogonally to the maximum horizontal stress σ_H. These fracture-like breakouts originate at the points of highest stress concentration at the borehole wall, along the σ_h spring line, and extend in the same direction for distances several times the borehole diameter, depending on the far-field stress condition.

Micrographs of the breakout tip in high-porosity Berea sandstone show the development of an apparent compaction band in the form of a narrow layer within which grains have been compacted normal to σ_H, creating a zone of low porosity. Some grain splitting is evident, a result of loosened grains crowding each other in the σ_H direction. The mechanism leading to fracture-like breakouts may be directly related to the formation of the compaction band, which is the result of sutured grain contact loosening and grain debonding. This process enables the drilling fluid to flush out the compacted loose grains. As the breakout tip advances, the stress concentration ahead of it persists, extending the compaction band, which in turn leads to breakout lengthening. It is envisaged that in field situations this process may continue for large distances, leading to substantial sand production.

Acknowledgements

This work is supported by the Geosciences Research Program, Office of Basic Energy Sciences, U.S. Department of Energy, grant no. DE-FGO2-98ER14850. Graduate Students I. Song and J. Kovacich, and A. Klaetsch conducted the experiments described in the paper. Their contribution is deeply appreciated.

References

Bell, J. S. and Gough, D. I. (1979), *Northeast-southwest Compressive Stress in Alberta: Evidence from Oil Wells*, Earth and Planet. Sci. Lett. *45*, 475–482.

Bessinger, B. A., Liu, Z., Cook, N. G. W., and Myer, L. R. (1997), *A New Fracturing Mechanism for Granular Media*, Geophys. Res. Lett. *24*, 2065–2608.

Brudy, M., Zoback, M. D., Fuchs, K., Rummel, F., and Baumgärtner, J. (1997), *Estimation of the Complete Stress Tensor to 8 km in the KTB Scientific Drill Holes: Implications for Crustal Strength*, J. Geophys. Res. *102*, 18,453–18,457.

Cundall, P. A., Damjanac, B., and Pestman, B. J. (1999), *Compression Cracks Observed in Numerical Simulations of a Thick Wall Cylinder in Porous Sandstone*, EOS Trans. Am. Geophys. Union *80*, F1068.

Haimson, B. C. (1999), *Borehole Breakouts, Compaction Bands, and Potential Sand Production in Berea Sandstone*, EOS Trans. Am. Geophys. Union *80*, F1067.

Haimson, B. (2001), *Fracture-like Borehole Breakouts in High-Porosity Sandstone: Are They Caused by Compaction Bands*? Physics and Chemistry of the Earth, Part A, *26*, 15–20.

Haimson, B. C. and Herrick, C. G., *Borehole breakouts – A new tool for estimating in situ stress*? In *Rock Stresses and Rock Stress Measurement, Proc. Int. Symp.* (ed. Stephansson O.) (Centek Pub., Lulea 1986) pp. 271–281.

HAIMSON, B. C. and SONG, I. (1993), *Laboratory Study of Borehole Breakouts in Cordova Cream: A Case of Shear Failure*, Int. J. Rock Mech. Min. Sci. and Geomech. Abstr. *30,* 1047–1056.

HAIMSON, B. C. and SONG, I. (1998), *Borehole breakouts in Berea sandstone: Two porosity-dependent distinct shapes and mechanisms of formation.* In *Rock Mechanics in Petroleum Engineering* (Soc. of Pet. Eng., Richardson, Texas) pp. 229–238.

HERRICK, C. G. and HAIMSON, B. C., *Modelling of episodic failure leading to borehole breakouts in Alabama limestone.* In *Rock Mechanics: Models and Measurements* (eds. Nelson P. and Laubach S.) (Balkema, Rotterdam 1994) pp. 217–224.

ISSEN, K. A. and RUDNICKI, J. W. (1999), *Conditions for Compaction Bands in Porous Rock*, EOS Trans. Am. Geophys. Union *80*, F1068.

JAEGER, J.C. and COOK, N.G.W. (1979), *Fundamentals of Rock Mechanics*, 3rd ed., Chapman and Hall, New York, 593 pp.

KOVACICH, J. R. and HAIMSON, B. C. (2000), *Factors affecting borehole breakout dimensions, and the potential for sand production in high-porosity Berea sandstone.* In *Pacific Rocks 2000* (eds. Garard J., Liebman M., Breeds C. and Doe T.) (Balkeema, Rotterdam 2000) pp. 271–277.

LABUZ, J.F. and BRIDELL, J.M. (1993), *Reducing frictional constraints in compression testing through lubrication*, Int. J. Rock Mech. Min. Sci. Geomech. Abstr., *30*, 451–455.

LEE, M.Y. and HAIMSON, B. C. (1993), *Laboratory Study of Borehole Breakouts in Lac du Bonnet Granite: A Case of Extensile Failure Mechanism.* Int. J. Rock Mech. Min. Sci. and Geomech. Abstr. *30*, 1839–1845.

MOLLEMA, P. N. and ANTONELLINI, M. A. (1996), *Compaction Bands: A Structural Analog for Anti-mode I Cracks in Aeolian Sandstone*, Techtonophysics *267*, 209–228.

OLSSON, W. A. (1999), *Theoretical and Experimental Investigation of Compaction Bands in Porous Rock*, J. Geophys. Res. *104*, 7219–7228.

RUDNICKI, J. W. and RICE, J. R. (1975), *Conditions for the Localization of Deformation in Pressure-sensitive Dilatant Materials*, J. Mech. Phys. Solids. *23*, 371–394.

SHAMIR, G. and ZOBACK, M. D. (1992), *Stress Orientation Profile to 3.5 km Depth Near the San Andreas Fault at Cajon Pass, California.* J. Geophys. Res. *97*, 5059–5080.

SONG, I. and HAIMSON, B. C. (1997), *Polyaxial Strength Criteria and their Use in Estimating in situ Stress Magnitudes from Borehole Breakout Dimensions*, Intl. J. Rock Mech. and Mining Sci, *34, 3–4, 116.*

VERNIK, L. and ZOBACK, M. D. (1992), *Estimation of Maximum Horizontal Principal Stress Magnitude from Stress-induced Well bore Breakouts in the Cajon Pass Scientific Research Borehole*, J. Geophys. Res. *97*, 5109–5119.

(Received February 2, 2001, revised June 6, 2001, accepted August 1, 2001)

Pure appl. geophys. 160 (2003) 833–849
0033–4553/03/060833–17

Pure and Applied Geophysics

A Model for the Mechanical Behaviour of Bentheim Sandstone in the Brittle Regime

EMMANUELLE KLEIN[1] and THIERRY REUSCHLÉ[1]

Abstract—The mechanical behaviour of Bentheim sandstone, a homogeneous quartz-rich sandstone with porosity of 22.8%, was investigated by triaxial compression tests conducted on dry samples. At confining pressures up to 35 MPa, the failure mode was characterized by a typical brittle deformation regime, as the samples showed dilatancy and failed by strain softening and brittle faulting. Previous studies have shown that the mechanical behaviour and failure mode of brittle porous granular rocks are governed by the time-dependent growth of microcracks. We analyse this process using the "Pore Crack Model" based on fracture mechanics analysis. It is consistent with the microstructure of porous granular rocks since it considers the growth of axial cracks from cylindrical holes in two dimensions. These cracks grow when their stress intensity factors reach the subcritical crack growth limit. Interaction between neighbouring cracks is introduced by calculating the stress intensity factor as the sum of two terms: a component for an isolated crack and an interaction term computed using the method of successive approximations. It depends on crack length, pore radius, pore density, and applied stresses. The simulation of crack growth from cylindrical holes, associated with a failure criterion based on the coalescence of interacting cracks, is used to compare the theoretical stress at the onset of dilatancy and at macroscopic rupture to the experimental determined values. Our approach gives theoretical results in good agreement with experimental data when microstructural parameters consistent with observations are introduced.

Key words: Fracture mechanics, micromechanical model, brittle deformation, crack interaction, subcritical crack growth.

1. Introduction

The understanding of a wide range of geophysical problems requires the knowledge of the mechanical properties of rocks. An important goal of rock mechanics is to provide useful methods for predicting failure mode, mechanical strength, porosity evolution and elastic moduli. These predictions are of great importance in the oil industry, for example, where the production-induced change in reservoir fluid pressure and total stress lead to reservoir compaction and wellbore instability, each with important consequences like subsidence, seismicity and sand production (e.g., MARTIN and SERDENGECTI, 1984; VEEKEN *et al.*, 1989; TEUFEL *et al.*, 1991).

[1] Institut de Physique de Globe de Strasbourg (CNRS-ULP UMR 7516), 5, rue René Descartes, 67084 Strasbourg Cedex, France. E-mail: Emmanuelle.Klein@eost.u-strasbg.fr
Corresponding author: Emmanuelle Klein

The basic mechanical data are on stress as a function of strain, which is intimately related to the mode of failure. Traditionally, rock failure in compression is categorized in terms of two end members. On the one hand, a sample may show dilatancy and fail by strain softening and brittle faulting under relatively low confining pressure. On the other hand, it may show pervasive (i.e., delocalized) compaction and fail by strain hardening under elevated confining pressure (PATERSON, 1978). At intermediate pressures, a transitional regime is sometimes observed with failure modes involving complex localized features such as conjugate shear bands. Although the two main regimes are very different on the macroscopic scale, they both involve brittle micromechanical processes (LOCKNER *et al.*, 1992; WONG *et al.*, 1992; READ *et al.*, 1995). In the brittle field, coalescence of microcracks leads to shear localization, whereas the ductile regime is a manifestation of homogeneously distributed microcracking (MENÉNDEZ *et al.*, 1996; KLEIN *et al.*, 2001).

Our laboratory decided to build a conventional triaxial testing apparatus in order to identify and to study the parameters controlling the compaction mechanisms. In the first part of this paper, we rapidly describe the main features of this new apparatus, focusing on its flexibility, and present the mechanical data obtained on Bentheim sandstone in the brittle regime. In the second part, we develop a model based on fracture mechanics that analyses the time-dependent growth of axial cracks from cylindrical holes in two dimensions and takes into account interaction between neighbouring cracks. A simple failure criterion, based on the coalescence of the interacting pore cracks, is then used to compare the theoretical critical stress to the experimental peak stress.

2. Experimental Procedure

2.1 The Testing Material

Bentheim sandstone occurs as well-known outcrop sandstone from the Gildehausen quarry, near the village of Bentheim (Germany). It is the same formation as the reservoir rock from the Schoonebeek oil field, a few tens of km north of the quarry. However, in terms of microstructure, petrophysical properties and composition, they are very different. The outcrop rock is very strong and quartz-rich, whereas the reservoir rock is much weaker and contains less quartz, and less quartz cement. Our tests have been done on the outcrop rock. It is relatively homogeneous, with mean composition of 95% quartz, 3% kaolinite, and 2% orthoclase (VOS, 1990).

Cylindrical samples (20 mm in diameter, 40 mm in length) were cored out of a single block, with an average porosity of 22.8%. Thin sections ($\sim$50 μm thick) of two intact samples impregnated with red epoxy were prepared and observed by optical microscopy. Quartz grains were found to be round to subround (Fig. 1) as observed by VOS (1990), and the distributions of the microstructural parameters (length of contact between two neighbouring grains, number of contacts per grain, pore size,

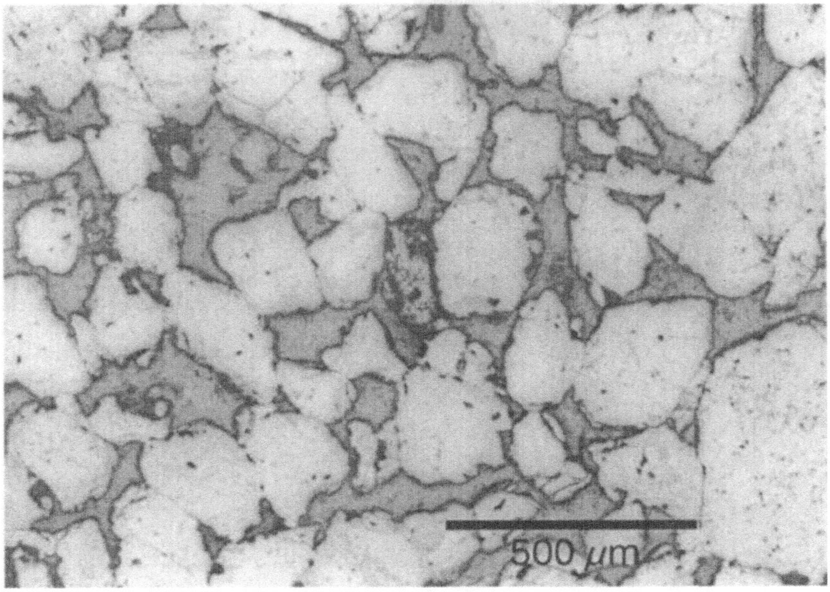

Figure 1
Optical micrograph of an intact sample of Bentheim Sandstone.

and grain size) determined by using stereological techniques (UNDERWOOD, 1970) were found to be very similar for both thin sections. The diameter of the quartz grains ranges from 50 μm to 550 μm as found by SCHUTJENS *et al.* (1995), and the mean grain size is ~210 μm. Because pores have various complex shapes, equivalent pore diameters were deduced from pore area measurements, using the cylindrical pore approximation (Fig. 2). In this case, the pore size ranges from 7 μm to 400 μm, with a mean size of ~80 μm. The pore density is about 21 pores/mm^2.

The other intact samples were dried and equipped with longitudinal and transverse strain gauges. To circumvent the problem of breakage of the strain gauges during loading, a thin layer of fast cured epoxy was first applied to fill up the surface pores. The strain gauges were then mounted on this layer of epoxy, in orthogonal directions, and on the central portion of the samples. During the experiments, jackets were used to isolate the samples from the confining medium. All tests were performed at room temperature on dry samples, at a strain rate of 1.25×10^{-5} s^{-1}.

2.2 The Conventional Triaxial Apparatus

A schematic diagram of the conventional triaxial testing apparatus is shown in Figure 3. The pressure vessel is designed for a maximum confining pressure of 200 MPa, a value sufficient to cover common reservoir conditions. During a typical experiment, the sample is positioned in the confining cell between two steel end plugs,

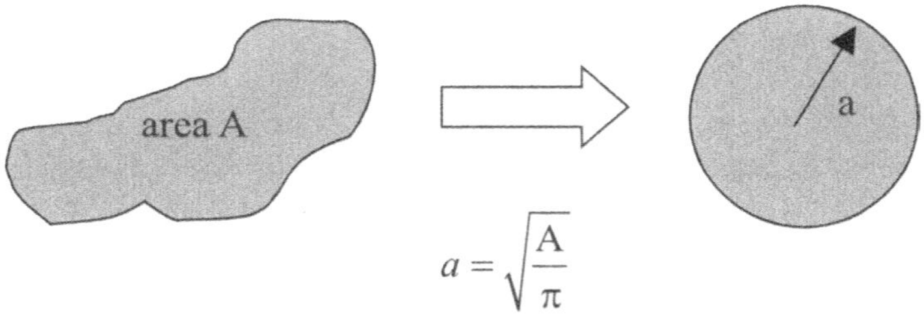

Figure 2
Schematic illustration of the cylindrical pore approximation. The equivalent pore radius a is deduced from pore area measurement.

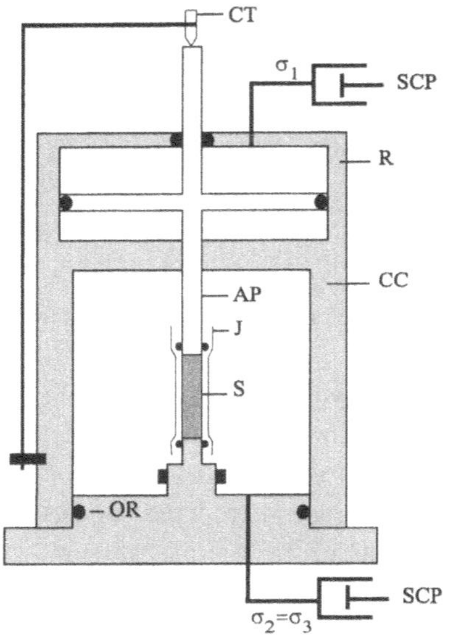

Figure 3
Schematic diagram of the conventional triaxial apparatus: CT = capacity transducer; SCP = servo–control stepping pump; R = ram; CC = confining cell; J = jacket; S = sample; OR = O-ring; AP = axial piston; σ_1 = axial stress; $\sigma_2 = \sigma_3$ = confining stress.

and a jacket is used to isolate it from the confining medium (kerdane). A pneumatic pump is first used to fill the confining cell and to increase progressively the confining pressure. Once the desired pressure is reached, a computer–controlled stepping motor connected to a pressure transducer monitors the confining pressure with a resolution of ± 0.03 MPa. The axial stress is applied by means of a piston, controlled by a second computer–controlled stepping motor, with a resolution of ± 0.15 MPa. The displacement of the piston is measured outside the pressure vessel with a capacity transducer mounted between the moving piston and the confining cell with a resolution of ± 0.2 μm. Calibration tests conducted on a steel sample were performed to estimate the friction and the intrinsic deformation of the apparatus; they allowed us to define the appropriate corrections.

This triaxial press has two appreciable features. First, the program enables us to record the applied stresses and the strains rapidly and accurately, and its graphic interface allows to draw stress-strain curves directly and continuously during an experiment. Second, the press is entirely servo-controlled (the operator just enters the parameters of the experiment) and the interactivity of the control/acquisition program gives enough flexibility to implement complex stress paths. Small stress excursions can easily be performed to investigate the mechanical behaviour of rocks and four different types of experiment can be realised (hydrostatic tests, triaxial tests at fixed strain rate, tests at constant mean stress with fixed strain rate, and tests at fixed axial stress and confining pressure).

2.3 Experimental Data

In the following, we adopt the convention that compressive stresses and compactive strains are positive. The maximum and minimum (compressive) principal stresses are denoted by σ_1 and σ_3, respectively. The differential stress is denoted by $Q = \sigma_1 - \sigma_3$, and the mean stress by $P = (\sigma_1 + 2\sigma_3)/3$. Volumetric strain was calculated from the strain gauge data by summing the axial strain and twice the transverse strain.

Under triaxial loading, the failure mode of Bentheim sandstone is characterized by a typical Mohr brittle deformation regime at confining pressures up to 35 MPa (Fig. 4). The differential stress attains a peak beyond which strain softening occurs, accompanied by a rapid stress drop to a residual level (Fig. 4a). The peak stress increases with confining pressure, which is typical of brittle failure of Mohr type (PATERSON, 1978). The volume initially decreases, probably due to crack closure and elastic grain contact deformation, however near the peak stress it reverses, indicating the inception of dilatancy, denoted here by C' (Fig. 4b). Visual inspection of these samples confirmed that each had failed by shear localization along an inclined macroscopic shear band cutting across the sample (Fig. 5). For all the samples, we determined the peak stress and C' (Table 1), the stress at the onset of dilatancy which corresponds to the departure from linearity of the mean stress — volumetric strain curve (WONG *et al.*, 1997).

3. Fracture Mechanics in the Brittle Regime: The Pore Crack Model

It is now established that the mechanical behaviour of rocks is closely related to the presence of heterogeneities such as pores, inclusions or small cracks. The nucleation process and microcrack growth are dominant mechanisms in the failure of brittle rock under compressive load (e.g., KRANTZ, 1979; COSTIN, 1983; AUBERTIN

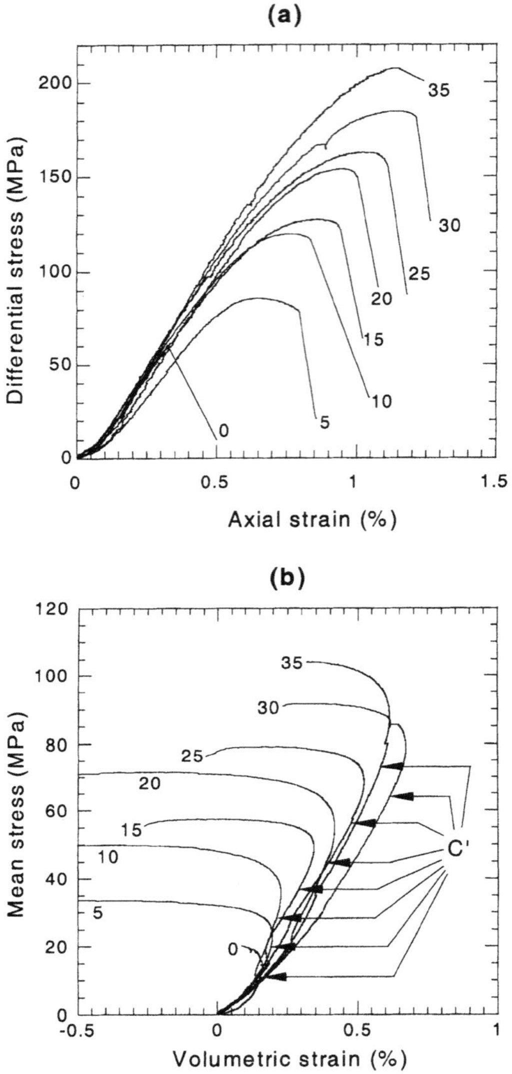

Figure 4
(a) Differential stress versus axial strain; and (b) mean stress $(\sigma_1 + 2\sigma_3)/3$ versus volumetric strain for Bentheim sandstone under triaxial tests at low confining pressures (labelled in MPa near the end of each curve). C' corresponds to the stress at the onset of dilatancy.

and SIMON, 1997). Microcracking begins at the onset of dilatancy C', and the number and length of microcracks increase as the applied stress is increased further, until the cracks link to form one or several macroscopic fractures.

Several attempts have been made to model the mechanics of compressive brittle fracture. A model widely used is the Wing Crack Model (e.g., ASHBY and HALLAM,

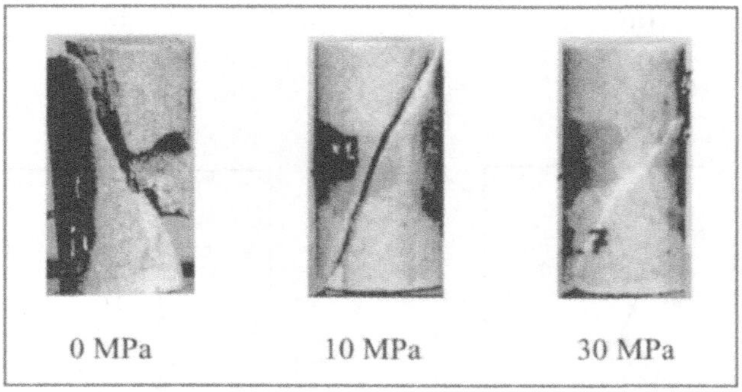

Figure 5
Picture of the samples deformed at 0, 10 and 30 MPa confining pressure.

Table 1

Compilation of mechanical data for Bentheim sandstone in the brittle regime

Confining Pressure σ_3 (MPa)	Onset of Dilatancy, C′ σ_1 (MPa)	Peak Stress σ_1 (MPa)
0	32	60.6
5	45	90.6
10	60	125.5
	52	129.6
15	79	142.4
20	95	174.1
25	118	187.9
30	130	214.8
	132	215.6
35	150	242.5

1986; HORII and NEMAT-NASSER, 1986; COSTIN, 1983), which considers the growth of wing cracks that initiate at the tips of pre-existing sliding cracks and propagate towards the direction of axial stress (Fig. 6a). This model is useful in the case of low-porosity materials such as crystalline rocks and metals, but remains inappropriate in the case of porous granular rocks in which sharp cracks are rare: small cracks initiate at pores or small inclusions. SAMMIS and ASHBY (1986) presented a model that is more consistent with the microstructure of porous materials. They analysed crack growth and crack-crack interaction from cylindrical holes in two dimensions and spherical holes in three dimensions, by means of analytical results of fracture mechanics and beam theory. Nevertheless, their approach is not fully satisfactory, as they did not consider the case of biaxial compression. They studied the case of uniaxial compression, and the case of a constant σ_3/σ_1 ratio. However, during a typical biaxial compression test, this ratio changes continuously as the applied stress σ_1 increases. In the following, we consider the same two-dimensional geometry as the one proposed by SAMMIS and ASHBY (1986), which is the growth of cracks from cylindrical holes (Fig. 6b). However our approach

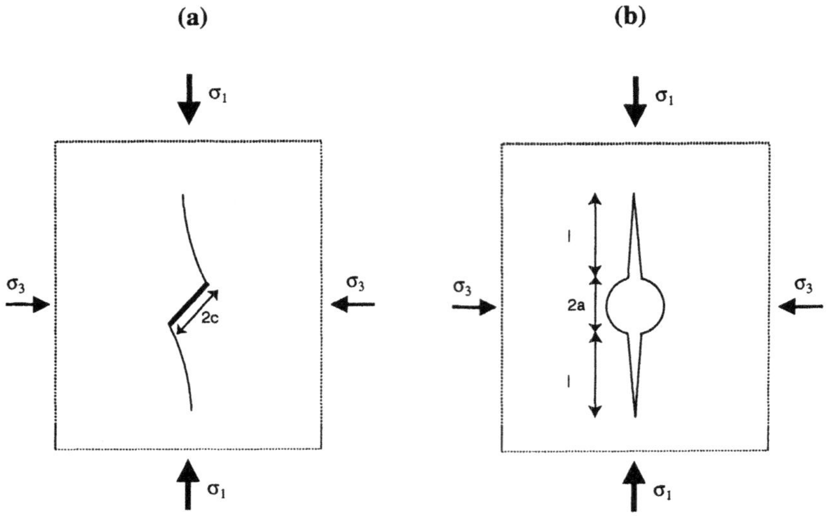

Figure 6
Crack geometry (a) for the Wing Crack Model; (b) for the Pore Crack Model.

differs from the one proposed by these authors in two aspects. Firstly, we do not base our analysis of crack-crack interaction on beam theory but on an iterative procedure previously used by SEGALL and POLLARD (1980) and BAUD and REUSCHLÉ (1997), and secondly, we include time-dependent crack growth, a phenomenon also known as subcritical crack growth (COSTIN, 1983; ATKINSON, 1984). As a matter of fact, we consider that due to the sufficiently low experimental strain rate, significant time-dependent crack extension can occur at values of stress intensity factor lower than the fracture toughness of the material.

3.1 Crack Growth from an Isolated Pore

Before modelling crack-crack interaction we first look at the crack growth from an isolated cylindrical pore in an infinite plate subjected to a biaxial state of stress (Fig. 7a). JAEGER and COOK (1979) calculated the components of the stress field around a hole of radius a. In the coordinate system defined by:

$$\begin{cases} X = x/a - 1 \\ Z = z/a \end{cases} \quad . \tag{1}$$

The normal stress along the axial plane aligned with the x axis is given by:

$$\sigma_{ZZ}(X, 0) = \sigma_1 \left[\lambda + \frac{(1+\lambda)}{2} \left(\frac{1}{1+X} \right)^2 - 3 \frac{(1-\lambda)}{2} \left(\frac{1}{1+X} \right)^4 \right] \tag{2}$$

where $\lambda = \sigma_3/\sigma_1$.

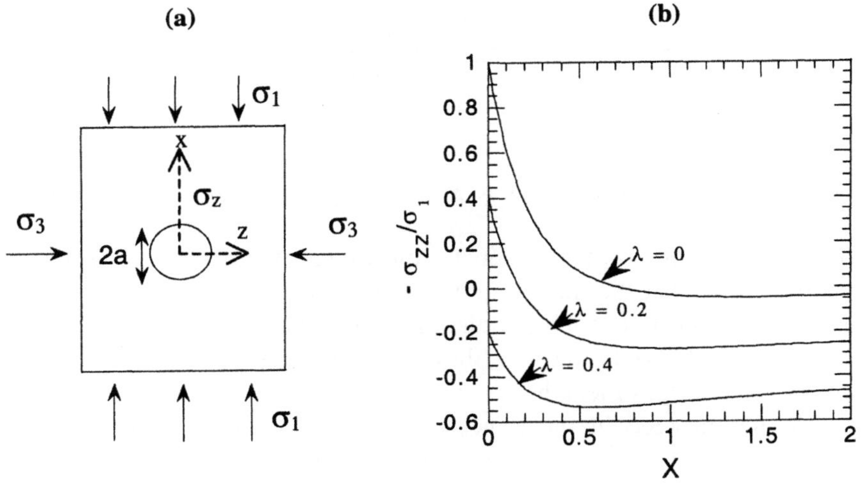

Figure 7
(a) Pore crack geometry and coordinate system; (b) Stress $\sigma_{ZZ}(X, 0)$ normalised by axial stress σ_1 versus the distance X to the hole.

Under uniaxial compression, the stresses are concentrated by the hole so that the stress is tensile and equal in magnitude to $|\sigma_1|$ at the top and bottom of the hole. This tensile stress rapidly decreases with the distance to the hole and becomes compressive before vanishing at infinity. Under biaxial compression, the local stress field is less tensile and above a threshold ratio σ_3/σ_1 no tensile stress appears along the x axis (Fig. 7b).

A pair of cracks of length l emanating from the top and bottom of the hole is subjected to mode I loading. A stress intensity factor K_I^{iso} is associated with each crack. This factor K_I^{iso} can be calculated using standard methods involving an integral of the local stress field over the surface of the crack (SIH and LIEBOWITZ, 1968). According to SHAH (1976), when the radius of the hole a is small compared to the crack length l, the stress intensity factor can be approximated by:

$$K_I^{iso} = \frac{\sqrt{a}}{\sqrt{\pi L}} \int_{-L}^{L} \sigma_{ZZ}(X,0) \left[\frac{L+X}{L-X}\right]^{1/2} dX, \tag{3}$$

where $\sigma_{ZZ}(X, 0)$ is given by equation (2), and $L = l/a$.

Attention should be paid to the case of small crack length, since this approximation becomes inaccurate for $L \ll 1$. In this case a better approximation of the stress intensity factor is obtained by neglecting the curvature of the hole (SIH, 1973):

$$K_{\mathrm{I}}^{\mathrm{iso}} = \frac{2\sqrt{\pi l}}{\pi} \int_{0}^{L} \sigma_{ZZ}(X, 0) \left[\frac{1 + G(X/L)}{(L^2 - X^2)^{1/2}} \right] dX, \tag{4}$$

where

$$G(X/L) = \left[1 - (X/L)^2\right]\left[0.29 - 0.39(X/L)^2 + 0.77(X/L)^4 - 0.99(X/L)^6 + 0.51(X/L)^8\right].$$

To avoid the use of two distinct expressions of $K_{\mathrm{I}}^{\mathrm{iso}}$ for an isolated hole (depending on whether the crack length is short or long when compared to the pore radius) SAMMIS and ASHBY (1986) fitted the exact solution of this problem that was tabulated by SIH (1973). In this way, they found an analytical expression to describe $K_{\mathrm{I}}^{\mathrm{iso}}$ over a wide range of lengths L.

We do not proceed in the same way, and choose to replace the real pore crack (the hole of radius a + the small axial cracks of length l) by an equivalent straight crack. First, we compare the results given by equations (3) and (4) with the exact solution tabulated by SIH (1973). We notice that for small crack length ($L \ll 0.1$) the stress intensity factor approximated by equation (3) is within 10% of the exact solution. For longer crack length, there is a good agreement between equation (3) and the exact solution, whereas equation (4) totally fails (Fig. 8). Taking into account the fact that equation (3) is based on the approximation that the pore crack of total normalised length $2(L + 1)$ is replaced by a straight crack of length $2L$, we introduce an equivalent straight crack of length:

$$2c = 2 * (l + \alpha a). \tag{5}$$

The stress intensity factor of this crack, equivalent to the real crack, is given by:

$$K_{\mathrm{I}}^{\mathrm{iso}} = \frac{\sqrt{a}}{\sqrt{\pi(L + \alpha)}} \int_{-L-\alpha}^{L+\alpha} \sigma_{ZZ}(X, 0) \left[\frac{L + \alpha + X}{L + \alpha - X} \right]^{1/2} dX. \tag{6}$$

The best agreement with the exact solution tabulated by SIH (1973), for small and long crack length, is obtained for parameter $\alpha = 0.0107$. Again, the low value of parameter α indicates that the effect on the stress intensity factor of the pore itself is negligible compared to that of the axial cracks emanating from it.

3.2 Interaction between Neighbouring Pore Cracks

When the pore crack density increases, the interaction between cracks can no longer be neglected. We consider in a first step the interaction between two neighbouring pore cracks. The total stress intensity factor K_{I} at the crack tip is now given by the sum of two terms: the component $K_{\mathrm{I}}^{\mathrm{iso}}$ for an isolated crack and an interaction term $K_{\mathrm{I}}^{\mathrm{int}}$:

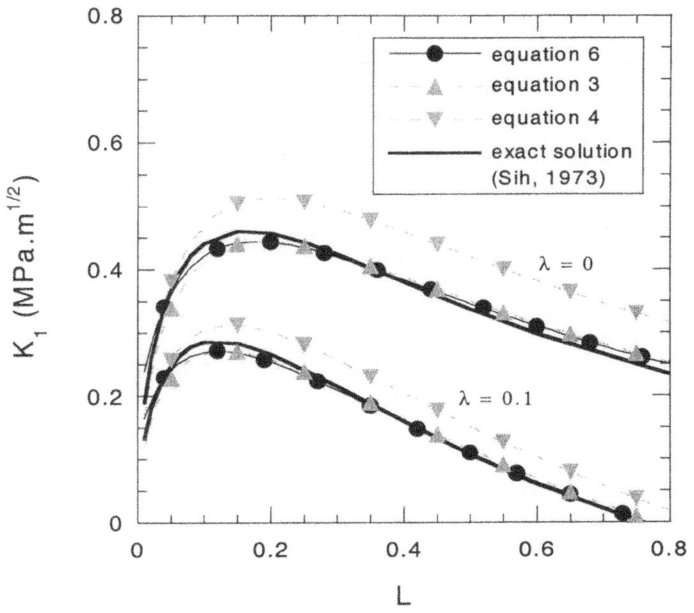

Figure 8
Stress intensity factor K_I for a crack of length l emanating from a pore of radius a ($L = l/a$). Comparison between equations (3), (4), (6) and the exact solution by SIH (1973). Parameter $\alpha = 0.011$.

$$K_I = K_I^{iso} + K_I^{int} \tag{7}$$

where K_I^{iso} is given by equation (6).

The great advantage of having introduced a straight crack equivalent to the real pore crack is that it will enable us to calculate the interaction component thanks to the method of successive approximations (SOKOLNIKOFF, 1956; MUSKHELISHVILI, 1977). This iterative method, which is based on the rule that at each iteration step the stress boundary conditions are met on all crack surfaces in the material, allows calculation of the complete stress field in the solid. A schematic diagram describing the procedure is proposed in Figure 9 (see BAUD and REUSCHLÉ (1997) for full details). At step (a), the pore crack 1 is isolated in an infinite medium subjected to a biaxial state of stress (σ_1, σ_3). The effect of the external stress field on the pore crack 1 creates a perturbation of the stress field $\Delta_1\sigma_{ww}$ on the pore crack 2. To ensure that the boundary conditions are fulfilled on pore crack 2, a correcting stress field $-\Delta_1\sigma_{ww}$ must be applied on it (step (b)). This correcting stress field leads to a perturbation $\Delta_1\sigma_{zz}$ of the stress field acting on the pore crack 1. Again, fulfillment of the boundary conditions requires that a correcting stress field $-\Delta_1\sigma_{zz}$ must be applied on pore crack 1 (step (c)), and so on. At each step the correcting stress field decreases until it becomes negligible. The final result of this procedure is the stress field in the solid containing the two interacting pore cracks. The interaction component K_I^{int} of the stress intensity factor for pore crack 1 is

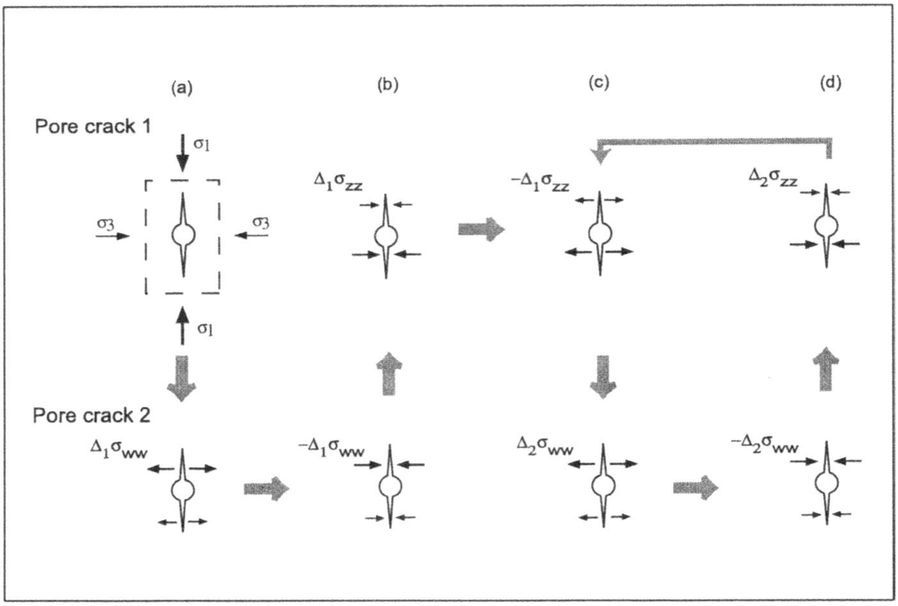

Figure 9
Schematic diagram describing the method of successive approximations.

derived from the sum of all stress perturbation terms $\Delta_i\sigma_{zz}$, and the total stress intensity factor K_I for pore crack 1 is then easily determined from equations (6) and (7). A similar result holds for pore crack 2.

Because the interaction effect creates non-uniform stress distributions on the cracks, we divide each crack into n segments (n $\sim$ 40), and calculate the stress perturbation on each segment. The interaction term of the stress intensity factor is then the cumulative effect of the stress perturbations acting on each individual segment.

The effect of interaction can be illustrated by considering two identical collinear pore cracks in an infinite plate. An axial compression is applied parallel to the pore cracks and the ratio K_I/K_I^{iso} is computed as a function of $b/2c$ (Fig. 10), where b is the centre-to-centre pore spacing and c is the equivalent crack half–length given by equation (5). As long as the spacing is more than twice the equivalent crack length, interaction is negligible. As the pores get closer, the interaction component becomes stronger and the total stress intensity factor increases rapidly.

Finally, once the stress intensity factor is calculated for both pore cracks, we have to introduce a propagation criterion to examine further crack growth. This may be done using Griffith's criterion, that is a crack propagates when $K_I \geq K_{IC}$, where K_{IC} is the fracture toughness of the material. However, previous experiments on rocks and other materials (e.g., ATKINSON, 1984 for a review) have shown that cracks may

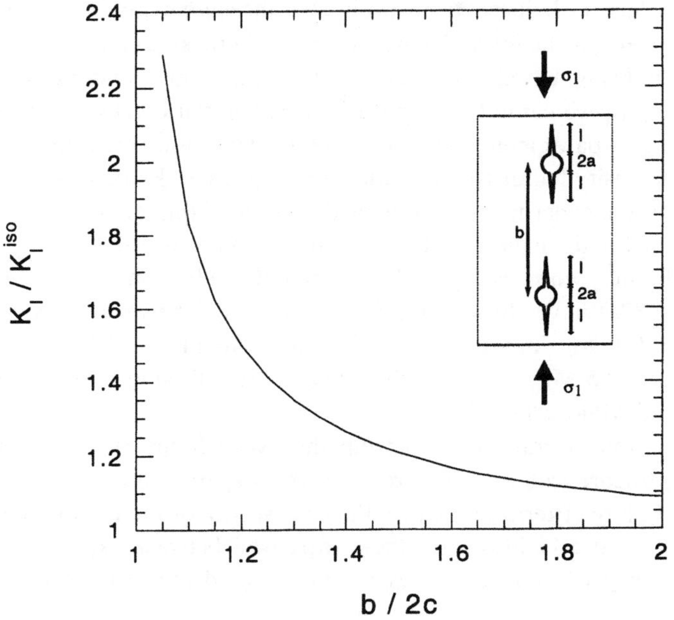

Figure 10
Stress intensity factor K_I for two collinear pore cracks under uniaxial compression as a function of centre-to-centre pore spacing b, where $c = l + a$, a is the pore radius, l is the crack length, and K_I^{iso} is the stress intensity factor for an isolated crack.

propagate under less critical conditions, that is to say for $K_I < K_{IC}$. Theoretical considerations have pointed out that the subcritical crack growth obeys an exponential law of the form (WIEDERHORN *et al.*, 1982):

$$v = v_0 e^{\beta(K_I - K_{I0})}, \tag{8}$$

where v is the crack velocity, K_{I0} is the subcritical crack lower limit, v_0 and β are constant material parameters determined experimentally. Subcritical crack growth occurs at velocities which decrease as K_I is lowered to a threshold value K_{I0}, below which no crack propagation is observed.

4. *Application of the Pore Crack Model to Bentheim Sandstone*

To apply this model to Bentheim sandstone, we simulate crack growth from two similar aligned holes in a material characterized by the same properties as the testing rock. We need different parameters to calculate the stress intensity factors at the crack tips. As seen previously, microstructural parameters are required. The centre-to-centre pore spacing is determined from the mean pore density: $b \sim 220$ μm; the mean pore radius a, deduced from pore area measurements, is ~40 μm. The last required parameter is the initial length of the axial cracks emanating from the holes. Our approach being based on linear fracture mechanics and not on a strength materials analysis, we cannot start without small initial cracks (INGRAFFEA, 1987). Finally, the major parameter that controls crack extension is the threshold value K_{I0} of the stress intensity factor for subcritical crack growth. Estimates for K_{I0} and l are constrained by the experimental data for the onset of dilatancy.

The best fit for the experimental data for the onset of dilatancy is obtained for $K_{I0} = 0.1$ MPa.m$^{1/2}$ and $l = 5$ μm. Experimental studies on rocks (ATKINSON and MEREDITH, 1989a) have shown that K_{I0} generally lies between $0.2K_{IC}$ and $0.5K_{IC}$. Our estimate for K_{I0} implies thus a fracture toughness of Bentheim sandstone between 0.2 and 0.5 MPa.m$^{1/2}$, which is consistent with values for other sandstones (ATKINSON and MEREDITH, 1989b).

The simulation of crack growth from the two cylindrical holes can be used to compare the theoretical critical stress to the experimental peak stress when a macroscopic failure criterion based on the coalescence of the interacting pore cracks is introduced. Figure 11 illustrates the comparison between experimental data (the onset of dilatancy C' and peak stress), and values deduced from our model. Best

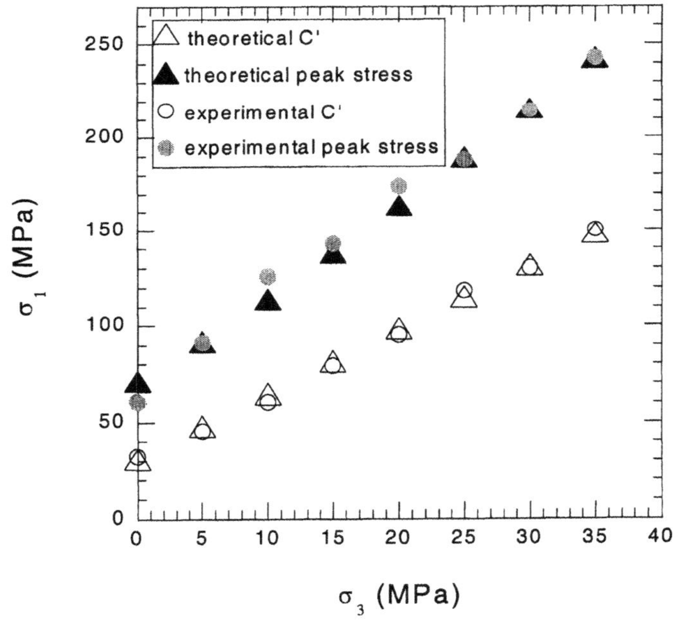

Figure 11
Comparison between the predictions of the pore crack model and experimental results obtained on Bentheim sandstone: full symbols correspond to the peak stress and open symbols correspond to the onset of dilatancy C'.

agreement between theoretical and experimental rupture is achieved when $\beta = 40\ \text{MPa}^{-1}.\text{m}^{-1/2}$ and $v_0 = 10^{-9}\ \text{m.s}^{-1}$. These parameters seem appropriate since they are comparable with experimental values obtained for Tennessee sandstone (ATKINSON, 1984).

Figure 11 shows that the pore crack is not only able to capture the general trend observed experimentally but also gives a quite good agreement between theoretical and experimental rupture. Certainly, this approach remains very simple in the sense that we consider that the coalescence of two pore cracks leads to the failure of the sample. However, such an assumption has been introduced by COSTIN (1983) for example who studied the time-dependent failure of crystalline rocks.

5. Conclusion

In this paper we presented the experimental data of Bentheim sandstone in the brittle regime. This rock is a typical porous granular rock, with porosity of about 23%. Considering that the brittle rupture is governed by the growth of microcracks, we developed a model based on fracture mechanics. It analyses the time-dependent growth of axial cracks from cylindrical holes in two dimensions and takes into account interaction between neighbouring cracks. The simulation of crack growth, associated with a simple failure criterion based on the coalescence of the interacting pore cracks, was used to compare the theoretical critical stress to the experimental peak stress. The results indicated that despite its simplicity, our model is able to capture the general trend observed experimentally and gives a quite good agreement between theoretical and experimental rupture when geometrical parameters deduced from microstructural observations are introduced.

Nevertheless adjustments are necessary. We plan to extend our procedure to account for cracks that are initially oriented at an angle to the direction of maximum principal stress. This evolution may explain the appearance of a shearing band as it occurs in the experiments. In its present form, our approach admits only failure by the development of vertical fractures. Another evolution will be to look at the generalisation to a population of pore cracks. The introduction of a damage parameter may be essential for calculating the evolution of the elastic moduli in order to build theoretical stress-strain curves. Concluding, we also plan to perform additional series of experiments, not only so as to better constrain our model, but also in order to test it on other rocks.

Acknowledgements

The authors thank R.W. Zimmerman and an anonymous reviewer for their helpful comments.

References

ASHBY, M. F. and HALLAM, S. D. (1986), *The Failure of Brittle Solids Containing Small Cracks under Compressive Stress States*, Acta Metall. *34*, 497–510.

ATKINSON, B. K. (1984), *Subcritical Crack Growth in Geological materials*, J. Geophys. Res. *89*, 4077–4114.

ATKINSON, B. K. and MEREDITH, P. G., *The theory of subcritical crack growth with applications to minerals and rocks*. In *Fracture Mechanics of Rocks* (ed. Atkinson B. K.) (Academic Press, London 1989a).

ATKINSON, B. K. and MEREDITH, P. G., *Experimental fracture mechanics data for rocks and minerals*, In *Fracture Mechanics of Rocks* (ed. Atkinson B. K.) (Academic Press, London 1989b) pp. 477–525.

AUBERTIN, M. and SIMON, R. (1997), *A Damage Initiation Criterion for Low Porosity Rocks*, Int. J. Rock Mech. and Min. Sci. *34*, 3–4.

BAUD, P. and REUSCHLÉ, T. (1997), *A Theoretical Approach to the Propagation of Interacting Cracks*, Geophys. J. Int. *130*, 460–469.

COSTIN, L. S. (1983), *A Microcrack Model for the Deformation and Failure of Brittle Rock*, J. Geophys. Res. *88*, 9485–9492.

HORII, H. and NEMAT-NASSER, S. (1986), *Brittle Failure in Compression: Splitting, Faulting and Brittle-ductile Transition*, Phil. Trans. R. Soc. Lond. *319*, 337–374.

INGRAFFEA, A. R., *Theory of crack initiation and propagation in rock*. In *Fracture Mechanics of Rocks* (ed. Atkinson B. K.) (Academic Press, London 1987) pp. 71–110.

JAEGER, J. C. and COOK, N. G. W., *Fundamentals of Rock Mechanics* (Chapman and Hall, London 1979).

KLEIN, E., BAUD, P., REUSCHLÉ, T., and WONG, T.-F. (2001), *Mechanical Behaviour and Failure Mode of Bentheim Sandstone under Triaxial Compression*, Phys. Chem. Earth (A). *26*, 21–25.

KRANTZ, R. L. (1979), *Crack-crack and Crack-pore Interactions in Stressed Granite*, Int. J. Rock Mech. Min. Sci. & Geomech. *Abstr. 16*, 37–47.

LOCKNER, D. A., BYERLEE, J. D., KUKSENKO, V., PONOMAREV, A., and SIDORIN, A., *Observation of quasistatic fault growth from acoustic emissions*. In *Fault Mechanics and Transport Properties of Rocks* (eds. Evans B. and Wong T.-F.) (Academic Press, San Diego 1992) pp. 1–31.

MARTIN, J. C. and SERDENGECTI, S. (1984), *Subsidence over Oil and Gas Fields*, Geol. Soc. Am. Rev. Eng. Geol. *6*, 23–24.

MENÉNDEZ, B., ZHU, W., and WONG, T.-F. (1996), *Micromechanics of Brittle Faulting and Cataclastic Flow in Berea Sandstone*, J. Struct. Geol. *18*, 1–16.

MUSKHELISHVILI, N. I., *Some Basic Problems of the Mathematical Theory of Elasticity* (Nooordhoff, Groningen 1977).

PATERSON, M. S., *Experimental Deformation — The Brittle Field* (Springer-Verlag, New York 1978).

READ, M. D., AYLING, M. R., MEREDITH, P. G., and MURRELL, S. A. F. (1995), *Microcracking During Triaxial Deformation of Porous Rocks Monitored by Changes in Rock Physical Properties. II. Pore Volumometry and Acoustic Emission Measurements on Water-saturated Rocks*, Tectonophys. *245*, 223–226.

SAMMIS, C. G. and ASHBY, M. F. (1986), *The Failure of Brittle Porous Solids under Compressive Stress States*, Acta Metall. *34*, 511–526.

SCHUTJENS, P. M. T. M., HAUSENBLAS, M., DIJKSHOORN, M., and VAN MUNSTER, J. G. (1995), *The Influence of Intergranular Microcracks on the Petrophysical Properties of Sandstone — Experiments to Quantify effects of core damage*, Proc. Int. Symp. Soc. Core Analysts.

SEGALL, P. and POLLARD, D. D. (1980), *Mechanics of Discontinous Faults*, J. Geophys. Res. *85*, 4337–4350.

SHAH, R. C., *Mechanics of crack growth*. In *Proc. Eighth Natn. Symp. on Fract. Mech.*) (ASTM Spec. Tech., 1976).

SIH, G. C., *Handbook of Stress Intensity Factors for Researchers and Engineers* (Lehigh Univ., Bethlehem, PA 1973).

SIH, G. C. and LIEBOWITZ, H., *Mathematical theories of brittle fracture*, In *Fracture, an Advanced treatise*. (ed. Liebowitz H.) (Academic Press, London 1968).

SOKOLNIKOFF, I. S., *Mathematical theory of elasticity* (McGraw-Hill, New York 1956).

TEUFEL, L. W., RHETT, D. W., and FARRELL, H. E. (1991), *Effect of Reservoir Depletion and Pore Pressure Drawdown on in situ Stress and Deformation in the Ekofish Field, North Sea.*, Proc. U.S. Rock Mech. Symp. *32*, 63–72.

UNDERWOOD, E. E., *Quantitative Stereology* (Addison-Wesley, Reading, 1970).

VEEKEN, C. A. M., WALTERS, J. V., KENTER, C. J., and DAVIS, D. R., *Use of plasticity models for predicting borehole stability*. In *Rocks at Great Depth* (eds. Maury V. and Fourmaintaux D.) (Balkema, Rotterdam 1989) pp. 835–844.

VOS, M. W. (1990). *Selection of Outcrop Samples for Acoustic Measurements on Reservoir rock*, Ph.D. Thesis, Delft University of Technology.

WIEDERHORN, S. M., FREIMAN, S. W., FULLER Jr, E. R., and SIMMONS, C. J. (1982), *Effects of Water and other Dielectrics on Crack Growth*, J. Mater. Sci. *17*, 3460–3478.

WONG, T.-F., DAVID, C., and ZHU, W. (1997), *The Transition from Brittle Faulting to Cataclastic Flow in Porous Sandstones: Mechanical Deformation*, J. Geophys. Res. *102*, 3009–3025.

WONG, T.-F., SZETO, H., and ZHANG, J. (1992), *Effect of Loading Path and Porosity on the Failure Mode of Porous Rocks*, Appl. Mech. Rev. *45*, 281–293.

(Received January 31, 2001, revised July 4, 2001, accepted August 1, 2001)

Pure appl. geophys. 160 (2003) 851–868
0033–4553/03/060851–18

Pure and Applied Geophysics

Failure Mode and Spatial Distribution of Damage in Rothbach Sandstone in the Brittle-ductile Transition

PIERRE BÉSUELLE,[1] PATRICK BAUD,[2] and TENG-FONG WONG[3]

Abstract—To elucidate the spatial complexity of damage and evolution of localized failure in the transitional regime from brittle faulting to cataclastic ductile flow in a porous sandstone, we performed a series of triaxial compression experiments on Rothbach sandstone (20% porosity). Quantitative microstructural analysis and X-ray computed tomography (CT) imaging were conducted on deformed samples. Localized failure was observed in samples at effective pressures ranging from 5 MPa to 130 MPa. In the brittle faulting regime, dilating shear bands were observed. The CT images and stereological measurements reveal the geometric complexity and spatial heterogeneity of damage in the failed samples. In the transitional regime (at effective pressures between 45 MPa and 130 MPa), compacting shear bands at high angles and compaction bands perpendicular to the maximum compression direction were observed. The laboratory results suggest that these complex localized features can be pervasive in sandstone formations, not just limited to the very porous aeolian sandstone in which they were first documented. The microstructural observations are in qualitative agreement with theoretical predictions of bifurcation analyses, except for the occurrence of compaction bands in the sample deformed at effective pressure of 130 MPa. The bifurcation analysis with the constitutive model used in this paper is nonadequate to predict compaction band formation, may be due to the neglect of bedding anisotropy of the rock and multiple yield mechanisms in the constitutive model.

Key words: Damage, sandstone, brittle-ductile transition, microscopy, X-ray computed tomography, bifurcation theory.

1. Introduction

In porous sandstones, a transition in failure mode from brittle faulting to cataclastic flow occurs as effective pressure increases. Under a relatively low pressure the differential stress attains a peak before it undergoes strain softening; the failure of sample occurs by shear localization. On the other hand, under high confinement the

[1] Ecole Normale Supérieure, Laboratoire de Géologie, Paris, France. Now, Laboratoire Sols, Solides, Structures (CNRS, UJF, INPG), BP53, 38041 Grenoble Cedex 9, France. E-mail: Pierre.Besuelle@inpg.fr

[2] E.O.S.T., Laboratoire de Physique des Roches, 5 rue René Descartes, 67084 Strasbourg Cedex, France.

[3] State University of New York at Stony Brook, Department of Geosciences, Stony Brook, NY 11794-2100, U.S.A.

sample strain hardens and fails by delocalized cataclastic flow (PATERSON, 1978; WONG *et al.*, 1997).

Some intriguing field observations and new theoretical analysis have focused on the mechanical behavior in the *transitional* regime: historically the failure mode in this regime is described as a mosaic of conjugate shear bands and yet recent field and experimental observations indicate the development of discrete compaction bands orthogonal to the maximum compressive stress σ_1 (MOLLEMA and ANTONELLINI, 1996; OLSSON, 1999). This agrees with recent bifurcation analyses (ISSEN and RUDNICKI, 2000) which predict a continuous transition of the localization mode from extension bands (parallel to σ_1) to compaction bands as a function of the constitutive parameters that characterize porosity change and frictional dependence of plastic yield. In the transitional regime between these two failure modes, localization develops by shear bands (inclined with respect to σ_1 at angles ranging from 0° to 90°) which may be dilating or compacting (BÉSUELLE, 2001).

Since they can act as barriers to fluid transport and influence the stress field and strain partitioning in sedimentary formations, it is important to gain a fundamental understanding of the extent of and conditions under which these localized compactant structures would occur. Preliminary microstructural studies (DIGIOVANNI *et al.*, 2000; WONG *et al.*, 2001) have shown that such complex development of localization can occur in sandstones with porosities ranging from 13% to 28%. Acoustic emission locations (OLSSON and HOLCOMB, 2000) have also mapped out the three-dimensional complexity of the localization development in the Castlegate sandstone (with 28% porosity). In this study we systematically conducted mechanical tests on the Rothbach sandstone (with a porosity of 20%), and characterized the spatial heterogeneity in porosity change and damage using X-ray computed tomography (CT) imaging and stereological measurements on optical micrographs, respectively. These two types of complementary measurements provide useful insights into the geometric complexity and micromechanics associated with the development of compaction and shear bands in the brittle-ductile transition.

2. *Experimental Procedure*

Rothbach sandstone from the Vosges mountains in eastern France was used is this study. Our samples came from the same block studied by DAVID *et al.* (1994) and WONG *et al.* (1997), who have provided the petrophysical description. Cylindrical samples (of diameter 18.1 mm and length 38.1 mm) were cored perpendicular to the sedimentary bedding. The arithmetic mean of connected porosities (measured on each sample) was 20.3%.

The samples were saturated with distilled water and deformed in conventional triaxial conditions following procedures described in detail by BAUD *et al.* (2000). The samples were deformed at a nominal strain rate of 1.3×10^{-5} s^{-1} under fully drained conditions at a fixed pore pressure of 10 MPa. Each end of the specimen was in contact with a stainless steel spacer with a central hole for fluid inlet. To uniformly distribute the pore fluid, several shallow grooves were made on the spacer surface that was touching the sample. The interfaces with the sample ends were lubricated with a mixture of stearic acid and vaseline (LABUZ and BRIDEL, 1993). Axial strain, differential stress, confining pressure, porosity change as well as acoustic emission activity were monitored during the tests. Axial strain was computed from the displacement of the cell piston measured by an external transducer. Adjustment of a pressure generator kept the pore pressure constant, and the pore volume change was recorded by monitoring the piston displacement of the pressure generator with a displacement transducer. The porosity change was calculated from the ratio of the pore volume change to the initial bulk volume of the sample.

Samples at different stages of deformation were unloaded and retrieved from the pressure vessel for X-ray CT imaging conducted at the Institut Français du Pétrole (Rueil-Malmaison, France). The CT-scanner measures the X-ray attenuation coefficient, which increases with increasing bulk density and atomic number (WELLINGTON and VINEGAR, 1987; ANTONELLINI *et al.*, 1994). In a homogeneous sample the attenuation coefficient is proportional to the bulk density. X-ray radiographs taken from different angles provide a cross-sectional image, and a multiplicity of such slices can be combined to visualize the three-dimensional distribution of density in the deformed specimen. The CT apparatus measured the average X-ray attenuation value in a voxel of $0.2 \times 0.2 \times 1.5$ mm^3. CT imaging has been used to characterize the three-dimensional development of strain localization in sandstone (VINEGAR *et al.*, 1991; BÉSUELLE *et al.*, 2000) and granite (RAYNAUD *et al.*, 1989; KAWAKATA *et al.*, 1999).

Petrographic thin sections of selected samples were also prepared. The deformed samples were first impregnated with epoxy and then sawed along a plane parallel to the axial direction. Using optical microscopy the damage states in the thin sections were characterized quantitatively. For each sample, the spatial distribution of crack density over a total area of 16.3×35.4 mm^2 was characterized. The area centrally located in a thin section was further divided into 10×29 subregions, each of which had an area of 1.63×1.22 mm^2. The reflected images were all acquired at a magnification of 100x. Using stereological techniques (MENÉNDEZ *et al.*, 1996; WU *et al.*, 1999), we counted the number of crack intersections with a test array of 5 parallel lines (spaced at 0.33 mm or 0.24 mm apart) in two orthogonal directions parallel and perpendicular to σ_1, respectively. We denote the linear intercept density (number of crack intersections per unit length) for the array oriented parallel to σ_1 by $P_L^{\parallel}$, and that for the perpendicular array by $P_L^{\perp}$.

For each sample, 290 pairs of stereological parameters ($P_L^{\parallel}$ and $P_L^{\perp}$) were measured to map out the spatial evolution of damage and stress-induced anisotropy.

Since the spatial distribution of damage is approximately axisymmetric in a triaxially compressed sample, the crack surface area per unit volume (S_v) is given by

$$S_v = \frac{\pi}{2} P_L^{\perp} + \left(2 - \frac{\pi}{2}\right) P_L^{\|} \tag{1}$$

and the anisotropy of crack distribution (UNDERWOOD, 1970) can be characterized by the parameter

$$\Omega_{23} = \frac{P_L^{\perp} - P_L^{\|}}{P_L^{\perp} + (4/\pi - 1) P_L^{\|}} \tag{2}$$

that represents the ratio between the surface area of cracks parallel to σ_1 and the total crack surface area.

3. Mechanical Data

The compressive stresses and compactive strains will be taken positive. The maximum and minimum (compressive) principal stresses will be denoted by σ_1 and σ_3, respectively. The pore pressure will be denoted by P_p, and the difference between the confining pressure ($P_c = \sigma_2 = \sigma_3$) and pore pressure will be referred to as "effective pressure."

The mechanical behavior and failure mode of porous sandstones have several common attributes (WONG *et al.*, 1997). At low pressures, failure involves dilatancy and faulting, with a failure envelope described by the Coulomb surface, with positive pressure dependence. At high pressures, the samples fail by strain hardening and shear-enhanced compaction. The compactive yield is characterized by an elliptical cap that has negative pressure dependence. The brittle-ductile transition occurs in the regime where these two envelopes meet.

Guided by the mechanical data of WONG *et al.* (1997) on Rothbach sandstone, we performed 15 triaxial experiments at effective pressures ranging from 5 to 130 MPa to capture the evolution of the failure modes in the transitional regime. Representative mechanical data for selected experiments are presented in Figure 1. At effective pressure up to 20 MPa, the mechanical response and failure mode were typical of the brittle faulting regime. The differential stress ($\sigma_1 - \sigma_3$) attained a peak, beyond which strain softening was observed (Fig. 1a). The peak stress showed a positive correlation with the confining pressure and effective mean stress (Fig. 2), and macroscopic shear bands oriented at ~30° with respect to the σ_1 direction were observed in the failed samples. Whereas dilatancy initiated before the stress peak was attained at 5 MPa of effective pressure, continuous compaction was observed beyond about 20 MPa of effective pressure (Fig. 1b).

At effective pressure $P_{\text{eff}} \geq 40$ MPa, visual examination of the failed samples did not discern features that would suggest localized failure. All the samples showed

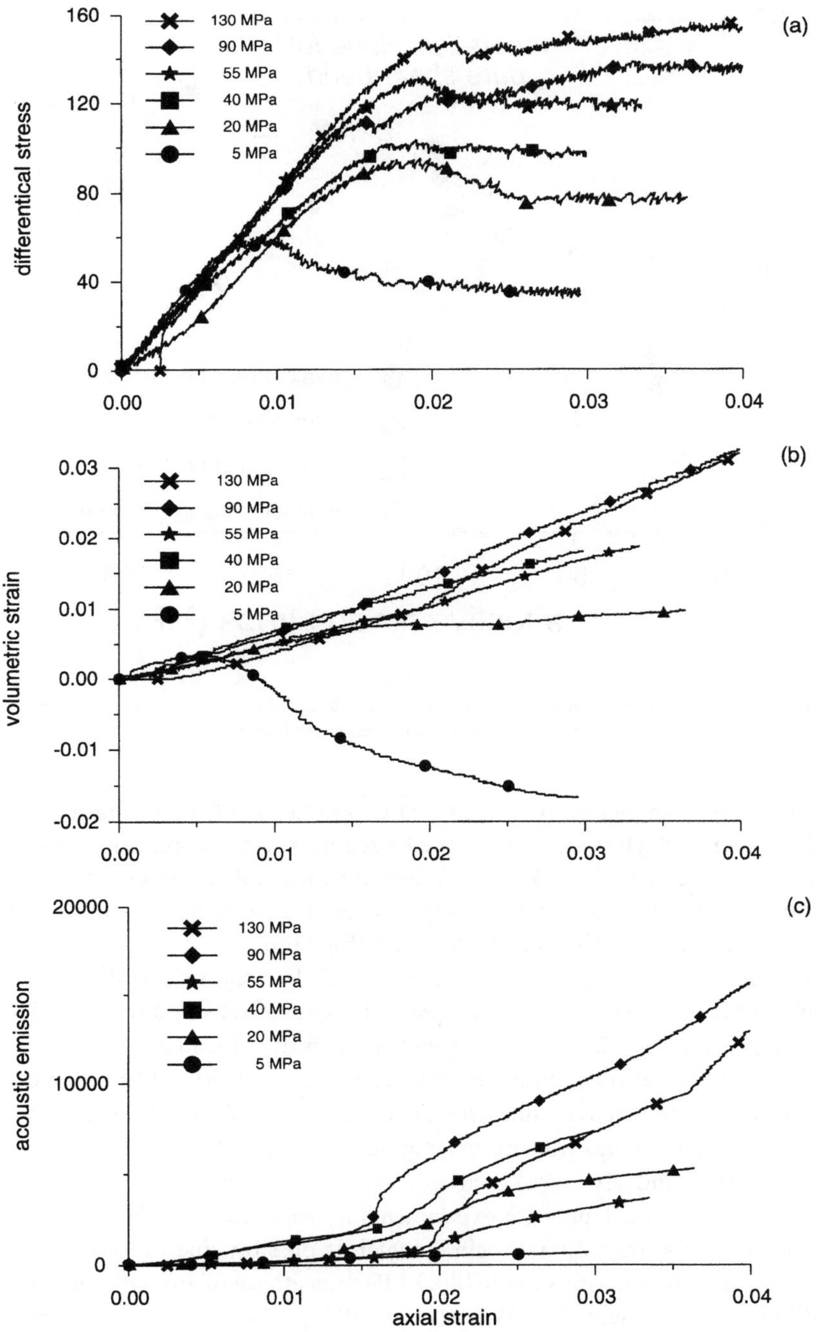

Figure 1
Mechanical data : Differential stress (a), volume strain (b) and acoustic emission (c) versus axial strain for effective pressure between 5 and 130 MPa.

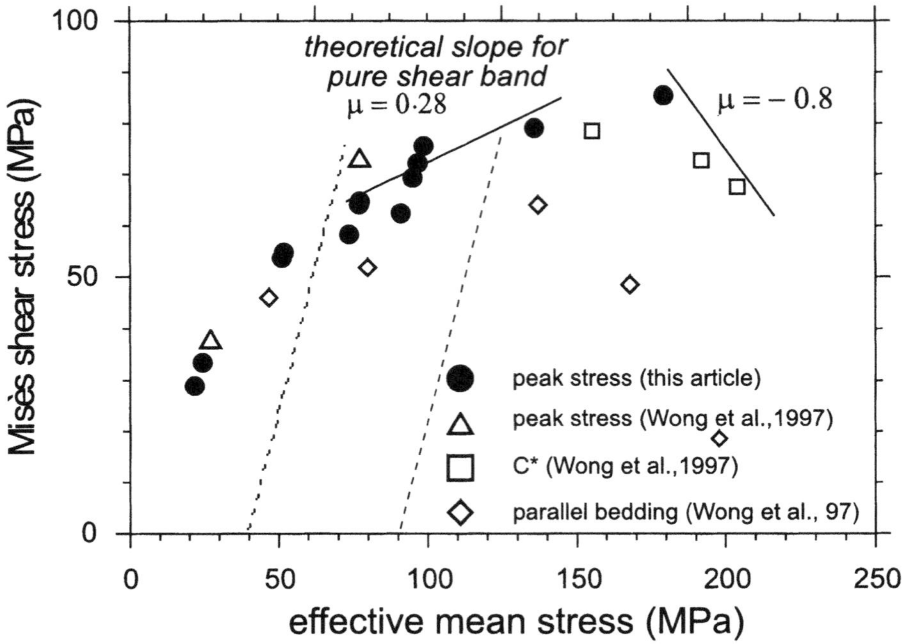

Figure 2
Peak stress and onset of shear enhanced compaction in the Misès shear stress versus effective mean stress plane. μ denotes the friction coefficient.

shear-enhanced compaction (CURRAN and CARROLL, 1979) that initiated at a stress level C^*, beyond which there was an accelerated decrease in volume in comparison to the hydrostat (Fig. 1b). A peak stress followed by a small amount of strain softening was still observed in most of the samples (Fig. 1a). A surge in acoustic emission activity was commonly observed beyond C^* (Fig. 1c).

We compiled the yield stress data in Figure 2. The equivalent Misès shear stress and effective mean stresses for a compressive axisymmetric stress state are defined by $(\sigma_1 - \sigma_3)/\sqrt{3}$ and $(\sigma_1 + 2\sigma_3) - P_p$, respectively. Our data (solid circles) correspond to the peak stresses at effective pressures between 5 and 130 MPa. Data of WONG *et al.* (1997) for the brittle strength and stresses at the onset of shear-enhanced compaction are also included. In the transitional regime (corresponding to effective pressures between 45 MPa and 130 MPa) dilatancy was inhibited even though the differential stress attained a peak. Duplicate experiments indicate appreciable variability in the peak stresses in the transitional regime, which is possibly due to local variations of the bedding direction in our sample block. The sensitivity of strength and yield stress to bedding was documented by WONG *et al.* (1997), who showed that samples cored parallel to bedding (open diamonds in Fig. 2) had significantly lower strengths and yield stresses.

4. Spatial Pattern of Strain Localization—CT Analysis

Altogether twelve deformed samples were prepared for CT imaging. Zones with higher attenuation coefficients are shown in lighter color. For all the specimens analyzed, a number of sub-parallel planar zones with somewhat higher attenuation than the average were observed (Fig. 3). These zones are almost perpendicular to the axial direction and probably associated with the natural sedimentary bedding. In a homogeneous material the brighter zones would correspond to denser materials. However, since the Rothbach sandstone (with modal composition of 68% quartz, 16% feldspar, ~12% clay, and 3% oxides and mica) is by no means monomineralic, the localized anomalies in attenuation may also arise from mineralogical heterogeneity.

At effective pressures of 5 and 20 MPa, the development of dilating shear bands was manifested as inclined planar zones of darker color in the CT images (Figs. 3 and 4). The geometric complexity of these shear bands is illustrated by the images of four circular slices perpendicular to the sample axis (Fig. 4). In each slice several dilatant zones were observed to radiate from an approximately circular zone. The diameter of the concentric zone and spacing between the radial zones both vary with the axial position of the slice.

Although we had expected to observe localized compactant features in samples deformed at effective pressure higher than 40 MPa, our CT images revealed no obvious signs of strain localization. In some specimens a cone of slightly higher density seems to have developed near the ends of the specimens (Fig. 5). Given the low resolution, it is difficult to unambiguously evaluate to what extent these represent strain localization features.

At 130 MPa effective pressure, if present, these localized zones are expected to be at very high angles to the sample axis and almost parallel to the bedding.

Figure 3

X-ray CT measurement of the whole sample at 20 MPa effective pressure. The imaged plane is parallel to the specimen axis. Shear band are inclined and dilating (darker than the average grey level) and sedimentary bedding is close to perpendicular of the specimen axis.

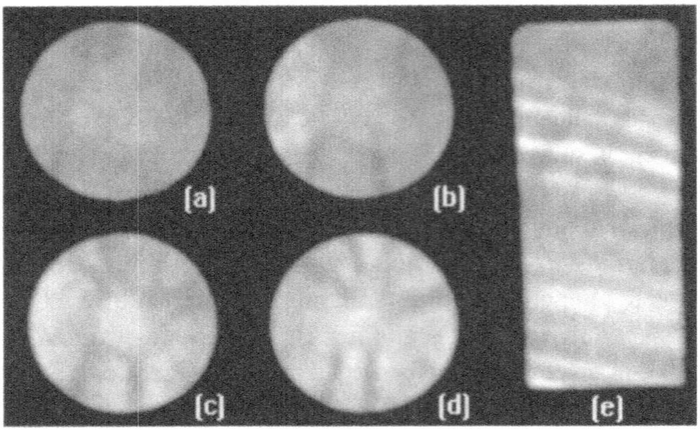

Figure 4
X-ray CT measurements of the whole sample at 5 MPa effective pressure : (a–d) in planes perpendicular to the specimen axis, the size and the spacing out of localisation traces (a circular zone and several radial bands) depend on the position in the specimen; (e) in a vertical mid-section of the specimen, the previous circular zone corresponds to a central V shape and the radial traces to several inclined bands.

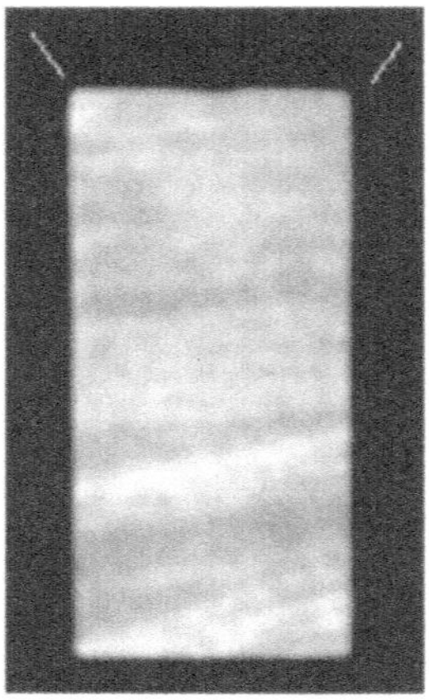

Figure 5
X-ray CT measurements of the whole sample at 55 MPa effective pressure. A cone of compactive strain (slightly brighter than the average grey level) is observed at the top of the specimen.

Consequently it is difficult to differentiate such localized features in a failed sample from the pre-existing bedding heterogeneities.

5. *Spatial Evolution of Anisotropic Damage—Microstructural Observation*

Microstructural observations were conducted to characterize the spatial distribution of cracking and damage localization in the deformed samples. For reference, we first studied a sample deformed in the brittle faulting regime at $P_{eff} = 20$ MPa : sample B1 was deformed to the post-failure stage (to about 3% of axial strain), where the differential stress reached a residual level (Fig. 1a). Shear localization developed from a corner of the sample along a planar zone inclined at an angle of ~40° to σ_1 (Figs. 3 and 6). The width of the shear band varied by a factor of ~3, with a minimum of ~ 0.5 mm (which is comparable to the average grain diameter of 0.6 mm). Its central part was subject to intense grain crushing. In the periphery of the band extensive inter- and intra-granular cracking occurred. Crack density decreased quickly and transversely to the band. Beyond a distance of 4 or 5 grains, damage became very slight. A few intra-granular cracks preferentially aligned in an approximately axial direction were observed in grains neighboring the shear zone (Fig. 7). Damage anisotropy in Rothbach sandstone is qualitatively similar to what was observed in Berea sandstone (21% porosity) by MENÉNDEZ *et al.* (1996).

Stereological measurements were performed on four samples deformed in the transitional regime at effective pressures 40, 55, 90 and 130 MPa. Using equation (1) the crack surface area per unit volume (S_v) was inferred from linear intercept measurements along two orthogonal directions. The samples deformed at 40 and 55 MPa effective pressures showed similar spatial distributions of damage (Fig. 8). In

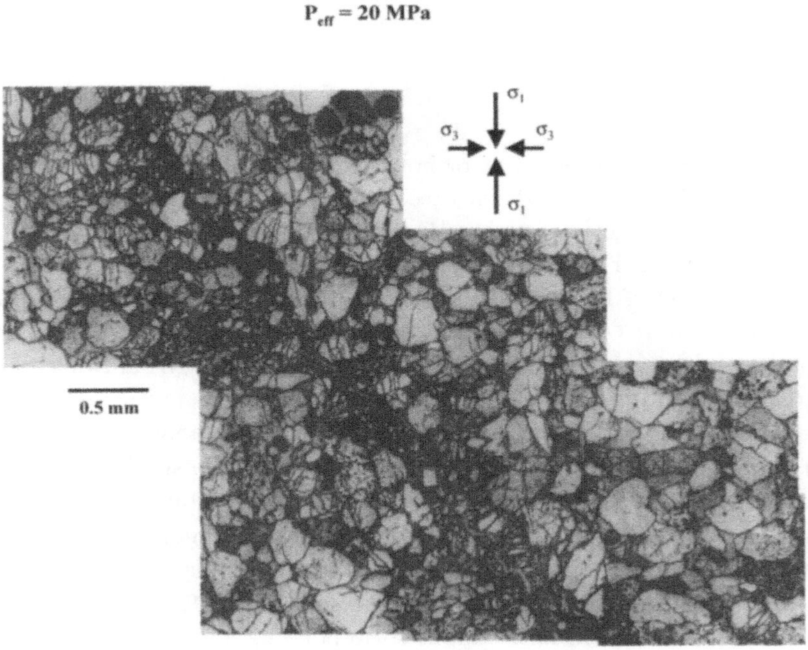

Figure 6
Mosaic of optical (reflexion) micrographs showing part of the shear band that developed in sample B_1 deformed in the brittle regime.

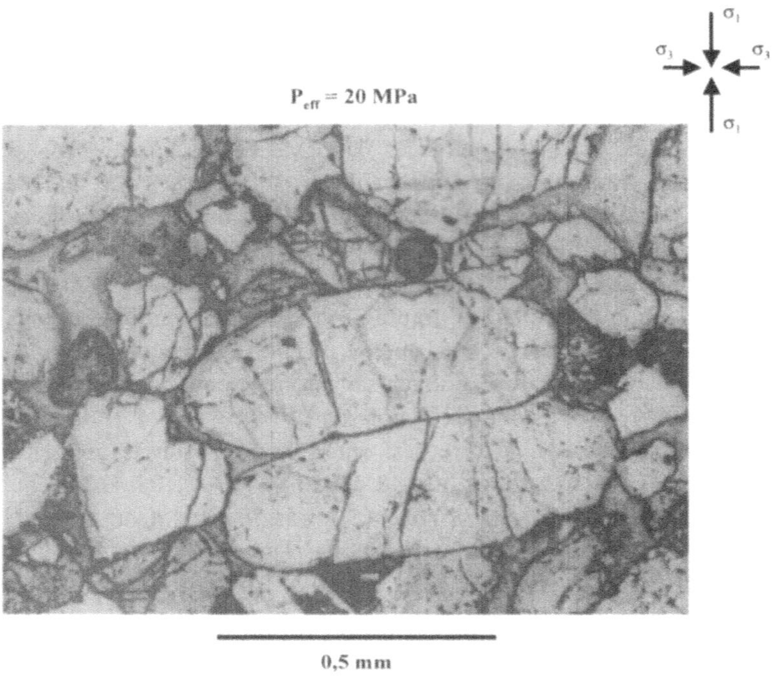

Figure 7
Intragranular cracks outside the shear band in sample B_1 deformed in the brittle regime.

all cases damage affected near one or both ends of the sample, and the cumulative area of the localized zones did not exceed 1/3 of the total surface of the sample. Within the high damage areas, the crack density is higher than the background by as much as a factor of 5. For example, we found in sample T_2 an average of 4.3 m^{-1} for S_v and a maximum of 14.6 m^{-1}. These values are in the range found by MENÉNDEZ *et al.* (1996) in shear compacted samples of Berea sandstone. Qualitatively similar distributions of damage were observed in samples T_2 and T_3 from two experiments conducted at effective pressure 55 MPa. High-angle conjugate zones initiated from the sample corners and intersected at a sub-horizontal tabular zone of variable length, where maximum crack density was measured in most cases. In sample T_2 deformed at 55 MPa effective pressure, the length of this tubular zone was ~4 mm. The thin section T_3 was taken from the sample shown in Figure 5. Locations of the high-angle conjugate zones in the thin section roughly correspond to those where an axi-symmetric compacting cone was vaguely resolved by CT.

In Figure 9, we present micrographs corresponding to different areas of sample T_3 (deformed to 3% of axial strain). Intensive grain crushing occured in the upper part of the sample (Figure 9a). Numerous "Hertzian fractures" radiating from grain contacts (MENÉNDEZ *et al.*, 1996; WU *et al.*, 1999) were observed. The damage

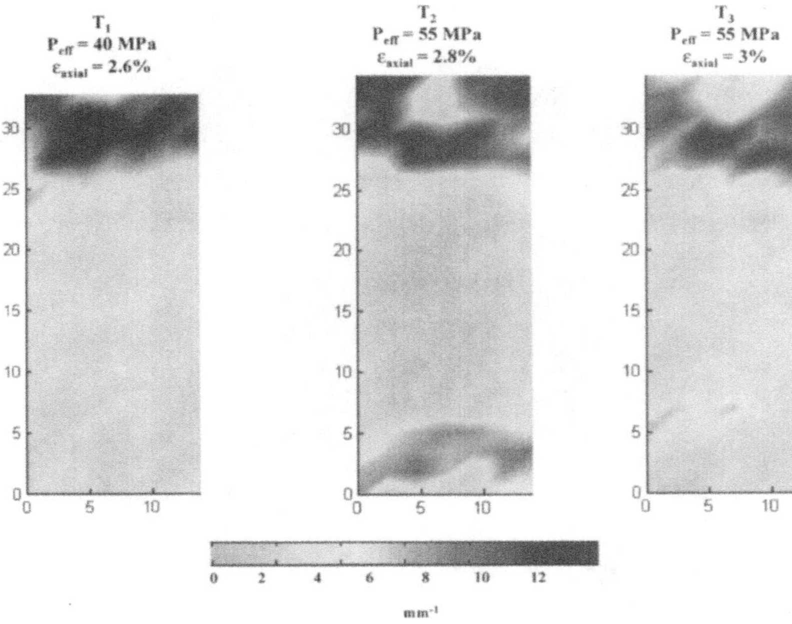

Figure 8
Spatial distribution of specific surface area in sample T_1 deformed at $P_{\text{eff}} = 40$ MPa (a), and in samples T_2 (b) and T_3 (c) deformed at $P_{\text{eff}} = 55$ MPa. The σ_1 direction is vertical.

anisotropy (evaluated using equation 2) was negligible in this sample, with $\Omega_{23} = 0.1$ within the localized zone and an average value $\sim$0 in the whole sample. In Figure 9b pore collapse was evident, and part of the pore space was filled with comminuted particles. This suggests an overall decrease of the porosity in the damaged part of the sample. Figures 9c and 9d depict two different regions of the sample that did not have intensive cracking. Minor damage was in fact observed in both areas, and some of the visible cracks were likely to be present prior to deformation. Figures 9c and 9d show the strong heterogeneity in grain size and porosity that is typical for different layers of the bedding in Rothbach sandstone. Bedding layers were found in most of the thin sections, except in the damaged zones where grain crushing "homogenized" the material. It is therefore unclear, at this point, if the bedding heterogeneity promotes or inhibits damage localization.

At 90 MPa effective pressure (Fig. 10), intensive cracking was observed in a larger portion of the sample (about 1/2), at both ends. The highest crack densities ($S_v = 14.6\,\text{mm}^{-1}$) were measured in association with grain crushing and pore collapse (Fig. 10b). In this area we found an anisotropy factor of about 0.1. In other regions, mostly close to the periphery of the damaged area, subvertical cracking was dominant (Fig. 10c). We consistently measured a maximum of $\Omega_{23} = 0.4$ in these zones. In the central part of the sample, significant differences

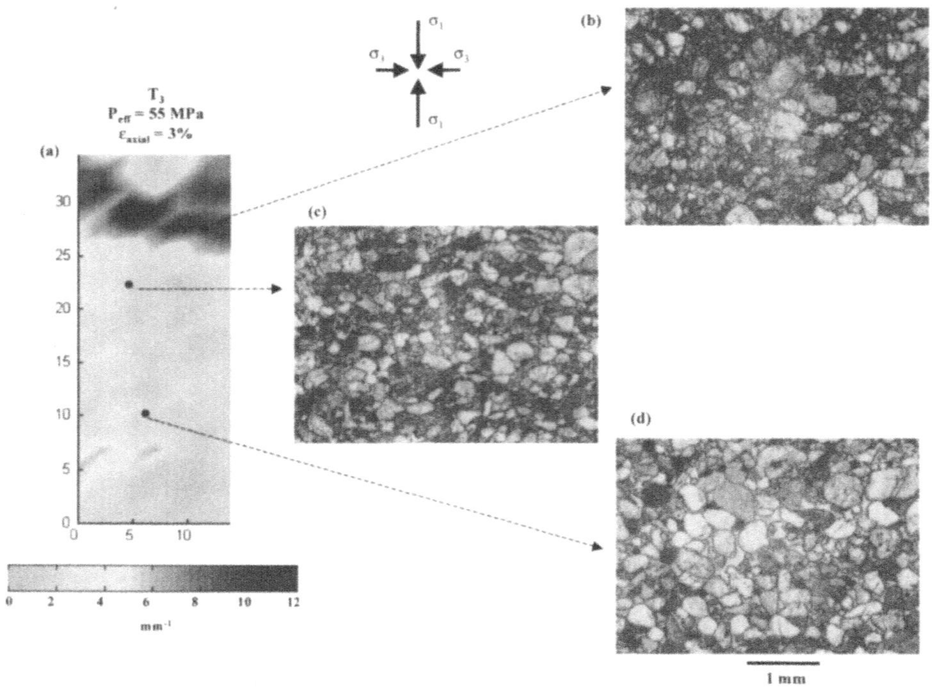

Figure 9
(a) Spatial distribution of specific surface area in sample T_3 deformed at P_{eff} = 55 MPa. (b) Extensive grain crushing where crack density is high. (c) No damage in a compact layer of the sample. (d) No damage in a more porous layer of the sample.

in porosity could be observed (Figs. 10d and 10e). Axial cracking such as observed in Figure 10c appeared to be predominant in the less compact layers of the sample.

At 130 MPa effective pressure, damage was spatially distributed in a clearly different way (Fig. 11). Intensive cracking was observed in about 2/3 of the sample, including the central area. Several localized and elongated clusters were identified (Fig. 11a). In all of them, grain crushing was evident (Fig. 11b). The orientations of these clusters were mainly sub-horizontal (perpendicular to σ_1), and the crack surface area reached a maximum of $S_v = 15.3\,\text{mm}^{-1}$. These localized structures can be interpreted as compaction bands or compacting shear bands at a relatively high angle of ~80° degrees with respect to the axial direction. As they are almost parallel to the bedding, it may have been difficult to resolve them in the CT images. We found a relatively low anisotropy factor of $\Omega_{23} = 0.02$ in these compaction bands. Damage level was appreciably less in grains outside the compaction bands (Fig. 11c) and the overall spatial distribution of damage remained anisotropic even at high effective pressure (Fig. 11d).

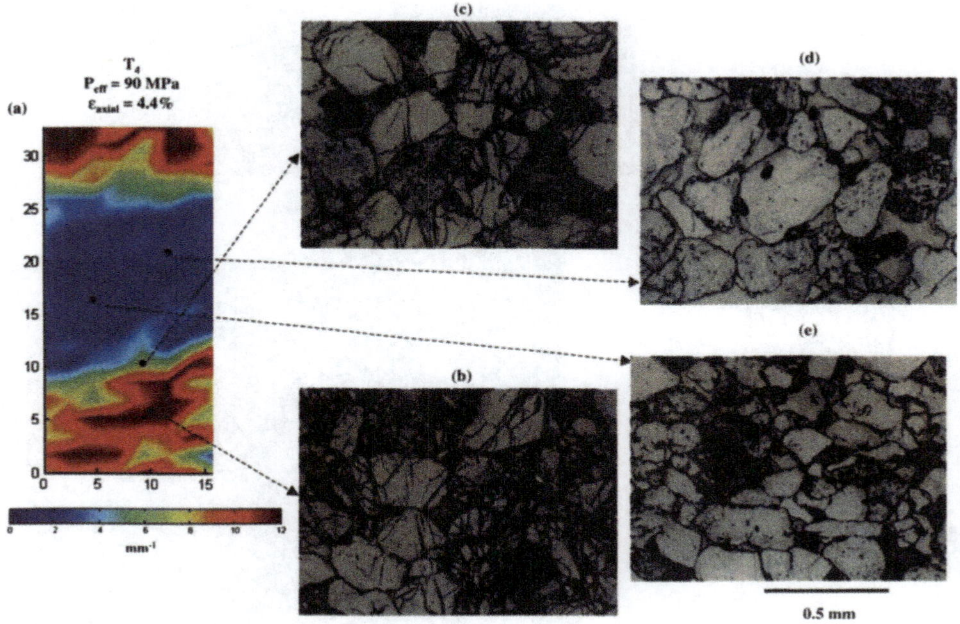

Figure 10
(a) Spatial distribution of specific surface area in sample T_4 deformed at $P_{eff} = 90$ MPa. (b) Extensive grain crushing in a localized cluster of intense damage. (c) Area where axial cracking is dominant. (d) No damage in a compact layer of the sample. (e) No damage in a more porous layer of the sample. The σ_1 direction is vertical.

6. Discussion and Conclusion

6.1. Strain Localization and Failure Mode in the Brittle-Ductile Transition

Our data on the Rothbach sandstone are in qualitative agreement with recent studies by OLSSON (1999) and WONG *et al.* (2001) who have documented the development of high-angle shear bands and compaction bands in sandstones with porosities ranging from 13% to 28%. The localized failure modes are associated with stress states in the transitional regime from brittle faulting to compactive cataclastic flow. These laboratory results suggest that such complex features can be pervasive in sandstone formations, not just in the very porous aeolian sandstone in which they were first documented (MOLLEMA and ANTONELLINI, 1996). Recognition of such structures in the field would elucidate the complex development of localization in sandstone formations (AYDIN and JOHNSON, 1978) and accretionary prisms (BYRNE *et al.*, 1993).

Our microstructural observations and stereological data underscore the spatial heterogeneity of damage associated with the development of high-angle shear and

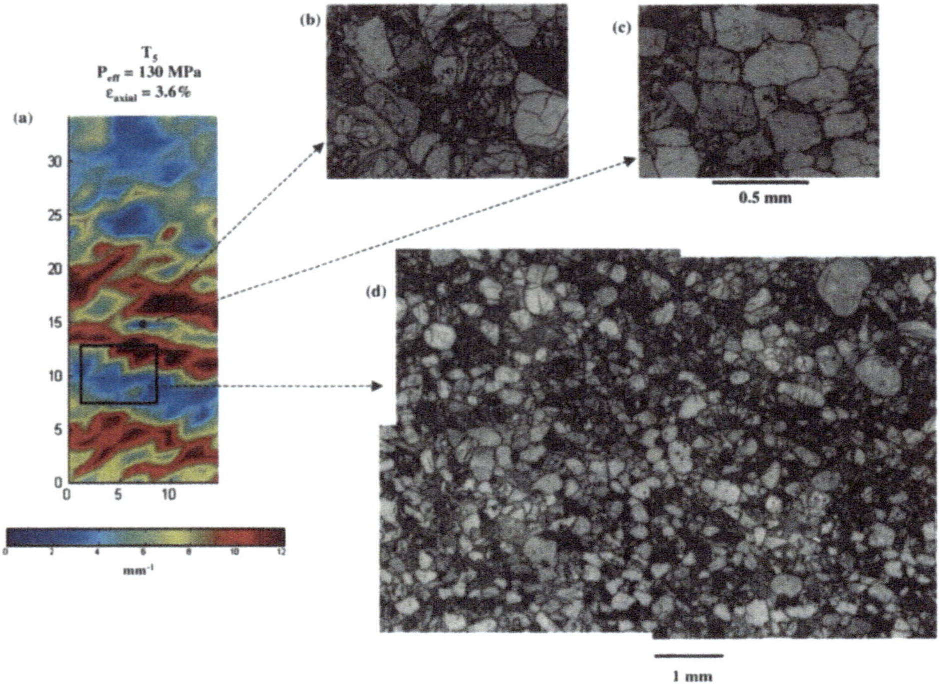

Figure 11
(a) Spatial distribution of specific surface area in sample T_5 deformed at P_{eff} = 130 MPa. (b) Detail of grain crushing in a compaction band. (c) No damage in an area between two compaction bands. (d) Mosaic of micrographs showing the transition between a compaction band and an undamaged zone. The σ_1 direction is vertical.

compaction bands. In the localized bands intensive grain crushing and pore collapse were observed. Similar micromechanical processes were documented by DIGIOVANNI *et al.* (2000) in laboratory deformed Castlegate sandstone and by MOLLEMA and ANTONELLINI (1996) in naturally deformed Navajo sandstone. Previous laboratory data show that bedding in Rothbach sandstone has significant influence on the failure and yield behavior. Our microstructural observations presented here suggest that the development of compaction band is promoted along planar zones parallel to the bedding.

The CT data have provided preliminary images of the three-dimensional geometric complexity associated with strain localization. At 5 MPa effective pressure, a complex pattern of localization was observed (Fig. 4). Axisymmetric loading and frictional constraints at the specimen ends appear to promote the development of such an axisymmetric pattern, and to inhibit the initiation of a single shear band (which is intrinsically a plane-strain mechanism). The pattern we observed is analogous to that in axisymmetrically compressed specimens of Hostun

RF sand, which was studied by DESRUES *et al.* (1996) as a combination of a cone centered on the axis of the specimen and on a set a plane strain mechanism associated with V-shape pairs.

Since compaction bands have an intrinsically axisymmetric geometry, their development should be promoted by axisymmetrical loading if the constitutive parameters attain the critical conditions for inception of such a localization mode. Since the compaction bands that developed at 130 MPa confining pressure were observed in the mid-section of the sample, we believe frictional end-effects to be minimal in our tests. However, the conical shape of the high-angle shear bands at effective pressures between 40 and 55 MPa may have been influenced by stress heterogeneity from frictional constraints, since they initiated in the proximity of the specimen ends.

6.2. Comparison with Localization Analysis

Bifurcation analyses (RUDNICKI and RICE, 1975; OLSSON, 1999; ISSEN and RUDNICKI, 2000; BÉSUELLE, 2001) specify the mechanical conditions under which such localized failure modes may develop. The simplest approach is to adopt a constitutive framework (RUDNICKI and RICE, 1975) whereby the yield envelope and inelastic volumetric change can be characterized by the pressure-sensitivity parameter μ and dilatancy factor β, respectively. The parameter μ can be inferred from the slope of the yield stress data as indicated in Figure 2. Our data together with those of WONG *et al.* (1997) for samples cored perpendicular to the bedding indicate that $-0.8 \leq \mu \leq 0.28$ in the transitional regime. For this range of μ values and Poisson ratio equal to 0.14 (inferred from our laboratory data), the bifurcation analyses predict that localized failure occurs by three possible modes :

1. A *pure shear* band (with negligible volumetric change throughout the sample) may develop for an associated model (i.e., $\mu = \beta$) for $\mu = 0.28$ (BÉSUELLE, 2001).
2. A *compaction* band perpendicular to σ_1 may develop for negative values of μ and β such that $\beta + \mu < -\sqrt{3}$ (ISSEN and RUDNICKI, 2000).
3. A *compacting shear* band at high angle may develop for intermediate values of μ and β such that $\beta + \mu \geq -\sqrt{3}$.

WONG *et al.* (2001) recently showed that the dilatancy factors for Bentheim, Berea and Darley Dale sandstones (deformed in the brittle-ductile transitional regime) fall in the range $-\sqrt{3}/2 < \beta < 0$. They argued that these limiting values arise from the plastic strain field in this regime which is characterized by axial shortening (probably dominated by pore collapse) and lateral expansion (probably induced by axial microcracking). Values of β for Rothbach sandstone in the transitional regime are expected to also fall within these limits, which imply that $\beta + \mu \geq -(\sqrt{3}/2 + 0.8) > -\sqrt{3}$. Accordingly, the bifurcation analysis would rule out the possibility of compaction band formation (mode 2) and predict the development of localization by either mode 1 or mode 3. These two localization modes correspond to the mosaics of high-angle shear bands revealed in our microstructural observations of samples deformed at effective pressures less than 130 MPa (Figs. 8–10). The bifurcation analysis also predicts that such localization modes develop during the strain softening stage (Fig. 1). To explore this further we should conduct observations on samples deformed to different stages to study the progressive development of localization.

Moreover, the bifurcation analysis (BÉSUELLE, 2001) predicts the development of dilating shear bands during the post-peak stage for $\mu > 0.28$ (if $\beta \approx \mu$). Constitutive parameters for Rothbach sandstone deformed at effective pressures of 5 MPa and 20 MPa satisfy these conditions. Our CT and microstructural observations indicate that the localized bands developed as dilating shear bands.

However, there is an apparent discrepancy in that our microstructural observations indicate formation of compaction bands in the sample failed at the effective pressure of 130 MPa (Fig. 11), even though the bifurcation analysis rules out this localization mode (mode 2). The discrepancy may arise from several intrinsic limitations in the constitutive framework of RUDNICKI and RICE (1975) that was adopted in the localization analyses. First, the plastic yield behavior is assumed to be isotropic, even though the mechanical data for Rothbach sandstone (WONG *et al.*, 1997; MILLIEN, 1993) and our microstructrual observations both indicate significant anisotropy due to the sedimentary bedding. RUDNICKI's (1977) theoretical analysis has shown that the onset of localization is sensitive to mechanical anisotropy. Second, the constitutive model does not adequately account for plastic yield and volume change induced by the mean stress (AYDIN and JOHNSON, 1983). ISSEN and RUDNICKI (2000) recently proposed a constitutive model that incorporates multiple yield surfaces and damage mechanisms, with predictions on compaction band formation that seem to be in better agreement with laboratory and microstructural observations (WONG *et al.*, 2001). More systematic studies are necessary before we can make quantitative comparisons with such a more elaborate model.

Acknowledgements

We are grateful to Maurice Boutéca who allowed us to use the X-ray apparatus at the Institut Français du Pétrole and to Camille Schlitter who performed the measurements. We thank Yves Guéguen who stimulated this work. We have benefited from discussions with John Rudnicki and Kathleen Issen. The research programs at Paris and Stony Brook were partially supported by joint program Centre National de la Recherche Scientifique (CNRS) and National Sciences Foundation (NSF) under grants EAR9805072, EAR0006932 and INT9815570.

References

ANTONELLINI, M., AYDIN, A., POLLARD, D. D., and D'ONFRO, P. (1994), *Petrophysical Study of Faults in Sandstone Using Petrographic Image Analysis and X-Ray Computerized Tomography*, Pure Appl. Geophys. *143*, 181–201.

AYDIN, A. and JOHNSON, A. M. (1978), *Development of Faults as Zones of Deformation Bands and as Slip Surfaces in Sandstone*, Pure Appl. Geophys. *116*, 931–942.

AYDIN, A. and JOHNSON, A. M. (1983), *Analysis of Faulting in Porous Sandstones*, J. Struct. Geol. *5*, 19–31.

BAUD, P., ZHU, W., and WONG, T.-f. (2000), *Failure Mode and Weakening Effect of Water on Sandstone*, J. Geophys. Res. *105*, 16371–16389.

BÉSUELLE, P. (2001), *Compacting and Dilating Shear Bands in Porous Rocks: Theoretical and Experimental Conditions*, J. Geophys. Res., 106, 13435–13442.

BÉSUELLE, P., DESRUES, J., and RAYNAUD, S. (2000), *Experimental Characterisation of the Localisation Phenomenon Inside a Vosges Sandstone in a Triaxial Cell*, Int. J. Rock Mech. Min. Sci. *37*, 1223–1237.

BYRNE, T., MALTMAN, A., STEPHENSON, E., SOH, W., and KNIPE, R. (1993), *Deformation Structures and Fluid Flow in the Toe Region of the Nankai Accretionary Prism*, Proc. of the ODP, Scientific Results *131*, 83–192.

CURRAN, J. H. and CARROLL, M. M. (1979), *Shear Stress Enhancement of Void Compaction*, J. Geophys. Res. *84*, 1105–1112.

DAVID, C., WONG T.-f., ZHU, W., and ZHANG, J. (1994), *Laboratory Measurement of Compaction-Induced Permeability Change in Porous Rock: Implications For the Generation and Maintenance of Pore Pressure Excess in the Crust*, Pure Appl. Geophys. *143*, 425–456.

DESRUES, J., CHAMBON, R., MOKNI, M., and MAZEROLLE, F. (1996), *Void Ratio Evolution Inside Shear Bands in Triaxial Sand Specimens Studied by Computed Tomography*, Géotechnique *46*, 529–546.

DIGIOVANNI, A. A., FREDRICH, J. T., HOLCOMB, D. J., and OLSSON, W. A. (2000), *Micromechanics of Compaction in an Analogue Reservoir Sandstone*, Proc. 4th North Am. Rock Mech. Symp. 1153–1160.

ISSEN, K. A. and RUDNICKI, R. J. (2000), *Conditions for Compaction Bands in Porous Rock*, J. Geophys. Res. *105*, 21529–21536.

KAWAKATA, H., CHO, A., KIYAMA, T., YANAGIDANI, T., KUSUNOSE, K., and SHIMADA, M. (1999), *Three-dimensional Observations of Faulting Process in Westerly Granite Under Uniaxial and Triaxial Conditions by X-Ray CT Scan*, Tectonophysics *313*, 293–305.

LABUZ, J. F. and BRIDEL, J. M. (1993), *Reducing Frictional Constraint in Compression Testing through Lubrication*, Int. J. Rock Mech. Min. Sci. *20*, 451–455.

MENÉNDEZ, B. ZHU, W., and WONG, T.-f. (1996), *Micromechanics of Brittle Faulting and Cataclastic Flow in Berea Sandstone*, J. Struct. Geol. *18*, 1–16.

MILLIEN, A. (1993), *Comportement anisotrope du grès des Vosges: élasto-plasticité, localisation, rupture*, Ph.D. Thesis, Univ. of Grenoble, Grenoble, France, 220 pp.

MOLLEMA, P. N. and ANTONELLINI, M. A. (1996), *Compaction Bands: A Structural Analog For Anti-Mode I Cracks in Aeolian Sandstone*, Tectonophysics *267*, 209–228.

OLSSON, W. A. (1999), *Theoretical and Experimental Investigation of Compaction Bands in Porous Rock*, J. Geophys. Res. *104*, 7219–7228.

OLSSON, W. A. and HOLCOMB, D. J. (2000), *Compaction Localization in Porous Rock*, Geophys. Res. Lett. *27*, 3537–3540.

PATERSON, M. S., *Experimental Rock Deformation-The Brittle Field* (Spinger-Verlag, New York 1978).

RAYNAUD, S., FABRE, D., MAZEROLLE, F., GERAUD, Y., and LATIERE, H. J. (1989), *Analysis of the Internal Structure of Rocks and Characterization of Mechanical Deformation by a Non-destructive Method: X-ray Tomodensitometry*, Tectonophysics *159*, 149–159.

RUDNICKI, J. W. (1977), *The effect of stress-induced anisotropy on a model of brittle rock failure as localization of deformation*. In *Energy Resources and Excavation Technology*, Proc. 18th U.S. Symp. on Rock Mechanics, pp. 1–8, Keystone, Colorado.

RUDNICKI, J. W. and RICE, J. R. (1975), *Conditions for the Localization of Deformation in Pressure Sensitive Dilatant Materials*, J. Mech. Phys. Solids *23*, 371–394.

UNDERWOOD, E. E., *Quantitative Stereology* (Addison Wesley, Reading 1970).

VINEGAR, H. J., DE WAAL, J. A., and WELLINGTON, S. L. (1991), *CT Studies of Brittle Failure in Castlegate Sandstone*, Int. J. Rock Mech. Min. Sci. *28*, 441–448.

WELLINGTON, S. L. and VINEGAR, H. J. (1987), *X-Ray Computerized Tomography*, J. Pet. Tech. *39*, 885–898.

WONG, T.-f., DAVID, C., and ZHU, W. (1997), *The Transition From Brittle Faulting To Cataclastic Flow in Porous Sandstones: Mechanical Deformation*, J. Geophys. Res. *102*, 3009–3025.

WONG, T.-f., BAUD, P., and KLEIN, E. (2001), *Localised Failure Modes in a Compactant Porous Rock*, Geophys. Res. Lett., *28*, 2521–2524.

WU, X. Y., BAUD, P., and WONG, T.-f. (1999), *Micromechanics of Compressive Failure and Spatial Evolution of Anisotropic Damage in Darley Dale Sandstone*, Int. J. Rock Mech. Min. Sci. *37*, 1–18.

(Received March 7, 2001, revised June 1, 2001, accepted August 1, 2001)

Pure appl. geophys. 160 (2003) 869–887
0033–4553/03/060869–19

Pure and Applied Geophysics

Velocity Measurements and Crack Density Determination During Wet Triaxial Experiments on Oshima and Toki Granites

A. SCHUBNEL,[1] O. NISHIZAWA,[2] K. MASUDA,[2]
X. J. LEI,[2] Z. XUE,[3] and Y. GUÉGUEN[1]

Abstract—A set of experiments on four samples of Oshima Granite at 15, 40 and 60 MPa confining pressure have been performed in order to investigate the damage behavior of granite submitted to deviatoric stress. In addition an experiment on one sample of Toki Granite at 40 MPa confining pressure was performed, in order to compare and elucidate the structural effects. Using acoustic emission data, strain measurements and elastic wave velocities allow to define consistently a damage domain in the stress space. In this domain, microcracking develops. The microcracking process is, in a first stage, homogeneous and, close to failure, localized. Elastic wave velocities decrease in the damage domain and elastic anisotropy develops. Using KACHANOV's model (1993), elastic wave velocities have been inverted to derive the full second-order crack density tensor and characterize the fluid saturation state from the fourth-order crack density tensor. Crack density is strongly anisotropic and the total crack density close to failure slightly above one. The results indicate that the rock is saturated in agreement with the experimental conditions. The model is thus shown to be very appropriate to infer from elastic wave velocities a complete quantitative characterization of the damaged rock.

Key words: Elastic velocities, granite, crack density tensor, acoustic emissions, effective medium.

Introduction

The development of a brittle fracture in an intact rock is preceded by the nucleation and propagation of cracks. Brittle fracture is attained at a small strain and is the result of shear localization. Before failure, microcracking can be detected through a variety of ways: (1) acoustic emission reflects dynamic increments of crack growth, (2) volumetric strain increases while dilatancy occurs as a result of crack opening, (3) elastic wave velocities decrease and exhibit anisotropy because crack nucleation, propagation, and opening affect the effective elastic properties of the rock.

Through a set of five experiments on two granitic rocks, we have defined a damage domain in the stress space where rock properties are strongly modified by

[1] Laboratoire de Géologie, Ecole Normale Supérieure, 24 rue Lhomond, 75005, Paris, France.
[2] Earthquake Research Dpt., Geological Survey of Japan, Tsukuba, Ibaraki 305-8567, Japan.
[3] Kisojiban Consultants Co. Ltd., Nara, Japan.

microcracking. Our purpose is (1) to define the damage domain in a consistent way using different data sets and (2) to model the observed variations of rock physical properties in the damage regime, and more specifically the elastic wave velocities.

Experimental Setup

The investigated rocks were Oshima granite (Western Japan) and Toki granite. All samples of Oshima granite originated from the same block. We used only one sample of Toki granite. This one was cored at 253 meters below the Toki Gifu prefecture, Japan. Petrographic descriptions of Oshima granite can be found in SANO (1981). Bulk density, crack porosity and atmospheric pressure wave velocities are given in SANO *et al.* (1992) for Oshima granite, in JNC Technical Report (1999) for Toki granite and are reported in Table 1. Five MPa effective pressure velocities were measured during our study. In the case of Oshima granite, one can see a decrease in wave anisotropy when pressure is applied. Unfortunately, in the case of Toki granite, the air velocity was only measured in one direction, however the initial anisotropy of the rock seemed to be much bigger as shown by the measurements under 5 MPa effective pressure. Granitic rocks have in general two sets of microcracks (DALE, 1923; OSBORNE, 1935; PENG and JOHNSON, 1972; SIMMONS *et al.*, 1975; SANO *et al.*, 1992). Most of the microcracks exist within the so-called "rift-plane." A secondary orientation of cracks is known as the "grain plane." The "hardway plane" is defined by the perpendicular plane to both the rift and the grain planes. Those are quarrymen's terms. The samples used in this study were cylindrical, 10 cm long and 5 cm diameter. All Oshima granite samples were oriented so that their height was normal to the hardway plane. In such case, the vertical, the minimum and maximum horizontal P velocities, respectively V_z, V_h, and V_H, corresponds respectively to V_H, V_G

Table 1

Elastic P-wave velocities and P-wave anisotropy at ambient air conditions and under 5 MPa effective confining pressure. Bulk density and crack porosity as measured for Oshima and Toki granites in SANO et al. (1992), JNC Technical Report (1999), and during this study

	Elastic P Wave Velocities (km/sec)				Anisotropy $(V_z - V_H)/V_z$		Bulk Density (g/cm^3)	Crack Porosity (%)
		V_z	V_h	V_H				
Oshima Granite	air*	4.685	4.390	3.635	air*	22%	2.652 ± 0.004*	0.21*
	5 MPa**	5.63	5.53	5.44	5 MPa**	3.3%		
Toki Granite	air***		4.9		air*	not determined	2.59***	–
	5 MPa**	4.85	5.35	5.65	5 MPa**	14%		

* SANO *et al.* (1992). ** This study, *** JNC Technical Report on Toki Gifu Borehole (1999).

and V_R, where H, G and R are the directions normal to hardway, grain and rift planes, respectively. Toki granite sample was oriented so that its height was normal to the rift plane. In such cases, V_z, V_h, and V_H corresponds respectively to V_R, V_G and V_H. Figure 1a shows an optical microscope photograph of an intact Oshima granite sample. Oshima granite is roughly composed of 50% quartz, 30% alcali feldspar, and 20% biotite. No preferred orientation could be observed in the mineralization of the rock, which looked very homogeneous. We therefore assumed here that the initial anisotropy in Oshima granite was entirely crack related, i.e. due to the initial preferred orientation of microcracks. The porosity of Oshima granite was found previously (SANO *et al.*, 1992) to be 0.2%, which is consistent with the crack porosity we could infer from stress-strain data measurement during this study (cf. Fig. 3). Thirty-eight Piezo Electric Transducers (PZT) were first mounted on the curvilinear surface of each sample, 18 PZT of 2 MHz eigenfrequency were compressional ones, the diameter of which varied from 3 mm to 5 mm. The 20 others were shear wave PZT of 1 MHz eigenfrequency, oriented either horizontally or vertically. Each PZT was numbered and accurately located on the sample as shown in Figure 1b. Accuracy on transducers location was $\pm 0,5$ mm. The PZT transducers were used as recorders, in order to locate the Acoustic Emissions (AE) events. They were also used as acoustic sources for measuring the P and S waves velocities during the experiment. Velocities were all measured simultaneously and online, during the experiment. This system enabled us to measure P and S velocities along many different paths. In particular, three horizontal paths along diameters and two diagonal ones at an angle of approximately 30° with the vertical axis, for each P, SV and SH waves were selected. Moreover, strain was locally measured thanks to 12 strain gages, 6 axial and

a) Optical micrograph of intact Oshima granite b) PZT sensors and strain gages location map

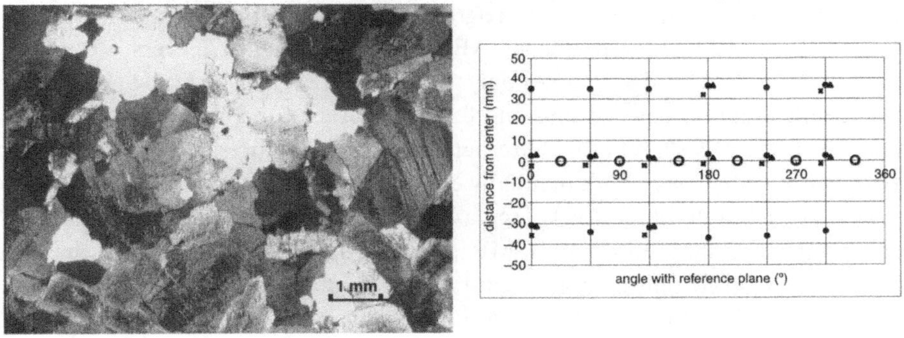

Figure 1
a) Optical micrograph of an intact Oshima granite sample in polarized light. Note the apparent homogeneity of the rock. Scale is shown on the figure and average grain size is equal to 2 mm. b) Planar projection of a sample cylindrical surface. Location of P wave PZT (crosses), *SV* wave PZT (triangles) and *SH* wave PZT (plain circles) as well as the location of the six double strain gages (open circles) is shown on the figure. The two vertical *P* wave PZT are not shown. Note that horizontal paths as well as diagonal ones are easily measurable with such an array.

6 radial ones pasted on the surface of the sample, one pair every 60°. Each strain gages pair was assumed to measure surface deformation along a line perpendicular to the diametric line joining the gage midpoints. Thus, the axial and radial strains were measured very locally.

The sensors were mounted on the sample and stainless steel pieces were attached to each rock extremity. These metal pieces contained holes and roves allowing the fluid access to the sample. Two compressional PZT sensors were mounted on those steel pieces, enabling us to measure the vertical *P*-wave velocity. A 5 mm thick silicon rubber layer was then pasted all over the sample in order to isolate it from the confining oil. The sample was then connected and put inside the pressure vessel.

Experimental Procedure

Many experiments, aimed at investigating failure zone development and fault growth under triaxial loading, have been performed in recent years. Following the technological advances made in AE location (LOCKNER *et al.*, 1991; MOORE and LOCKNER, 1995; KUSUNOSE, 1995); LEI *et al.* (1998, 2000) and JOUNIAUX *et al.* (2001) have developed an AE localization technique (for details see LEI *et al.*, 1998). Figure 2 presents a schematic view of the sample-piston-data recording arrangement. Two recording systems are used. The first one enables recording of the waveform of the acoustic emission and/or of the recordings by each PZT of the pulse generated by the source PZT, thus allowing measurement of wave velocity. This "analogic signal" requires considerable memory, and due to the huge flux of information, buffers are inevitable. The second one only detects peaks in real-time, and counts the total number of AE since the beginning of the experiment. This one is not subjected to the buffer processing. Thanks to recent progress in electronics and computer science, it was possible to record high frequency AE waveforms of 32 PZT every 50 μs. The 8 remaining PZT were used as sources only. Time arrivals could be measured with 0.15 nanoseconds accuracy for *P* waves and 0.3 ns accuracy for *S* waves.

Distilled water was used as the internal fluid. Confining pressure system as well as pore pressure system were both manually controlled. When inside the pressure vessel, the rock was submitted to a slowly increasing confining pressure, up to 15 MPa. The pore pressure was then raised to 10 MPa. Several days were needed to saturate the sample. We assumed that complete saturation was reached when the pore pressure remained stable for at least 10 hours. Axial stress was applied through a piston. The axial stress was servo-controlled and increased until failure. The load rate was kept constant and in general approximately equal to 10^5 Pa.sec^{-1}. In order to record the maximum number of AE, a step loading was applied. For each experiment the total AE number exceeded 50,000. Yet, only large ones could be recorded and located, which reduces that number to approximately 5000. The estimated location error for most AE events is <2 mm (LEI *et al.*, 1992).

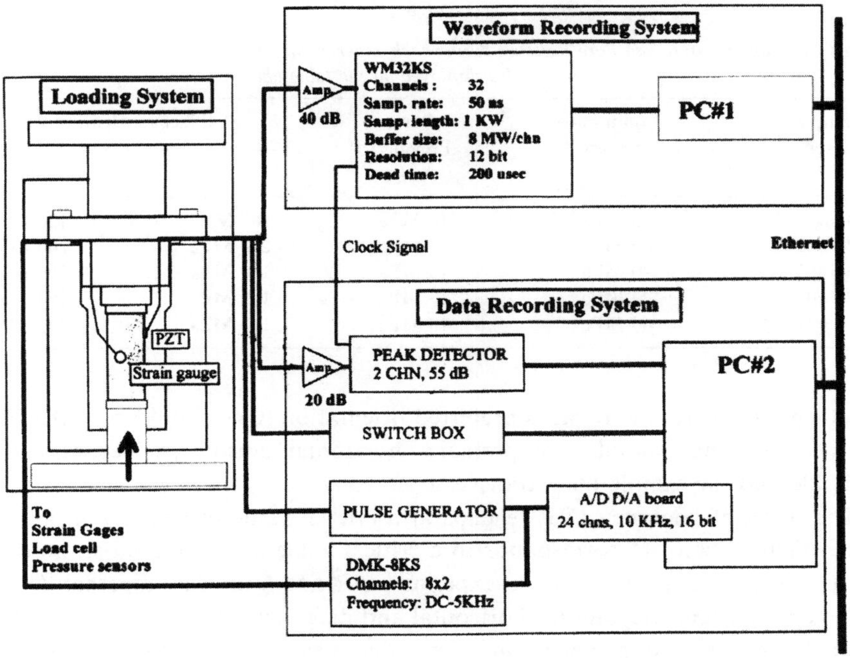

Figure 2
A block diagram of the experimental system used in this study (from LEI *et al.*, 2000). Ten piezoelectric transducers (PZT) were mounted on the surface on each test sample (see Fig. 1b for details). The signal was preamplified by 40 dB before feeding into a fast waveform recording system which has a maximum sampling rate reaching 40 ns. An automatic switch box was used for switching some sensors between AE and velocity measurements during the experiment. Stress, strain, pore pressure and confining pressure were recorded using two 16-bit analog to digital (A/D) boards. For the sake of simplicity, the pore pressure system was not displayed on the figure.

Experimental Results

Elastic Wave Velocities

All experiments were performed at the Earthquake Research Department in Tsukuba, Geological Survey of Japan (Table 2). Four experiments investigated Oshima granite samples at confining pressure of 15, 40 and 60 MPa, pore pressure being constant and equal to 10 MPa. One experiment was performed on Toki granite with $\sigma_3 = P_c = 40$ MPa and $Pp = 10$ MPa, in order to compare both rocks mechanical and physical behaviors. Figure 3 shows P, SV and SH wave-velocity measurements versus effective mean stress P' where $P' = ((\sigma_1 + 2\sigma_3)/3 - P_p)$ in the case of OG#03. In this experiment, confining pressure was equal to 40 MPa. One notices an increase in all velocities as the confining pressure was raised to 40 MPa. Anisotropy becomes lower than 5% (see Fig. 3d) as the cracks are being closed with effective mean stress, which shows again that in Oshima granite, most of the initial

Table 2

Applied Confining pressure and Pore Pressure during each experiment. Effective Mean Stress at C′ and Peak stress are also displayed on the table above

Sample Reference	Confining Pressure	Pore Pressure	Effective Mean Stress	
			at C′	at Peak
OG#01	15 MPa	10 MPa	80 MPa	120 MPa
OG#02	60 MPa	10 MPa	200 MPa	240 MPa
OG#03	40 MPa	10 MPa	150 MPa	200 MPa
OG#05	40 MPa	10 MPa	130 MPa	190 MPa
TK#01	40 MPa	10 MPa	155 MPa	185 MPa

anisotropy is due to microcracks preferred orientation. One should note that in the intermediate range of effective pressure, horizontal anisotropy becomes almost negligible and the sample fits a transverse isotropy.

The onset of dilatancy C' is reached at a critical value of effective mean stress. This critical value of P' corresponds to a critical value of differential stress Q since $P' = Q/3 + (P_c - P_p)$, so that in our experiments P' and Q increase together, $(P_c - P_p)$ being kept constant. Beyond C', horizontal and diagonal velocities decrease because of crack growth and propagation. Vertically measured P velocity stays nearly constant. The P' and Q increase corresponds also to an increase in acoustic emission rate. In this experiment the critical value of P' at point C' is 155 MPa and the associated value of Q is therefore 375 MPa. Load was slowed down as the AE rate was increasing and was stopped right before failure of the sample. Peak effective mean stress was equal to 195 MPa. Vertical P velocities were constant, showing a slight decrease before failure. This suggests that most of the stress induced microcracks are vertical, and thus invisible to vertical P wave.

Local axial and radial strain were also measured during deformation of the sample. Figure 4 shows local volumetric strain in the case of the experiment OG#01. In this experiment confining pressure was equal to 15 MPa. Local volumetric strain was calculated using $\epsilon_v = \epsilon_{\text{axial}} + 2 * \epsilon_{\text{radial}}$. As one can see on Figure 3, the six different strain gage pairs record six different volumetric strains, especially during creep deformation. Thus, the strain pattern is elliptical, direction 1 and 6 showing maximum deformation, 3 and 4 showing minimum deformation, as previously observed by HADLEY (1975). Although ϵ_1 shows dilatancy from the beginning of the experiment, we believe that this is an artefact. Indeed, the strain gage signals might have been polluted by air bubbles, or very inhomogeneous deformation. On figure 3, the onset of dilatancy C' can be measured when the local volumetric strain curves become nonlinear. In this experiment C' was equal to approximately 80 MPa. The high strain anisotropy in this experiment can be explained by the low effective confining pressure (5 MPa), and thus very inhomogeneous deformation, as expected. Indeed, we saw that during this experiment, horizontal velocity anisotropy, as well as

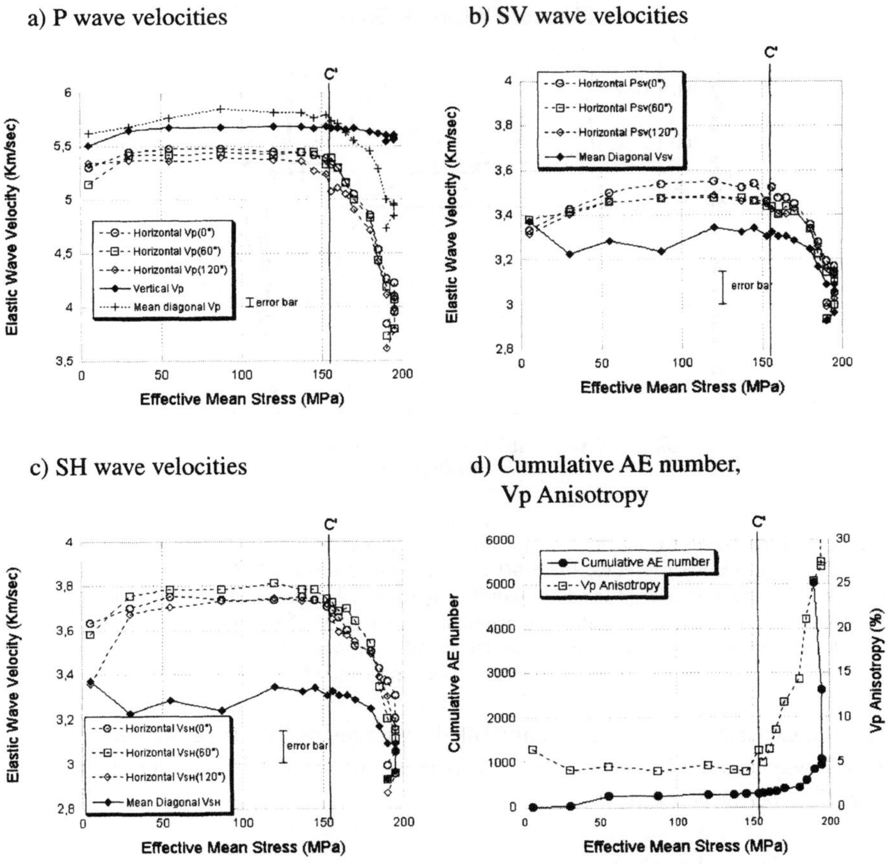

Figure 3

Wave-velocity measurements versus Effective Mean Stress $P' = [\frac{(\sigma_1+2\sigma_3)}{3} - P_p]$. Confining Pressure σ_3 was equal to 40 MPa and Pore Pressure P_p equal to 10 MPa. The relationship between Differential Stress $Q = \sigma_1 - \sigma_3$ and Effective Mean Stress P' is $P' = Q/3 + (\sigma_3 - P_p)$ so that the increase in P' is due to an increase in differential stress Q. Confining Pressure $(\sigma_3 - P_p)$ being kept constant. a) P-wave velocities. Error bar on P-wave measurements is ± 0.1 km/sec. Plain diamonds show vertical P-wave velocity, crosses mean diagonal P velocity. Open circles, squares and diamonds show horizontal P wave at 0, 60 and 120 degrees from reference plane, respectively. b) SV wave velocities. Error bar on SV-wave measurements is ± 0.15 km/sec. Plain diamonds show mean diagonal SV velocity. Open circles, squares and diamonds show horizontal SV wave at 0, 60 and 120 degrees from reference plane, respectively. c) SH wave velocities. Symbols and error bar are the same as in c. d) Plain line with plain circles corresponds to cumulative AE number. The first increase corresponds to the point where the piston touches the sample, and thus to when differential stress starts being applied. The second one, considerably more dramatic, corresponds to the onset of dilatancy C', shown on the figure. Dashed line and open squares show the maximum V_p anisotropy in percent and defined as $An = (V_z - V_h)/V_z$.

radial strain anisotropy were larger than in other experiments. We believe this can be easily explained by the relatively low effective pressure of this experiment. Peak effective mean stress was equal to 118 MPa. Moreover, axial strain was very linear,

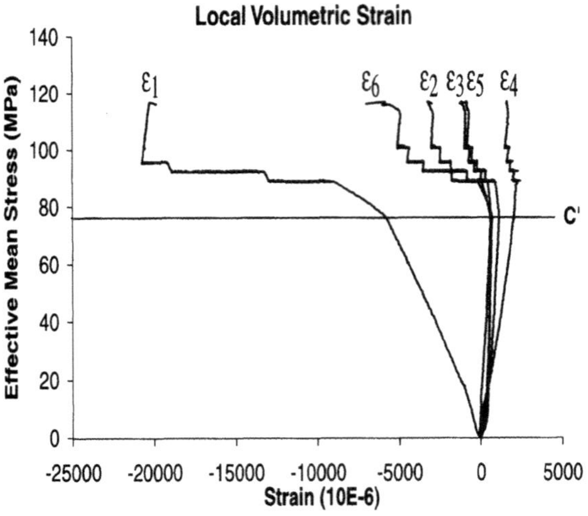

Figure 4

Calculated local volumetric strain ($\epsilon_v = \epsilon_{axial} + 2 * \epsilon_{radial}$) versus effective mean strain for Oshima sample OG#01. Confining pressure was equal to 15 MPa and Pore Pressure to 10 MPa in this experiment which explains the high inhomogeneity of the deformation. Numbers 1, 2, 3, 4, 5, and 6 correspond to an angle of 30, 90, 150, 210, 270, 330 degrees with reference plane respectively. C' was reached at approximately 80 MPa effective Mean stress (e.g., 225 Differential stress) and is represented by a horizontal line on the figure. 10,000 on strain scale represents 1%.

though the experiment was stress controlled, whereas radial strain became nonlinear when C' was reached. Again, this is consistent with microcracks growth and propagation being vertical.

Acoustic Emissions Localization

Figure 5 shows the total Acoustic Emissions distribution for each experiment. One can see on figure 5a that the AE distribution for OG#01 is very inhomogeneous, which is coherent with the local volumetric strain distribution seen in Figure 4 and with the experimental conditions (low effective confining pressure). For all other experiments, acoustic emissions were at first rather homogeneously distributed in the sample, and started localizing when close to failure. Whether experiments were stopped right before or right after brittle fracture, one can clearly see on Figure 5 that AE are localized in swarms along a fault plane. Thereafter, localization occurred and fracture was initiated along a plane shown on Figure 5. AE were starting at the onset of dilatancy, and AE rate increases further beyond. The AE-data hint that cracks are localized and not randomly oriented. However one should keep in mind first that only up to 5000 AE were localized from 50,000 recorded. Moreover many small events were probably below the recording threshold. Second, looking at AE distribution in time, one notices that more than 80% of the events we were able to localize occur in the very pre-peak and post-peak state. AE localization gives us an

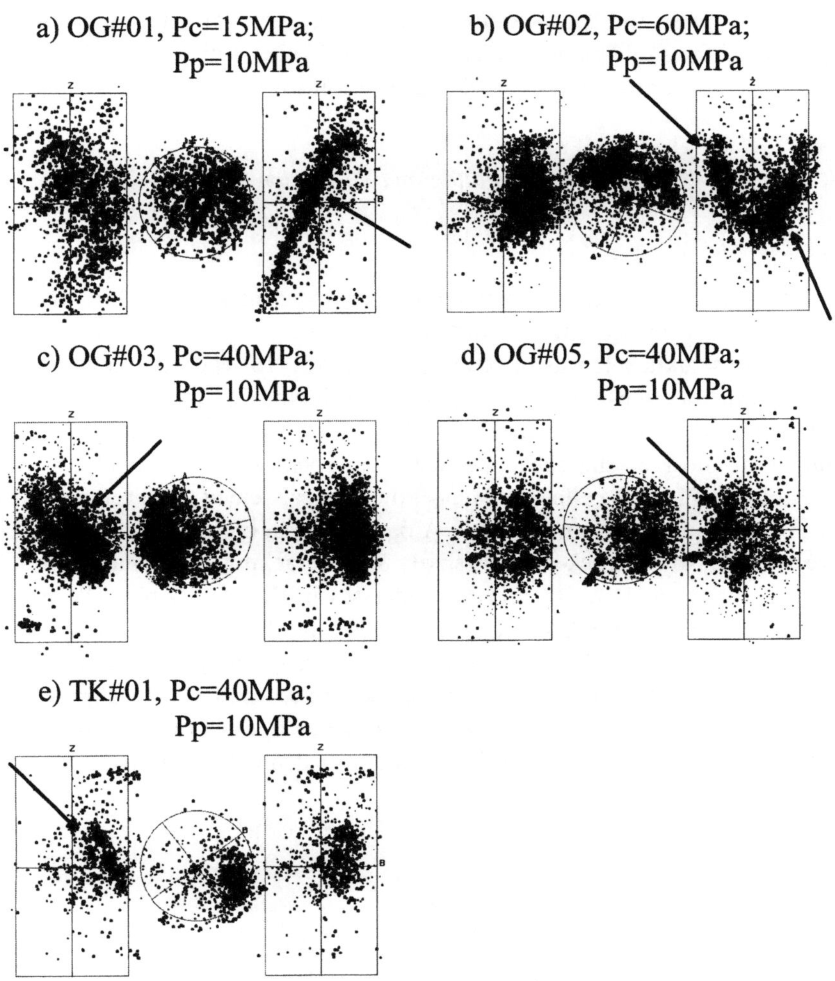

Figure 5

Acoustic emission distribution. Error on localization is 2 mm. Projection shows two perpendicular vertical planes as well as one horizontal plane. Total acoustic emission number was generally over 50,000, but only 2,000 to 5,000 could be localized for each sample. Fracture planes are indicated by arrows. a) OG#01 deformed under 15 MPa confining pressure and 10 MPa pore pressure. AE were nonhomogeneously distributed from the beginning of the experiment and sample eventually failed. b) OG#02 deformed under 60 MPa confining pressure and 10 MPa pore pressure. Experiment was stopped before failure, but localization was taking place along two potential fracture planes. c) OG#03 deformed under 40 MPa confining pressure and 10 MPa pore pressure. Sample eventually failed. Localization appeared close to failure and fault plane is situated inside the AE cloud. d) OG#05 deformed under 40 MPa confining pressure and 10 MPa pore pressure. Experiment was stopped before failure, but localization was taking place and a fracture was initiated from the free surface. e) Toki granite TK#01 deformed under 40 MPa confining pressure and 10 MPa pore pressure. Experiment was stopped before failure, but localization was observed. A fracture has been initiated from the free surface.

image of the pre- and post-failure localization, i.e., on the nucleation and propagation of the fault planes indicated on the figures with arrows, but provides sparse information on the actual crack distribution in the rock before those stages.

Interpretation and Discussion

Damage Domain

C' and peak stresses were determined for each experiment using AE, stress-strain data and velocity measurements and are reported in Table 2. The agreement between all those data sets was in general very good. However, one must remember that each parameter sensitivity to the onset of dilatancy C' depends on the spatial orientation and the measuring direction with respect to the crack network. Indeed, assuming that most of the shear enhanced microcracks are vertical, the vertical compressibility will not be affected. Hence, the vertical velocity will remain constant, as we have previously seen on Figure 3. A damage area, corresponding to an area in the stress space where rock was intensely microcracking and dilating, could be determined. Figure 6 shows C' and peak stresses in the $P' - Q$ plot (WONG *et al.*, 1997; ZHU and WONG 1997; BAUD *et al.*, 2000). In this figure, P' corresponds to effective mean stress calculated as previously showed, and Q to differential stress. One can clearly see that in such a plot, peak stresses and C' are aligned on two parallels. The domain in between those two lines corresponds to that where the rock starts microcracking and dilating before it fails. We defined this domain as a damage

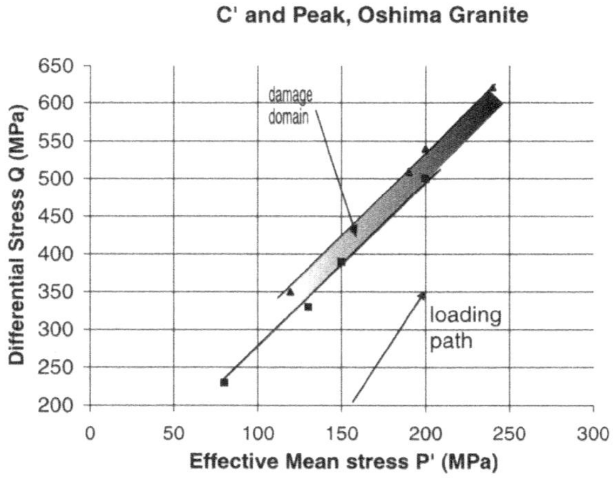

Figure 6
Onset of dilatancy C' (squares) and peak stresses (triangles) in the $P' - Q$ plot for Oshima granite, as defined by WONG *et al.* (1997). P' is effective mean stress, Q is differential stress. Peak stresses and C' are aligned on two parallels. In between is the damage domain. The loading path corresponds to $Q = 3P' - 3(\sigma_3 - P_p)$, where σ_3 is the confining pressure and P_p the pore pressure.

area. The loading path in this plot is not vertical but has a slope of 3 since $Q = 3P' - 3(P_c - P_p)$. Experiments at constant effective pressure would be interesting in order to see the stress path and history effect. One should note that the damaging area is relatively narrow.

Elastic Stiffnesses Determination

From velocity and bulk density measurements, elastic constants could be inverted. The model used is, as previously mentioned, a transverse isotropic one. In such a model, both horizontal axis x_1 and x_2 are equivalent whereas vertical axis x_3 is not. For a medium with transversely isotropic (hexagonal) symmetry, the elastic stiffness tensor C_{ij} has only five independent constants:

$$\begin{pmatrix} C_{11} & C_{12} & C_{13} & 0 & 0 & 0 \\ C_{12} & C_{11} & C_{13} & 0 & 0 & 0 \\ C_{13} & C_{13} & C_{33} & 0 & 0 & 0 \\ 0 & 0 & 0 & C_{44} & 0 & 0 \\ 0 & 0 & 0 & 0 & C_{44} & 0 \\ 0 & 0 & 0 & 0 & 0 & C_{66} \end{pmatrix} \cdot C_{66} = \frac{C_{11} - C_{12}}{2} \; .$$

In this case, relationships between elastic wave velocities and elastic stiffness tensor C_{ij} are the following (MAVKO *et al.*, 1998):

quasi-longitudinal mode:

$$V_p = \left(C_{11}\sin^2\theta + C_{33}\cos^2\theta + C_{44} + \sqrt{M}\right)^{1/2}(2\rho)^{-1/2} \tag{1}$$

quasi-shear mode:

$$V_{Sv} = \left(C_{11}\sin^2\theta + C_{33}\cos^2\theta + C_{44} - \sqrt{M}\right)^{1/2}(2\rho)^{-1/2} \tag{2}$$

pure-shear mode:

$$V_{Sh} = \frac{(C_{66}\sin^2\theta + C_{44}\cos^2\theta)^{1/2}}{\rho} \tag{3}$$

where

$$M = [(C_{11} - C_{44})\sin^2\theta - (C_{33} - C_{44})\cos^2\theta)]^2 + (C_{13} + C_{44})\sin^2 2\theta$$

and θ is the angle between the wave vector and the axis of symmetry. This axis is the vertical one in our case. The five components of the stiffness tensor were obtained from five velocity measurements: $V_p(0°)$ $(= V_z)$, $V_p(90°)$ $(= V_h = V_H)$, $V_p(30°)$, $V_{Sh}(90°)$, $V_{Sv}(30°)$.

Figure 7 shows the results for all experiments. One notices that damage affects the different constants in different ways. Compressibility along axis 1, C_{11} decreases

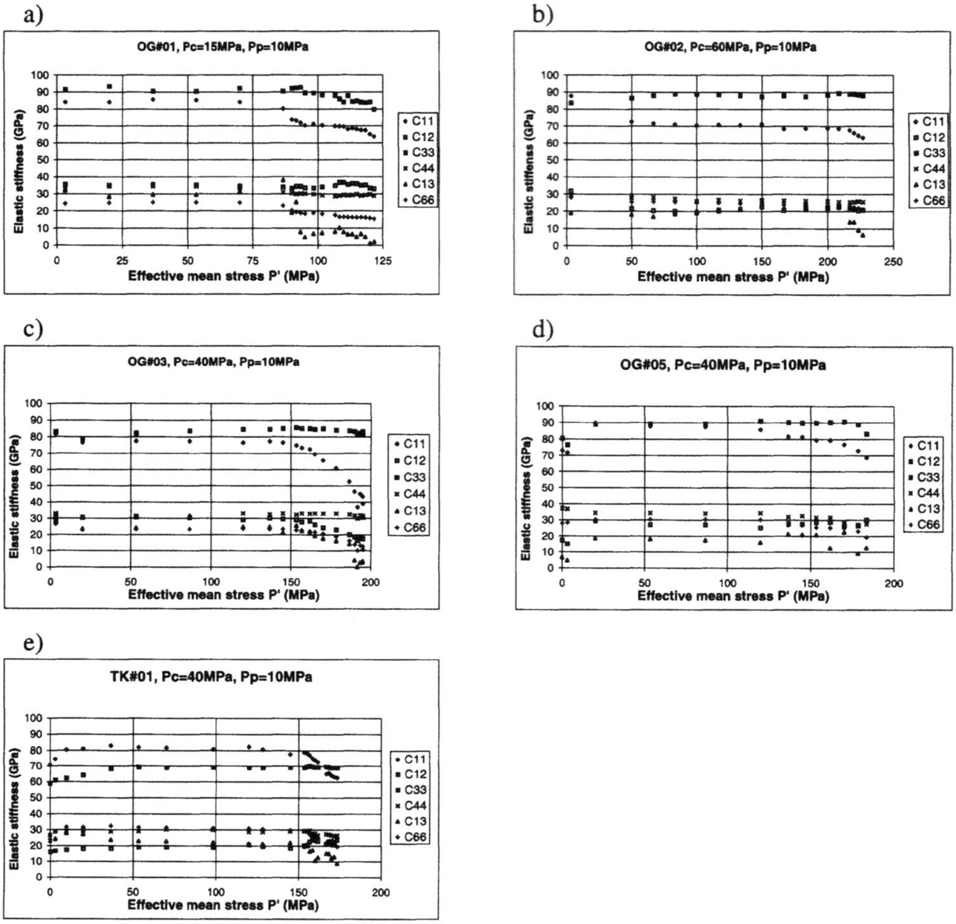

Figure 7

Elastic stiffnesses as inferred from velocity measurements. *P* vertical (*P*0°), *P* and *SH* horizontal (*P* and *SH* 90°) as well as *P*(30°) and *SV*(30°) were used. Angle corresponds to the angle between the wave vector and the vertical axis. a) OG#01 deformed under 15 MPa confining pressure and 10 MPa pore pressure. Sample eventually failed. b) OG#02 deformed under 60 MPa confining pressure and 10 MPa pore pressure. Experiment was stopped before failure. c) OG#03 deformed under 40 MPa confining pressure and 10 MPa pore pressure. Sample eventually failed. d) OG#05 deformed under 40 MPa confining pressure and 10 MPa pore pressure. Experiment was stopped before failure. e) Toki granite TK#01 deformed under 40 MPa confining pressure and 10 MPa pore pressure. Experiment was stopped before failure.

substantially (50%) whereas compressibility along axis 3, C_{33}, the vertical one, is almost constant. Again, this suggests that the cracks are vertical. C_{44} is not affected, whereas vertical shear modulus C_{66} is strongly affected. C_{13} is also affected, which is again consistent with vertical microcracking, as C_{13} represents the stress-strain coupling between the vertical and horizontal axes.

Crack Density Tensor

The average macroscopic strain $\bar{\epsilon_{ij}}$ in an elastically homogeneous volume element V containing N cracks with surfaces S_r (where $r = 1, \dots, N$) can be represented as the sum of two terms (KACHANOV, 1993):

$$\bar{\epsilon_{ij}} \equiv (S^0_{ijkl} + \Delta S_{ijkl})\bar{\sigma_{kl}} \tag{4}$$

where $\bar{\sigma_{kl}}$ is the average stress, S^0_{ijkl} is the fourth rank compliance tensor of the rock and ΔS_{ijkl} is the crack compliance tensor. Assuming that crack interactions can be neglected, SAYERS *et al.* (1995) showed that ΔS_{ijkl} can be expressed by:

$$\Delta S_{ijkl}\bar{\sigma_{kl}} = \frac{1}{2V}\sum_r \left[B_T(\bar{\sigma_{iq}}n_j n_q + \bar{\sigma_{jq}}n_i n_q) + 2(B_N - B_T)\bar{\sigma_{pq}}n_i n_j n_p n_q\right]^r S_r \tag{5}$$

where n_i is the i-th component of the crack normal, B_N and B_T are respectively the normal and shear term of the crack compliance tensor B_{ij}. The cracks are assumed to be flat with a area S_r. This is clearly an approximation; the simplest one indeed. Defining a second rank tensor α_{ij} by

$$\alpha_{ij} = \frac{1}{V}\sum_r [B_T n_i n_j]^r S_r \tag{6}$$

and a fourth rank tensor β_{ijkl} by:

$$\beta_{ijkl} = \frac{1}{V}\sum_r [(B_N - B_T)n_i n_j n_k n_l]^r S_r \ . \tag{7}$$

It follows that ΔS_{ijkl} (SAYERS *et al.*, 1995) is equal to

$$\Delta S_{ijkl} = \frac{1}{4}(\delta_{ik}\alpha_{jl} + \delta_{il}\alpha_{jk} + \delta_{jk}\alpha_{il} + \delta_{jl}\alpha_{ik}) + \beta_{ijkl} \ . \tag{8}$$

For a general transversely isotropic orientation distribution of cracks, $\alpha_{11} = \alpha_{22}$, $\beta_{1111} = \beta_{2222}$, $\beta_{1212} = \beta_{1122} = \beta_{1111}/3$ and equation (8) gives

$$\begin{aligned}
\Delta S_{1111} = \Delta S_{2222} &= \alpha_{11} + \beta_{1111} \\
\Delta S_{3333} &= \alpha_{33} + \beta_{3333} \\
\Delta S_{1212} &= \frac{\alpha_{11}}{2} + \frac{\beta_{1111}}{3} \\
\Delta S_{2323} = \Delta S_{3131} &= \frac{\alpha_{11} + \alpha_{33}}{4} + \beta_{1133} \\
\Delta S_{1122} &= \frac{\beta_{1111}}{3} \\
\Delta S_{2233} = \Delta S_{3311} &= \beta_{1133} \ .
\end{aligned} \tag{9}$$

Remembering $(S_{ij}) = (C_{ij})^{-1}$ where (S_{ij}) and (C_{ij}) are respectively the compliance tensor and the stiffness tensor using Voigt notation, and using equations (8)–(9), we obtain the tensors α and β for all our experiments. Results are shown in Figure 8. Note that on Figure 8, α and β are normalized by a normalizing factor h equal to

$$h = \frac{3E_0(2 - \nu_0)}{32(1 - \nu_0^2)} \tag{10}$$

so that

$$(\alpha^*_{ij}, \beta^*_{ijkl}) = (h\alpha_{ij}, h\beta_{ijkl}) \ , \tag{11}$$

where E_0 and ν_0 are the initial Young's modulus and Poisson ratio of the rock, respectively. Remembering that for open penny-shaped with radius a in a homogeneous background medium, the crack shear compliance B_T is equal to (KACHANOV, 1993)

$$B_T = \frac{32(1 - \nu_0^2)a}{3\pi E_0(2 - \nu_0)} = \frac{a}{\pi h} \tag{12}$$

and using equation (6), one can easily see that α^*_{ij} is therefore the normalized crack density

$$\alpha^*_{ij} = \frac{1}{V}\sum_N a^3 n_i n_j \ . \tag{13}$$

For Oshima and Toki granites, initial Young's modulus E_0 and initial Poisson ν_0 were taken directly as inferred from velocities at 100 MPa effective mean stress during the experiment and were respectively equal to 70 ± 5 GPa and $0,27 \pm 0,02$. Those values are in close agreement with the hydrostatic moduli observed previously by SANO *et al.* (1992). In the same way, initial elastic compliances of the rock were needed to perform the inversion shown by equation (9). The values we used in order to complete this inversion were systematically the plateau values at 100 MPa effective mean stress. Indeed, we considered that at such a stress state, initial cracks were mostly closed, and because we were far from C',

▶

Figure 8
Normalized crack density tensors α^*_{ij} and β^*_{ijkl} versus effective mean stress. α^*_{ij} represents the normal crack density tensor. Total crack density is equal to tr (α^*_{ij}). For Oshima and Toki granites, initial Young's modulus $E_0 = 70 \pm 5$ GPa and initial Poisson ratio $\nu_0 = 0.27 \pm 0.02$. a) OG#01 deformed under 15 MPa confining pressure and 10 MPa pore pressure. Sample eventually failed. b) OG#02 deformed under 60 MPa confining pressure and 10 MPa pore pressure. Experiment was stopped before failure. c) OG#03 deformed under 40 MPa confining pressure and 10 MPa pore pressure. Sample eventually failed. d) OG#05 deformed under 40 MPa confining pressure and 10 MPa pore pressure. Experiment was stopped before failure. e) Toki granite TK#01 deformed under 40 MPa confining pressure and 10 MPa pore pressure. Experiment was stopped before failure. Note that initial anisotropy is reversed as expected.

no additional shear-enhanced cracks had yet appeared in the rock. This hypothesis, valid for large effective confining pressures, is obviously not for low ones. This might explain the results of the inversion for OG#01. Indeed, horizontal cracks seem to close (i.e., α_{33} decrease) with applied differential stress, which is expected.

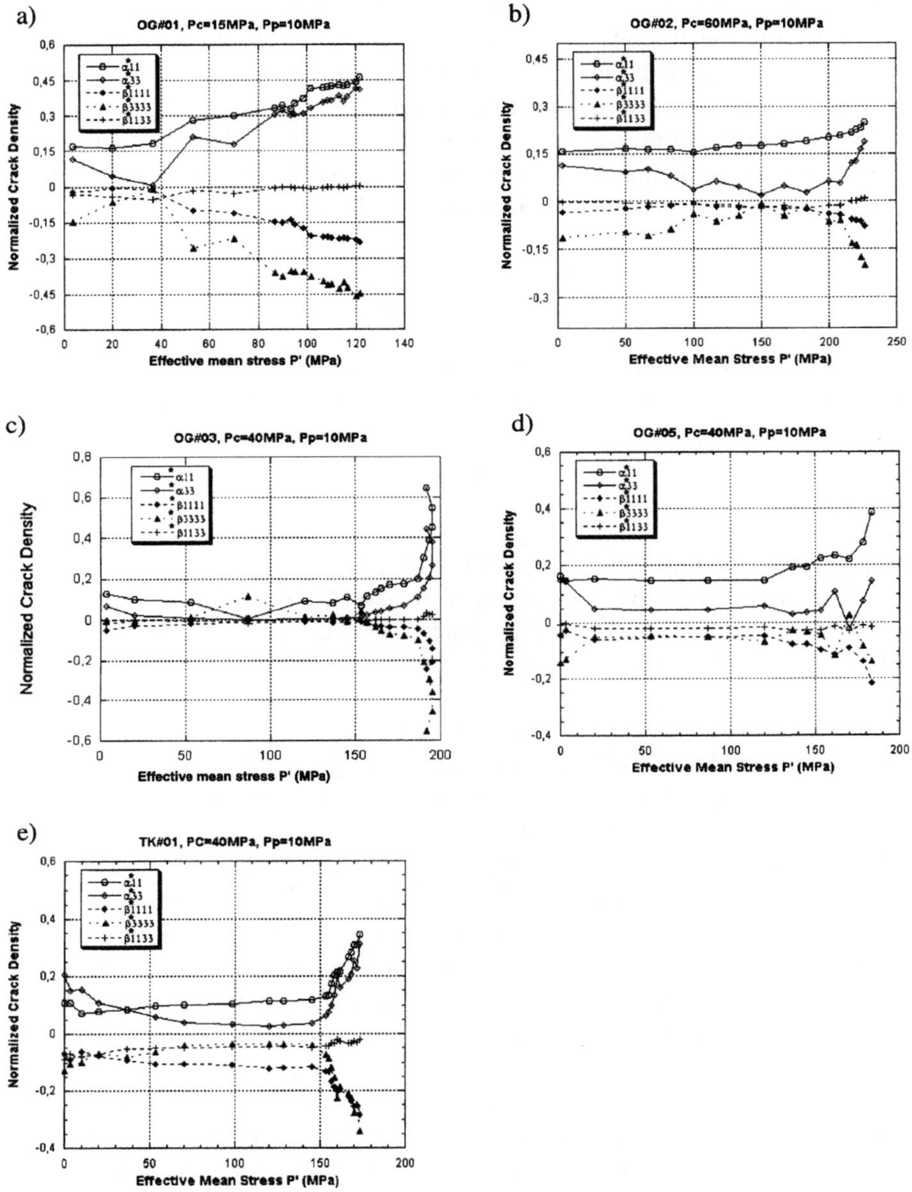

However, vertical crack density seems to increase throughout the experiment, and in such conditions it is difficult to point out the onset of dilatancy C' on the plot. For other experiments, the increase of the crack density clearly occurs when C' is reached. Total crack density is $\mathrm{tr}(\alpha^*_{ij}) = \frac{Na^3}{V} = \rho$. For Oshima and Toki granites, fracture is reached for a ρ value slightly larger than 1. Note that such a value is expected from 3-D percolation models (GUÉGUEN *et al.*, 1997). Indeed, there are two different percolation thresholds for ρ: the first one is the permeability threshold at $\rho \sim 0.15$ and the second one a fracture threshold at $\rho \sim 1$. The confusion one can sometimes find in the literature probably stems from the fact that in a 2-D percolation model, those two thresholds merge.

The components of the tensor β are negative, and diagonal terms β_{iiii} turn out to become large close to failure. Indeed, for cracks filled up with fluids, the crack compliance tensor B_{ij} is such that (KACHANOV, 1993)

$$\frac{B_N}{B_T} = \left(1 - \frac{v_0}{2}\right)\frac{\delta}{1+\delta} \ , \tag{14}$$

where δ characterizes the coupling between the stress and the fluid pressure, and determines the impact of the fluid on the effective crack compliance:

$$\delta = \beta^{-1}\frac{E_0\zeta_0}{K_f} \ , \tag{15}$$

where K_f is the fluid bulk modulus, ζ_0 is the average aspect ratio of the crack defined as $\zeta_0 = \frac{\langle w \rangle}{a}$, where $\langle w \rangle$ is the average crack aperture, and β a numerical factor: $\beta = 16\frac{1-v_0^2}{3\pi}$. In the case of an incompressible fluid such as water $K_f = 2.10^9$ Pa and with $E_0 = 7.10^{10}$, $\delta \sim 20\zeta_0$, which is negligible if ζ_0 is small ($< 10^{-3}$) as expected for a crack.

Therefore, $B_N/B_T \sim 0$ and the diagonal terms $\beta_{iiii} \sim -B_T \sim -\alpha_{ii}$. This theoretical prediction for fluid-saturated cracks is in perfect agreement with results shown on Figure 8, which suggests strongly that in the damage regime where ρ is high enough, cracks are interconnected and saturated.

Finally, comparing the results shown on Figures 8c, d and e is interesting to validate our model. Experiments OG#03 and OG#05 were performed in the same stress conditions, except that experiment OG#05 was stopped before failure. Examining the figures, one can see that initial crack densities are the same, and C' is somewhat reached at the same effective stress. This indicates that our results were very reproducible, as well as our inversion. Comparing Oshima granite sample OG#03 and Toki granite sample TK#01 is also of great interest. Both rocks were deformed in identical stress conditions as well. The main difference between those two rocks lies in the fact that initial anisotropy was reversed for Toki granite ($\alpha_{33} > \alpha_{11}$), as explained previously. This initial anisotropy vanishes as the horizontal cracks are being closed by the applied differential stress, and the α_{33} and α_{11} cross over one another. At the conclusion of the experiment, the residual anisotropy is reversed compared to the initial one. Again this shows that our results were very reproducible, and that initial crack distribution weakly influenced final crack density distribution, in the case of a homogeneous intact granite.

Conclusion

Experimental data from acoustic emissions, elastic wave velocities and strain measurements are in close mutual agreement and allow the definition of a domain in the stress space where the rock is mechanically damaged. Multitudes of acoustic emisions taking place in this domain could also be relocated, enabling observation of localization and final fracturing. A straightforward interpretation is that this domain is the domain where microcracking is active. Microcracks can account for all the reported observations. This interpretation fits well to a quantitative model using theoretical results on elastic solids with many cracks. KACHANOV (1993) has calculated the elastic constants of a cracked solid in terms of two diffferent tensors. Both tensors are initially obtained here from inversion of the experimental elastic wave velocities, measurements and asuming a tranversely isotropic symmetry. Oshima and Toki granites seem to reach a total crack density slightly larger than one when close to fracture. Total vertical crack density $\alpha_{11} + \alpha_{22}$ is at least twice larger than horizontal one α_{33} as expected. Moreover, the results are in good agreement with the theory of cracks replete with fluids, and the method allows qualitative quantification of the saturation of the rock.

We showed that those results were highly reproducible, and that although AE localization is rather inhomogeneous, a transverse isotropic model for crack distribution appears valid up to the very pre-peak state for mechanically damaged rocks. This indicates that the above analysis could widely be used to obtain informations on the fluid saturation of a medium as well as on its crack anisotropic distribution. A further step would be to use the crack densities parameters which can be inferred from such models to predict how permeability is expected to vary in the damaged regime. The recent model of SIMPSON *et al.* (2001), is appropriate to develop such a calculation.

Acknowledgements

We would like to thank Xiachin Li and Chidaram Sivaji for their assistance while experimenting at Geological Survey of Japan, Pierre Bésuelle, Teng-Fong Wong, Till

Popp and an anonymous reviewer for their constructive comments on this paper. This research was made possible thanks to an exchange agreement (JITA) between the research groups directed by Osamu Nishizawa at Geological Survey of Japan and Yves Guéguen at Ecole Normale Supérieure.

References

BAUD, P., SCHUBNEL, A., and WONG, T-f. (2000), *Dilatancy, Compaction and Failure Mode in Solnhofen Limestone*, JGR *105*, 19,289–19,320.

DALE, T. N. (1923), *The Commercial Granites of New England*, US. Geol. Surv. Bull. *738*, 1–97.

HADLEY, K. (1975), *Azimuthal Variation of Dilatancy*, JGR *80*, 4845–4850.

GUÉGUEN, Y., CHELIDZE, T., and LE RAVALEC, M. (1997), *Microstructures, Percolation Thresholds and Rock Physical Properties*, Tectonophysics *279*, 23–35.

KACHANOV, M. (1992), *Continuum Model of Medium with Cracks*, J. Eng. Mech. Div. Am. Soc. Civ. Eng. *106*, 1039–1051.

KACHANOV, M. (1993), *Elastic Solids with Many Cracks and Related Problems*, Adv. Appl. Mech. *30*, 259–445.

KUSUNOSE, K. (1995), *Fracture Mechanics of Rocks*, J. Phys. Earth *43*, 479–504.

JNC TECHNICAL REPORT (1999), *Report on Toki Gifu Borehole* (in Japanese), JNC Tech. J. *7400*, 99–014, 33–41.

JOUNIAUX, L., MASUDA, K., LEI, X. L., NISHIZAWA, O., KUSUNOSE, K., LIQIANG, L., and MA, W. (2001), *Comparison of the Microfracture Localization in Granite between Fracturation and Slip of a Pre-existing Macroscopic Healed Joint by Acoustic Emission Measurements*, JGR *106*, 8687–8698.

LEI, X. L., NISHIZAWA, O., KUSUNOSE, K., and SATOH, T. (1992), *Fractal Structure of the Hypocenter Distribution and Focal Mechanism Solution of AE in Two Granites of Different Grain Size*, J. Phys. Earth *40*, 617–634.

LEI, X. L., SATOH, T., NISHIZAWA, O., MASUDA, K., and KUSUNOSE, K. (1998), *A Real-time AE Hypocenter Monitoring System for Laboratory Rock Fracture Experiment*, Bull. Geol. Surv. Japan *49*, 447–457.

LEI, X. L., KUSUNOSE, K., RAO, M. V. M. S., NISHIZAWA, O., and SATOH, T. (2000), *Quasi-static Fault Growth and Cracking in Homogeneous Brittle Rock under Triaxial Compression Using Acoustic Emission Monitoring*, JGR *105*, 6127–6139.

LOCKNER, D. A., BYERLEE, J. D., KUKSENKO, V., PONOMAREV, A., and SIDORIN, A. (1991), *Quasi-static Fault Growth and Shear Fracture Energy in Granite*, Nature *350*, 39–42.

MAVKO, G., MUKERJI, T., and DVORKIN, J., *The Rock Physics Handbook* (Cambridge University Press, 1998) pp. 329.

MOORE, D. E. and LOCKNER, D. A. (1995), *The Role of Cracking in Shear Fracture Propagation in Granite*, J. Struct. Geol. *17*, 95–114.

OSBORNE, F. F. (1935), *Rift, Grain, Hardway in Some pre-Cambrian Granites*, Québec, Econ. Geol. *30*, 540–551.

PENG, S. and JOHNSON, M. A. (1972), *Crack Growth and Faulting in Cylindrical Specimens of Chelmsford Granite*, Int. J. Rock. Mech. Min. Sci. *9*, 37–86.

SANO, O. (1981), *A Note on the Sources of Acoustic Emissions Associated with Subcritical Crack Growth*, Int. J. Rock Mech. Min. Sci. Geomech. Abst. *18*, 259–263.

SANO, O., KUDO, Y., and MIZUTA, Y. (1992), *Experimental Determination of Elastic Constants of Oshima Granite, Barre Granite, and Chelmsford Granite*, JGR *97*, 3367–3379.

SAYERS, C. M. and KACHANOV, M. (1995), *Microcrack Induced Elastic Wave Anisotropy of Brittle Rocks*, JGR *100*, 4149–4156.

SIMMONS, G., TODD, T., and BALRIDGE, W. S. (1975), *Toward a Quantitative Relationship between Elastic Properties and Cracks in Low Porosity Rocks*, AM. J. Sci. *275*, 318–345.

SIMPSON, G., GUÉGUEN, Y., and SCHNEIDER, F. (2001), *Permeability Enhancement Due to Microcrack Dilatancy in the Damage Regime*, JGR *106*, 3999–4016.

VAKULENKO, A. A. and KACHANOV, M. (1971), *Continuum Theory of Medium with Cracks*, Mechanika Tverd. Tela *6*, 159–166.

WONG, T-F., DAVID, C., and ZHU, W. (1997), *The Transition from Brittle Faulting to Cataclastic Flow in Porous Sandstones: Mechanical Deformation*, JGR *102*, 3009–3025.

ZHU, W. and WONG, T-F. (1997), *The Transition from Brittle Faulting to Cataclastic Flow: Permeability Evolution*, JGR *102*, 3027–3041.

(Received February 27, 2001, revised July 5, 2002, accepted August 1, 2002)

Pure appl. geophys. 160 (2003) 889–908
0033–4553/03/060889–20

Pure and Applied Geophysics

Permeability of Triaxially Compressed Sandstone: Influence of Deformation and Strain-rate on Permeability

JULIANE HEILAND[1]

Abstract—The influence of differential stress on the permeability of a Lower Permian sandstone was investigated. Rock cylinders of 50 mm in diameter and 100 mm length of a fine-grained (mean grain size 0.2 mm), low-porosity (6–9%) sandstone were used to study the relation between differential stress, rock deformation, rock failure and hydraulic properties, with a focus on the changes of hydraulic properties in the pre-failure and failure region of triaxial rock deformation. The experiments were conducted at confining pressures up to 20 MPa, and axial force was controlled by lateral strain with a rate ranging from 10^{-6} to 10^{-7} sec^{-1}. While deforming the samples, permeability was determined by steady-state technique with a pressure gradient of 1 MPa over the specimen length and a fluid pressure level between 40 and 90% of the confining pressure. The results show that permeability of low-porosity sandstones under increasing triaxial stress firstly decreases due to compaction and starts to increase after the onset of dilatancy. This kind of permeability evolution is similar to that of crystalline rocks. A significant dependence of permeability evolution on strain rate was found. Comparison of permeability to volumetric strain demonstrates that the permeability increase after the onset of dilatancy is not sufficient to regain the initial permeability up to failure of the specimen. The initial permeability, which was determined in advance of the experiments, usually was regained in the post-failure region. After the onset of dilatancy, the permeability increase displays a linear dependence on volumetric strain.

Key words: Sandstone, triaxial compression, permeability, stress dependency, strain rate.

Introduction

The investigations described in this paper were motivated by a scientific project on geothermal energy production in Germany. One task of this project is the laboratory examination of the hydro-mechanical coupling behavior of potential reservoir rocks. Conventional use of geothermal heat exploits reservoirs where the petrophysical properties allow the extraction of hot water with a flow rate higher than 50 m^3/h. Reservoir rocks with these properties and in depths with the necessary temperature can be found in Germany only in limited areas. An extension of the potential reservoirs to so-called jointy-porous rocks could lead to

[1] GeoForschungsZentrum Potsdam, Telegrafenberg, D – 14 472 Potsdam, Germany.
E-mail: jheiland@slb.com
Present address: Schlumberger Cambridge Research, High Cross, Madingley Road, Cambridge, CB3 0EL, England.

a much broader usage of geothermal energy. However, geologic units of these rock types require stimulation (e.g., by hydraulic fracturing) in order to achieve the flow rates needed for geothermal energy production. The results of the stimulation cannot be understood without a thorough petrophysical characterization of the rocks. As part of these petrophysical investigations we examined the coupling between mechanical deformation and hydraulic properties in a well-consolidated, low-porosity sandstone. According to the project task, we restrict our laboratory investigations to the brittle field and stress conditions as they are found in the upper few kilometres of the earth's crust.

One of the first publications on the permeability evolution under increasing differential stress was published by ZOBACK and BYERLEE (1975). They determined the permeability of granite under stepwise increased axial stress at effective confining pressures of 390 and 140 MPa. They loaded the rock specimens up to a maximum differential stress of approximately 80% of their peak strength.

More recently, ZHU and WONG (1997) published data on the permeability evolution of porous sandstones under high confining pressures (maximum effective confining pressures reaching 500 MPa). The focus of their investigation lay on the transition from brittle faulting to cataclastic flow. The porosity of all sandstones investigated by them exceeded 14% and the determination of the permeability was made under stepwise increased axial stress.

KEANEY *et al.* (1998) reported about experiments on Tennessee Sandstone, a very low-porosity (4.5–7.5%), quartz-rich sandstone with a permeability in the range of microDarcy (10^{-18} m^2) at 20 MPa effective confining pressure. Using transient pulse permeability measurements together with a triaxial deformation apparatus, they determined the permeability and specific storage characteristics of this rock when undergoing triaxial deformation. The rocks were deformed well beyond failure and into post-failure region, in order to determine the influence of shear bands on the permeability. KEANEY *et al.* (1998) identified five phases a rock is experiencing in triaxial deformation experiments and they linked these phases with the evolution of the pore space and the permeability: (1) During compaction the elastic closure of existing cracks and grain boundaries leads to a decrease of permeability and specific storage. The permeability decreases by a factor of about 2. (2) When dilatancy starts to dominate the deformation regime, the nucleation and growth of microcracks results in an increase of permeability and specific storage. They found that both properties increase to a value considerably higher than their initial values. (3) When approaching peak strength, developing microcracks are mainly aligned parallel to the maximum principle stress direction. During their experiments they found that brittle failure of the sample is accompanied by a transient permeability high, where permeability may increase for a short time extending a factor of 17 compared to the initial value. (4) After peak strength their samples showed strain softening and the deformation becomes highly localized, which eventually results in the formation of a macroscopic fracture. (5) Continuous straining results in frictional sliding on this fracture. At this stage permeability of their samples reaches a stable value, which is about 1.5 times higher than the initial value.

The above-mentioned investigations used the standard triaxial experimental setup with cylindrical specimens laterally constrained by a confining pressure and loaded in an axial direction. The permeability was measured along the cylinder axis. In an experiment, the confining pressure was first increased up to the desired level after which the specimen was deformed by increasing axial stress.

RUISTEN *et al.* (1999) investigated the influence of stress paths other than the standard triaxial deformation on the permeability of weakly cemented sandstones. Although they did not perform standard triaxial tests, they schematically describe the evolution of permeability under increasing differential stress in standard triaxial stress path: Firstly decreasing permeability due to compaction, and at higher stresses increasing permeability.

The above-mentioned publications either dealt with crystalline rocks or clastic rocks with high porosities (above 14%). KEANEY *et al.* (1998) were the only authors publishing data on a sandstone with a porosity lower than 10%. ZHU and WONG (1997) found that sandstones with a porosity above 14% exhibit a different permeability evolution in the brittle field than sandstones with a porosity below 14%. However, there is clearly a lack of investigation in the field of strongly cemented, low porosity ($< 15\%$) sandstones.

The so-called Flechtinger Bausandstein is a Permian (Rotliegend) sandstone consisting of 65% quartz, 9% illite, 13% feldspar and 4% carbonates (STAMMNITZ, 1992) with a porosity between 6 and 9% and is cemented by phyllosilicates. It crops out in northern Germany in the area of Magdeburg (SW of Berlin) in the tectonical uplift called "Flechtinger Scholle." It is a well consolidated, well-sorted, cross-bedded, and fine-grained red sandstone with a permeability in the order of 10^{-17} to 10^{-16} m^2. The mean grain size is between 0.1 and 0.4 mm and the grains are mostly subangular to rounded with concavo-convex grain contacts. Most of the quartz grains are enclosed by a thin phyllosilicate layer and quartz overgrowth is found between the grains. According to its texture and fabric this rock may be taken as a model rock for the expected reservoir rocks of the geothermal energy production project. Two parameters were investigated which control permeability in a triaxial stress field: (1) Difference between continuous and stepwise deformation and (2) dependence of permeability evolution on strain rate. Most authors (e.g, ZHU and WONG, 1997; ZOBACK and BYERLEE, 1975; KEANEY *et al.*, 1998) who investigated permeability under differential stress stopped the axial actuator movement while measuring permeability. However, after the onset of dilatancy, rocks exhibit stable crack growth even under constant load which leads to ongoing deformation (creep), especially in a lateral direction. With a fixed actuator, this causes a decrease of axial stress due to stress relaxation, which leads to increasing lateral strain, which again influences permeability.

KEANEY *et al.* (1998) stated that the stop of the actuator movement does not lead to a steady-state permeable microstructure in the rock, especially in the dilatant deformation phase. In order to achieve a steady-state condition, they reduced the load during the permeability measurements following a standardized procedure, but took care that each measurement was done at successively higher axial strains.

In order to examine the influence of actuator stops on the permeability measurements, we conducted an experiment with a stepwise loading path (exp. III). In this experiment we tried to suppress lateral deformation while the axial actuator was stopped, by controlling lateral strain through the confining pressure.

In another experiment (exp. II) we investigated the influence of the deformation rate on the permeability evolution. For this purpose the sample was deformed continuously with the strain rate varied between $3*10^{-6}$ and $6*10^{-7}$ sec^{-1}.

Experimental Procedures and Equipment

Rock cylinders of 50 mm diameter and 100 mm length were cut from rock blocks of Flechtinger Bausandstein, which had been sampled in a quarry in the region of Magdeburg southwest of Berlin. The end planes of the rock cylinders were ground plan-parallel and perpendicular to the cylinder axis.

The experiments were conducted in a triaxial testing system by MTS consisting of a stiff loading frame and a triaxial cell with a maximum confining pressure of 140 MPa. The experimental setup is schematically shown in Fig. 1. The triaxial cell is equipped with in-vessel sensors, which allow very precise measurement of axial force and sample deformation. Axial strain (ε_{ax}) is detected with two extensometers located 180° from each other on the sides of the specimen. They measure axial strain over a length of 50 mm in mid-height of the specimen. Lateral strain (ε_{lat}) is calculated from measurements of a circumferential extensometer. The readings of both, axial and lateral strain, can be used as a feedback signal for force control. Volumetric strain (ε_{vol}) is calculated from the strain measurements using the equation:

$$\varepsilon_{vol} = \varepsilon_{ax} + 2 \cdot \varepsilon_{lat}. \quad (1)$$

As it is convention in rock mechanics, axial strain (shortening of specimen length) is positive, and lateral strain (thickening of the specimen) is negative. Based on formula (1), increasing volumetric strain indicates that compaction is the dominating deformational mechanism, whereas decreasing volumetric strain indicates dilatancy to be the dominant mechanism.

Usually the experiments were done in lateral strain control with a strain rate between 10^{-6} and 10^{-7} sec^{-1}. We prefer lateral strain control, because in this control mode the failure process can be controlled more precisely than in axial strain control. FAIRHURST and HUDSON (1999) stress that, in order to determine the full stress-strain-curve of rock, one has to carefully choose the control mode. They suggest using lateral strain control when testing rocks with brittle failure behavior. From our experience there is no difference in the results of triaxial tests using axial strain control compared to

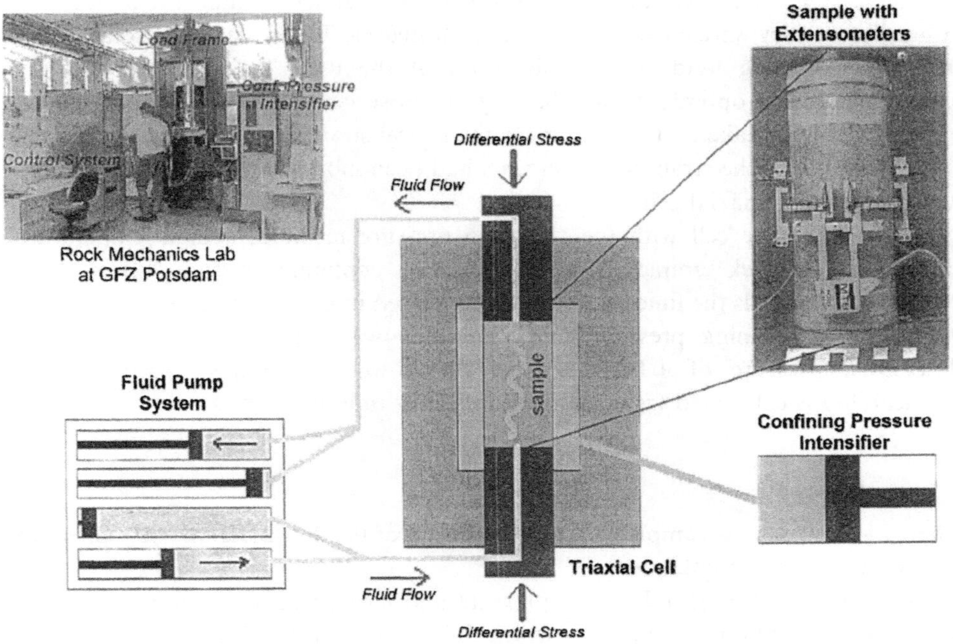

Figure 1
Sketch and photographs of the experimental equipment. The photograph of the rock mechanical lab on the left shows the electronic control system, the load frame with the closed triaxial cell in the center and on the right the confining pressure intensifier. The sample assembly shows the axial extensometers on the sides of the sample and the circumferential extensometer in the center.

tests in lateral strain control. We conducted tests in both control modes on brittle rock (limestone), as well as on relative ductile rocks (metamorphic anhydrate). The results showed no difference for rock strength, failure criterion and elastic properties.

Lateral strain control also has a technical advantage because in our system lateral strain is measured around the whole circumference of the sample and thus deformation is detected wherever the shear band develops. In contrast, the axial extensometers measure between two pins with 50 mm distance and are located at the side of the specimen cylinder. Depending on the location of the shear fracture, deformation in a post-failure region may not be detected by one or both axial extensometers.

Permeability is determined using the steady-state method. Hereby, water is pumped through the specimen in an axial direction using a pump system which allows continuous flow-through with a minimum pressure gradient over the sample length of 1 MPa and an absolute fluid pressure of up to 140 MPa (Fig. 1). The volume resolution of the fluid flow is $1.8 * 10^{-10}$ l. The experiments were conducted at room temperature (20 °C), which is not necessarily representative of the actual temperature in the geothermal well.

The samples were saturated with distilled water under vacuum for at least 24 hours, before they were covered with a thin, shrinkable Teflon sleeve to isolate them from the confining fluid. The combination of the used Teflon sleeve and the extensometers was optimized for achieving the most accurate results. The thickness of the sleeves was taken into account for the lateral strain determination. When the extensometers for the strain measurements had been mounted, the sample assembly was put into the triaxial cell.

After filling the cell with the confining pressure medium, a high-temperature silicone oil, the rock samples were subjected to a confining pressure (σ_3) of 10 or 20 MPa. Afterwards the fluid pressure was increased to a total value between 40 and 90% of the confining pressure, and fluid flow was established by applying a differential pressure of 1 MPa between the lower and upper specimen end. Permeability is calculated from the measured flow rates as follows:

$$k = \frac{q \cdot l \cdot v}{\Delta P \cdot A} \tag{2}$$

where q—flow rate, l—sample length, v—viscosity of the flow medium, ΔP—pressure difference and A—sample area.

Applying a constant differential pressure and measuring flow rate resembles in our opinion the natural conditions, where fluid flow is assumed to be controlled by pressure differences.

When stable hydraulic conditions (i.e., equal and constant flow rates between fluid in- and outlet) were reached, the sample was deformed in axial direction, using lateral strain for force control. The initial permeability under hydrostatic conditions was determined in advance of each deformation experiment. Three different axial loading paths were applied which will be referred to as experiment I to III:

Experiment I: Continuous deformation with a lateral strain rate of $3*10^{-6}$ sec^{-1}.

Experiment II: Continuous deformation with varied lateral strain rate from $6*10^{-7}$ to $3*10^{-6}$ sec^{-1}. The strain rate variations took place in the pre-peak region, whereas after peak strength the strain rate was constant, although lower than in experiment I.

Experiment III: Stepwise deformation of the sample. During the stops, the axial actuator was fixed in order to keep axial deformation constant. By adapting the confining pressure according to changes in lateral deformation, it was attempted to suppress lateral deformation. The reason for these measures was to "freeze" sample deformation, in order to facilitate steady-state conditions for the permeability measurements. Between the stops the specimen was deformed with a lateral strain rate of $6*10^{-7}$ sec^{-1} in the beginning of the experiment, and after the onset of dilatancy the strain rate was increased to $3*10^{-6}$ sec^{-1}. The first two stops took place before the onset of dilatancy, the 3rd stop was exactly at the onset of dilatancy and the 4th stop was shortly after reaching the peak strength, but before a shear band developed.

Results

Experiment I

We conducted four single experiments of this type, which are presented here. The experimental conditions and petrophysical properties of all presented specimens are given in Table 1.

Figure 2 shows the stress-strain curves for all experiments (types I, II and III). The different stages of a triaxial deformation experiment can be seen in these curves: Under increasing differential stress, rocks first exhibit elastic deformation, as indicated by the linear portion of the axial strain curve. The onset of dilatancy, that is the highest volumetric strain reached during an experiment, is marked by large dots and is taking place between 50% and 65% of peak strength. During this phase the axial stress—axial strain curve is still linear (indicating elastic deformation), although the dilatant volumetric strain curve indicates stable crack growth. At about 90% of the peak strength the axial strain curve starts to deviate from linear, which marks the end of elastic deformation and the beginning of plastic deformation. Around peak strength the samples display major volumetric strain changes, whereas axial strain remains relative slight. A sharp drop of differential stress in the post-peak region marks the development of a shear band. As can be seen in Figure 2, the axial strain decreases during the shear band development, which indicates brittle failure behavior. This kind of post-peak behavior cannot be detected when using axial strain as a control parameter. Axial strain control in an experimental situation described above would result in excessively high deformation rates during the shear band evolution, which would result in catastrophic failure of the sample. After the development of a shear band, the differential stress is stable, indicating the residual strength of the rock at this confining pressure. The volumetric strain reached at residual strength is at least 1% (FB 3.5 p), but normally more than 5%. The axial strain on the contrary is normally less than 1%. The axial stress—axial strain curve of samples FB 2.2 p, FB 1.4 and FB 3.6 p show no residual strength, because in these

Table 1

Petrophysical properties and experimental details

Specimen	FB 1.3	FB 2.2 p	FB 3.6	FB 3.5 p	FB 1.4	FB 3.6 p
Initial Permeability [10^{-15} m^2]	0.53	0.017	0.58	0.18	0.04	0.26
Porosity [%]	8.33	–	8.26	–	7.85	6.16
Grain Density [g/cm^3]	2.66	–	2.66	–	2.66	2.59
Confining Pressure [Mpa]	20	10	10	10	20	10
Pore Pressure (at fluid inlet) [Mpa]	19	8	9	8	10	7
λ = Pore Pressure / Confining Pressure	0.9	0.8	0.9	0.8	0.5	0.7
Experiment Type	I	I	I	I	II	III
Lateral strain rate			$3*10^{-6}$ sec^{-1}			varied

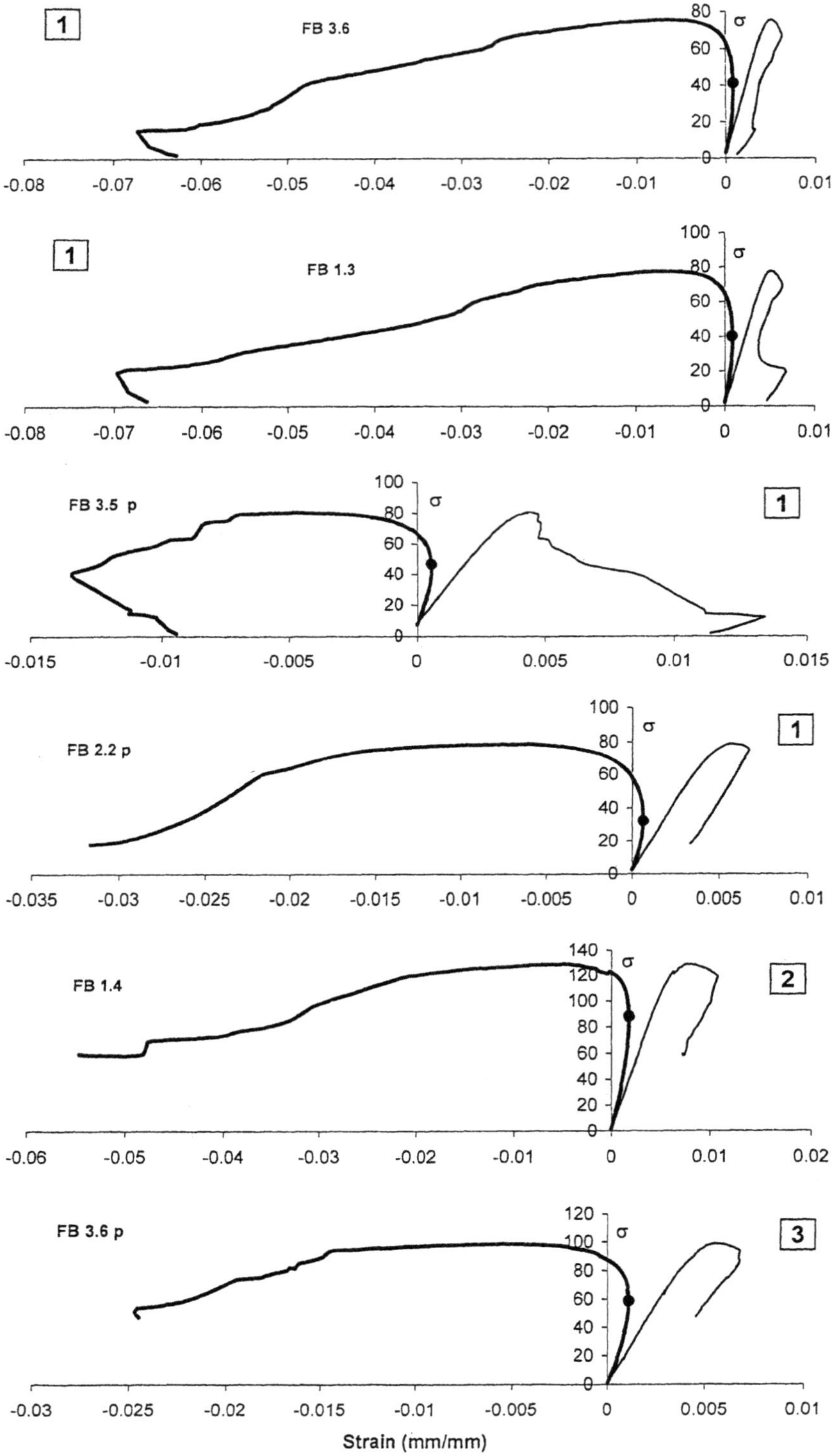
1
FB 3.6
σ
1
FB 1.3
σ
FB 3.5 p
σ
1
FB 2.2 p
σ
1
FB 1.4
σ
2
FB 3.6 p
σ
3
Strain (mm/mm)

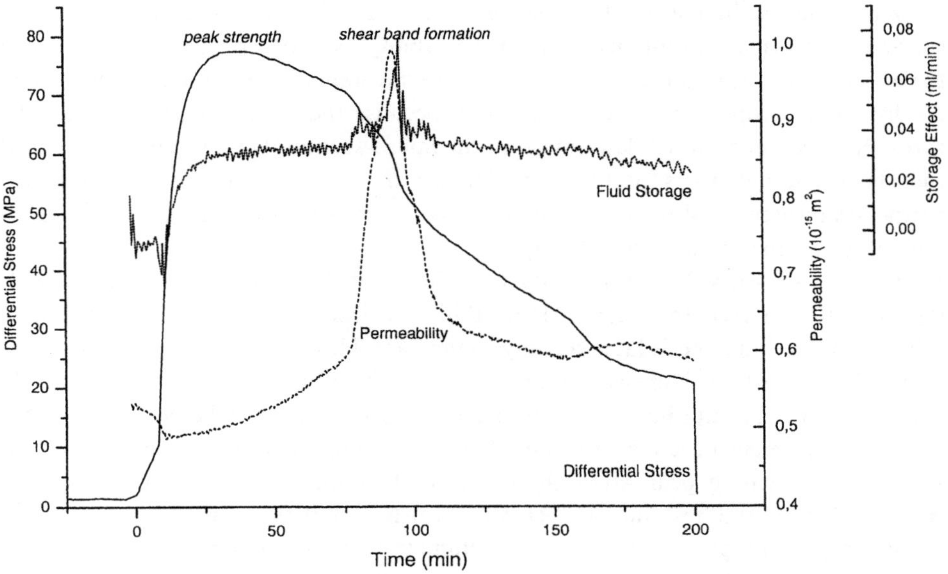

Figure 3
Temporal development of differential stress, permeability and fluid storage of specimen FB 1.3. The graph covers the whole experiment, which was conducted with a confining pressure of 20 MPa and a fluid pressure of 19 MPa ($\lambda = 0,9$).

experiments the shear band developed in such a orientation within the samples that it could not be detected by the axial extensometers.

Figure 3 shows the temporal development of axial stress, permeability and fluid storage of sample FB 1.3. The features shown in this graph are similar for all experiments I. Due to the ongoing deformation of the rock, water is stored in induced microfractures. This leads to slightly larger in-going than out-going flow-rates. The flow-rate difference is a measure for the stored fluid volume. The permeability as calculated from the in-going flow-rates decreases at the beginning of the deformational phase and starts to increase well before the peak strength is reached. In the post-peak region, a sudden stress drop and simultaneously a large peak in permeability mark the development of a shear band. The development of the shear band leads to a sudden increase in pore space, which obviously causes a transient disturbance of the fluid-flow expressed by the permeability peak in Figure 3, as was also described by KEANEY *et al.* (1998).

◀

Figure 2
Plots of differential stress (σ) vs. axial and volumetric strain for all experiments. The onset of dilatancy is marked by dots. The boxed numbers indicate the experiment type. Thin line: axial strain, thick line: volumetric strain.

When the residual strength is reached, e.g., the axial stress is constant and the specimen deforms continuously, the permeability is more or less constant and the experiment was terminated. The residual permeability is very much influenced by the hydraulic properties of the shear band, because the band is cutting diagonally through the sample. In the description of the micro-sections (Fig. 6) one can see that the shear band consists partly of mylonized material which acts as a seal for permeating water, resulting in a residual permeability which was in the same order of magnitude as the initial permeability.

Figure 4 presents the permeability vs. differential stress plots for all experiments (types I, II and III). In general, permeability is decreasing in the early stages of differential loading, and is later starting to increase. However, the point at which the permeability starts to increase cannot clearly be linked to a specific stress state. In experiment FB 3.6 the increase starts beyond peak strength, whereas in the other plots of experiment I the increase starts before reaching the peak strength. However, also before reaching peak strength the start of permeability increase takes place between 25% (FB 3.5 p) and 50% (FB 2.2 p, FB 1.3) of the peak strength.

When plotting permeability against volumetric strain, the dependence of permeability from sample deformation becomes much clearer. The permeability evolution during triaxial deformation is clearly linked to changes in volumetric strain as can be seen from permeability vs. volumetric strain plots (Fig. 5). As long as compaction is the dominating mechanism (as indicated by positive volumetric strain), permeability decreases. After the onset of dilatancy (i.e., when dilating sample deformation is getting the dominant mechanism, which is marked by the point of inflection towards negative volumetric strain), permeability starts to increase. The point of inflection towards negative volumetric strain is the point with the largest volumetric strain (Fig. 5). However, when "zero" volumetric strain is reached again (i.e., where the zero-volumetric-strain-line is crossed after the onset of dilatancy), the permeability reduction, which took place during the compactional phase, is not recovered.

For most experiments of type I (FB 1.3, FB 3.5 p and FB 2.2 p) and for experiment III (FB 3.6 p) the onset of permeability increase is identical with the onset of dilatancy. Sample FB 3.6 exhibits a slower change from permeability reduction to permeability increase, however ultimately (after reaching peak strength), the rate of permeability increase is similar to the other experiments.

In the post-failure region a linear dependence of permeability on volumetric strain can be detected which shows a similar derivative for experiments FB 3.6, FB

▶

Figure 4

Permeability vs. differential stress plot for all experiments: The upper four plots are the experiments of type I (marked by diamonds). The next plot is for experiments of type II (squares) and the last plot is for experiment type III (triangles). For this experiment, the phases with stopped deformation are marked by large dots. All plots contain the experimental data from the start of the differential loading up to the peak strength and beyond into post-failure region until the onset of shear band formation.

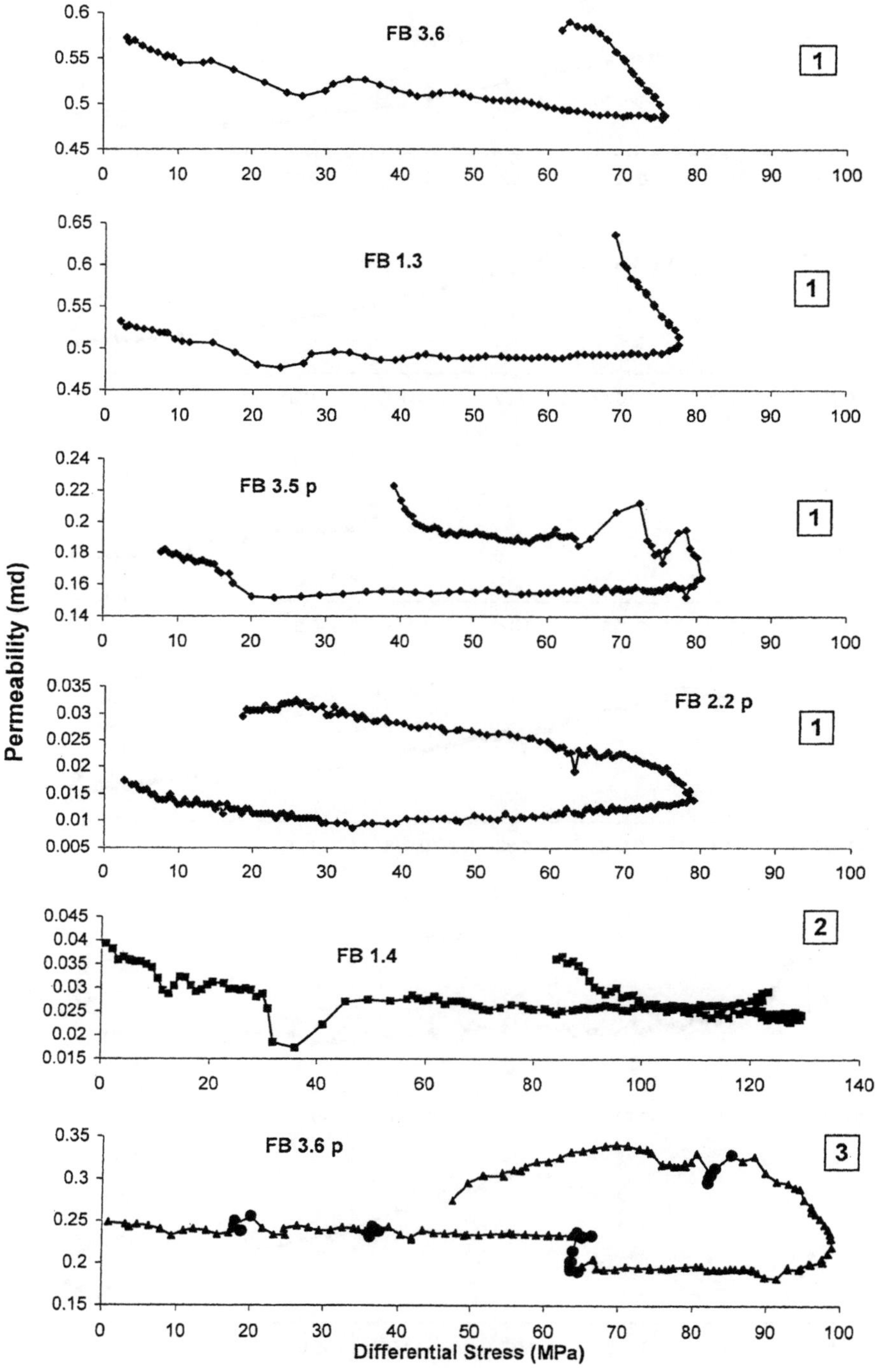
FB 3.6
1
FB 1.3
1
FB 3.5 p
1
FB 2.2 p
1
FB 1.4
2
FB 3.6 p
3
Permeability (md)
Differential Stress (MPa)

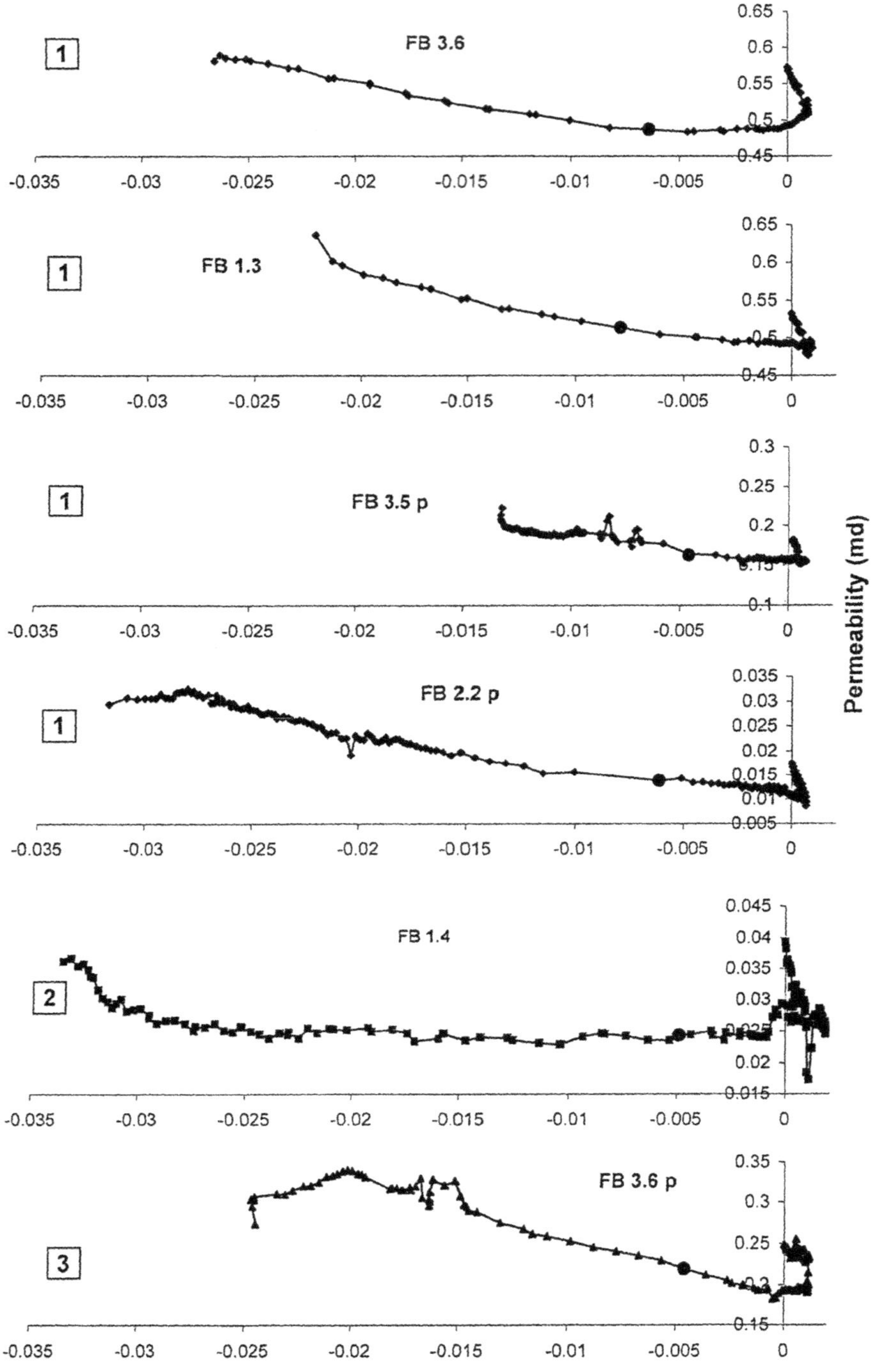
1
FB 3.6
1
FB 1.3
1
FB 3.5 p
1
FB 2.2 p
2
FB 1.4
3
FB 3.6 p
Permeability (md)
Volumetric Strain (mm/mm)
-0.035
-0.03
-0.025
-0.02
-0.015
-0.01
-0.005
0

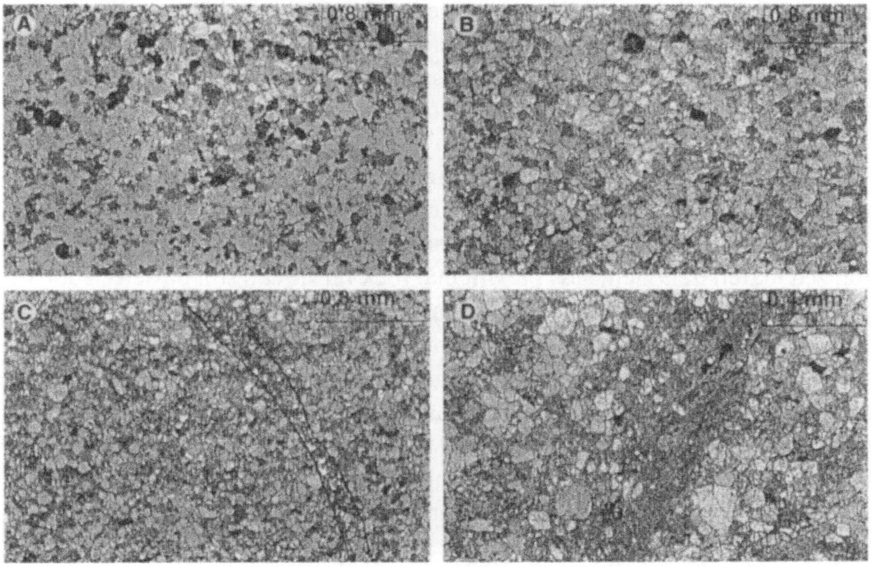

Figure 6
Micro-section photographs of A: undisturbed specimen, B–D: specimen deformed under triaxial stress conditions up to post-failure region. Picture A shows an undisturbed sample. Pictures B–D are from different locations within a tested sample. The loading axis is oriented vertically. In picture C thick lines mark the location of a conjugated shear fracture. For further details see text.

1.3, FB 3.5 p and FB 3.6 p. These samples also display similar initial permeabilities (depending on the effective pressure). Sample FB 2.2 p has a considerably lower initial permeability, which is reflected in a lower rate of permeability increase. Sample FB 1.4 shows no permeability increase, even after the peak strength, which is probably due to the lower strain rate (10^{-6} sec^{-1} instead of $3*10^{-6}$ sec^{-1}).

Figure 6 presents a series of thin sections from an undisturbed sample (picture A) and from a specimen which was deformed up to the residual strength (pictures B–D). Pictures B–D of Figure 6 are taken from one large thin section which was cut along the cylinder axis of a specimen in such a way that the shear band crosses the micro-section diagonally. Picture B was taken in one of the corners of the thin-section which is not crossed by the shear band. Picture C shows a secondary shear fracture and Picture D depicts the central part of the shear band. In this series of micro-section

◄

Figure 5
Permeability as a function of volumetric strain plotted for all experiments: The upper four plots are the experiments of type I (marked by diamonds). Experiment type II is marked by squares and experiment type III by triangles. Large dots mark the data points at peak strength. The plotted data contain the experimental phases from the start of differential loading, beyond peak strength and extending to the beginning of shear band formation (as in Fig. 4).

pictures the fabric changes can be examined which occur with increasing damage. In picture B (Fig. 6), only few micro-fractures are found and the fabric shows little change as compared to the undisturbed sample. The micro-fractures are randomly distributed, and no preferred orientation of the fractures can be detected.

The secondary shear fracture shown in picture C has a thickness of about 0,4 and 0,8 mm, and its margins are relatively sharp. This fracture is probably the onset of a conjugate shear band which started to develop but did not intersect the sample totally. The filling of this fracture consists mainly of crushed grains and some mylonized material, nonetheless the original rock fabric can still be detected. Crushed grains can frequently be found in this detail of the thin section, with a concentration of crushed grains in some parts of the picture leaving less deformed spots in-between.

The main shear band which crosses picture D of Figure 6 is about 1–2 mm thick and its margins are not very sharp. The shear band itself consists of mylonized material with few residuals of crushed grains "floating" in it. In general, the original rock fabric is completely destroyed within the shear band. The crushed grains in the vicinity of the shear band show a preferred orientation of the fractures parallel to the loading axis.

Experiment II

With this experiment conducted on one specimen, we investigated the influence of strain rate on the evolution of permeability. During the compactional phase of deformation fluid flow reacts very sensitively to strain rate changes (Fig. 7C), where the large negative permeability peak is connected to an increase of strain rate from $6*10^{-7}$ sec^{-1} to $3*10^{-6}$ sec^{-1}. After the onset of dilatancy a strain rate decrease triggers a change from permeability increase to decrease. However, after the onset of dilatancy, a change of strain rate does not cause a large flow rate disturbance (i.e., no spike in the permeability curve). This is probably due to a higher degree of hydraulic connectivity caused by the dilatant sample deformation. Figure 7B illustrates the permeability evolution for the whole experiment until the development of the shear band. After peak strength the strain rate was kept constant at 10^{-6} sec^{-1} (this is in between the minimum and maximum strain rate used in the earlier stages of the experiment). It can clearly be seen that permeability stays constant after the peak strength is reached and only increases shortly before the shear band develops.

This experiment demonstrates that the evolution of permeability is very sensitive to the deformation rate. The variation of the strain rate was less than one order of magnitude and still, after the onset of dilatancy, the permeability evolution changed between permeability increase at the higher strain rate and a constant permeability at lower strain rates.

Experiment III

The results of this experiment performed on one specimen can be divided into observations made during the elastic phase of the deformation and observations

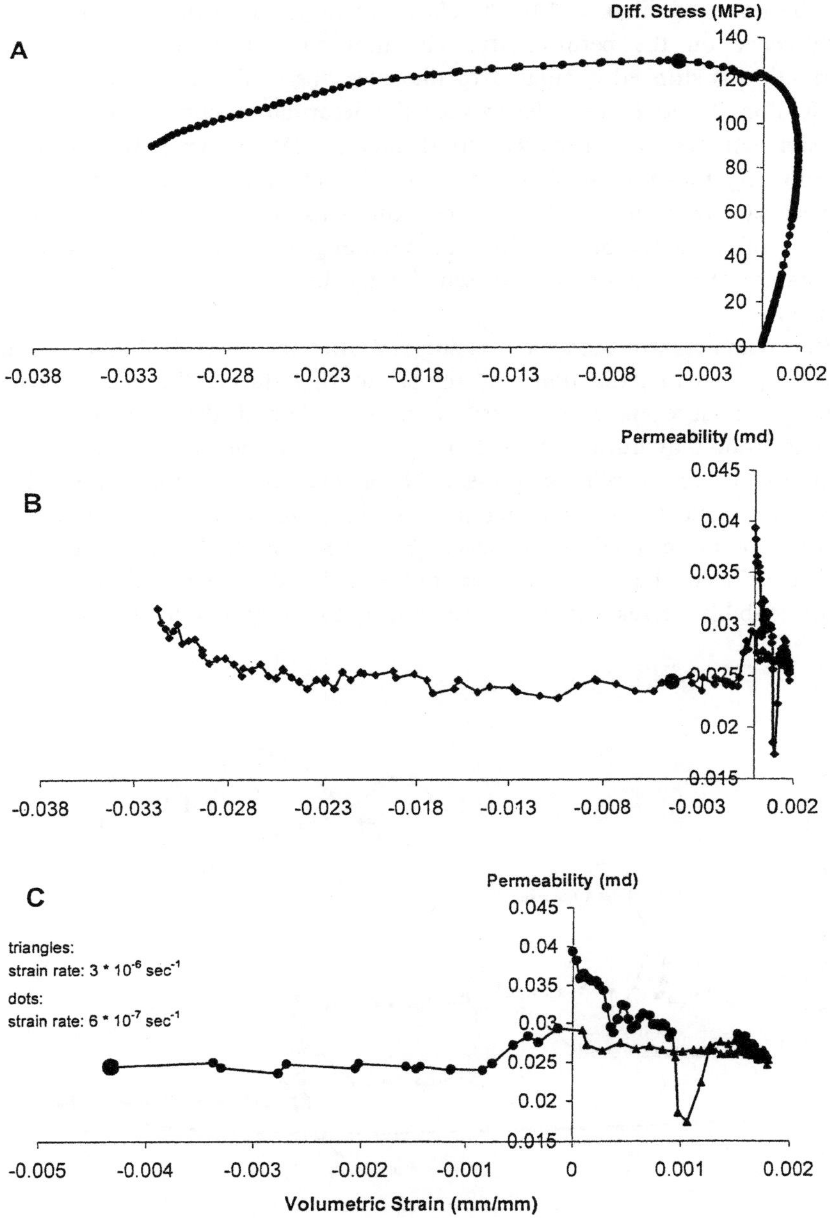

Figure 7
Experiment type II: specimen FB 1.4. A: Differential stress plotted vs. volumetric strain. B: Permeability vs. volumetric strain plot, peak strength marked by a large dot, data shown up to development of shear band. C: detail of plot B, data shown up to peak strength.

after the onset of dilatancy. In the elastic region, a stop of the deformation has no influence on the permeability. The flow rates remain constant when the deformation is stopped compared to the rates observed before and after the stop (Fig. 8, stops 1 and 2). In order to keep the deformation constant during the stop, the axial actuator movement was fixed, and the lateral strain was controlled by the confining pressure. A slight increase of confining pressure (about 0.5 MPa) was necessary to keep lateral strain constant. This increase happened directly after stoppage of the actuator and then the confining pressure (and the lateral strain) were stable. The axial stress decreased slightly due to stress relaxation in the rock specimen.

When the actuator movement is stopped after the onset of dilatancy, the axial stress drop due to relaxation was similar to that during the elastic phase. The confining pressure increase necessary to stop lateral deformation was barely different from that during stops 1 and 2, however the increase was not stable, which means the confining pressure kept increasing. Permeability decreased continuously due to the ongoing increase in confining pressure, and no stable flow-rate conditions could be achieved (Figure 8, stop 3). Before triaxial deformation is continued, the confining pressure was reduced to its initial value. However, the permeability decrease that occurred during this stop was not recovered (Fig. 8).

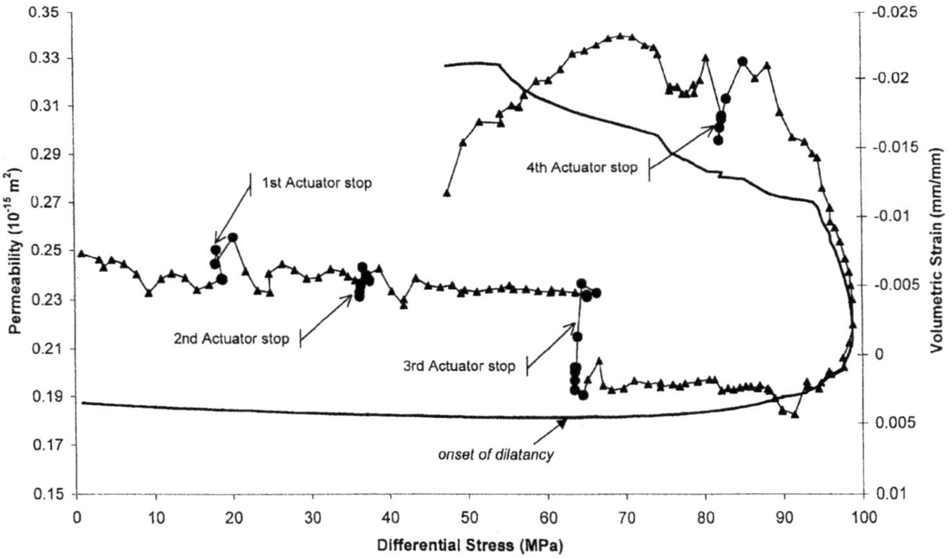

Figure 8

Experiment III (sample FB 3.6 p): Permeability (triangles) and volumetric strain (thick line) plotted against differential stress. Phases with "frozen" sample deformation are marked with large dots. Data shown up to peak strength.

During the 4th stop, which was in the post-peak region, the confining pressure displayed the same behavior as during stop 3 and the permeability decreased. However, when restarting the deformation, the permeability reduction was recovered (Figure 8, stop 4).

In general, the overall permeability evolution of this experiment was similar to that found in continuous deformation experiments. Permeability decreased due to compaction and started to increase after the onset of dilatancy (Fig. 4 where the data of all experiments are shown). The data of experiment type III perfectly fit to the data of the continuous experiments of type I.

Discussion and Conclusion

In the continuous experiments of type I, we found a pronounced reduction of the permeability during the compactional phase of the experiment. As presented in Figure 5, for six experiments this permeability reduction was not recovered until the peak strength of the rock was reached, but was recovered in the post-peak region before the forming of a shear band. ZHU and WONG (1997) found that high-porosity sandstones show no permeability increase, even in the brittle field. Only for Darly Dale sandstone, which has the lowest porosity (~14%) of their sandstones, they reported a small permeability increase after the peak strength. However, in all their experiments, the compaction-driven permeability was never recovered in the post-failure region.

In contrast, ZOBACK and BYERLEE (1975) showed that in triaxial compression tests, the permeability in crystalline rock first decreases due to compaction and starts to increase well before the peak strength is reached. They also found a connection between the onset of dilatancy and the turning point in permeability evolution from decrease to increase. In contrast to our results they reported that the compaction-driven decrease of permeability was recovered well before the peak strength was reached, since they stopped their experiments at about 80% of the peak strength. Depending on the effective pressure, the permeability increased in their experiments by a factor of about 2 to 3 between the initial permeability and the permeability under maximum load (80% of peak strength).

Comparison of the results of these previous studies with our results suggests that the permeability evolution of low-porosity sandstones (< 14%) can be described as a combination of characteristics known from crystalline rocks and from porous sandstones. The change from compaction-driven permeability reduction to dilatancy-induced increase of permeability in the pre-failure region is similar to the behavior of crystalline rock as described by ZOBACK and BYERLEE (1975). However, the permeability increase due to dilatancy is considerably smaller than in the case of crystalline rock, which is similar to the behavior of sandstones with a porosity of about 14% as described by ZHU and WONG (1997).

In experiments of type I a linear dependence between permeability and volumetric strain was established which shows a similar slope for all conducted experiments. ZHU and WONG (1997) replotted the data of Westerly Granite given by ZOBACK and BYERLEE (1975) as permeability vs. porosity plots. In these plots, one could assume a fairly linear dependence between these parameters, too, which would be comparable to our results, as volumetric strain includes the porosity change. In an earlier paper (HEILAND and RAAB, 2001), we were able to show that the overall specimen volume, which was calculated from extensometer readings during a continuous experiment, is similar to the fluid volume change which resembles the porosity change of the specimen. In contrast, the sandstones investigated by ZHU and WONG (1997) showed no linear trend on the permeability vs. porosity plots, which would again underline the difference in hydro-mechanical coupling behavior between high-porosity and low-porosity sandstones and the similarity between low-porosity sandstones and crystalline rocks.

In the experiment of type II, a strong dependence of permeability evolution even on small changes in strain rate was detected. At a low strain rate ($6*10^{-7}$ sec^{-1}), permeability did not increase, even after the onset of dilatancy. This means that although the specimen volume increased (i.e., volumetric strain decreased), the permeability did not change, and thus the dependence of permeability on volumetric strain has a different (i.e., zero) slope than in experiment I. This implies that the gradient of the volumetric strain – permeability curve not only depends on the petrophysical properties of the rock, but also on the experimental conditions. Increasing the strain rate by less than one order of magnitude to $3*10^{-6}$ sec^{-1} resulted in an immediate change towards the normal permeability evolution that is increasing permeability during dilatancy.

Combining the results of experiments I and II, two possible theories for the nature of the dependence of permeability on volumetric strain could be established:

1. The derivative is a combination of characteristic behavior of the rock and strain rate, i.e., each rock type shows a similar derivative only at one specific strain rate or in a specific range of strain rates.
2. The derivative depends only on strain rate and is similar for all low-porosity sandstones.

The permeability reduction during the compactional phase of a continuous experiment may be explained by an increase of the tortuosity of the pore space. If one assumes this increase of tortuosity to be time-dependent, the influence of the strain rate on the permeability evolution may be explained. The time-dependence may be caused by the effect that at low strain-rates the rock fabric has more time to reorganize and respond to the stresses than at higher strain rates.

From our experiments we can conclude that the determination of the evolution of permeability during compression experiments is best made in continuous deformation experiments, because each stop of the actuator movement causes additional disturbance of the hydro-mechanical system in the rock. A stop of the axial actuator movement causes a significant drop of the axial stress while the specimen continues to deform. Lateral deformation may be suppressed by controlling it with the confining pressure as was done in experiment III. In order to suppress lateral deformation, an increase of confining pressure is needed and an increase of less than 1 MPa confining pressure was enough to ensure "zero" sample deformation. However, even this small confining pressure increase resulted in a strong permeability decrease. This permeability reduction was not recovered when the triaxial compression continued under initial confining pressure conditions. It is widely known that under hydrostatic conditions permeability decreases with increasing effective pressure. This is due to the fact that increasing effective pressure leads to compaction of the pore space, which results in permeability reduction. However, in our experiment III the confining pressure increase was only intended to suppress increasing lateral strain and no compaction of the sample took place due to this confining pressure increase.

Micro-section analysis revealed that triaxial compression induced microcracks oriented parallel to the load axis, causing an increase of the permeability after the onset of dilatancy. The initiation of microcracking is generally assumed to coincide with the onset of dilatancy. Since the confining pressure acts perpendicularly to the orientation of these cracks, an increase of it results in a reduction of crack-opening which eventually leads to a permeability decrease as observed in experiment III.

With the combination of continuous triaxial deformation experiments and steady-state permeability measurements, one is able to investigate the permeability evolution at continuing rock deformation under well-defined states of stress. Combined with precise strain measurements at the specimen surface, these experiments generate well-constrained results for the understanding of the hydromechanical coupling in low-porosity sandstones.

Acknowledgements

The author would like to thank G. Borm for enlightening discussions concerning the geomechanical aspects of this work. S. Raab and E. Spangenberg provided valuable contributions to the hydraulical and petrophysical part of this work. An early review by D. Bruhn substantially enhanced the manuscript. L. Liebeskind and O. Ryll were very supportive in the laboratory. The experiments were conducted in the rock mechanical laboratory of GFZ Potsdam. The contribution of two anonymous reviewers is acknowledged.

References

Fairhurst, C. E. and Hudson, J. A. (1999), *Draft ISRM Suggested Method for the Complete Stress-strain Curve for Intact Rock in Uniaxial Compression*, Int. J. Rock Mech. and Min. Sci. *36*, 279–289.

HEILAND, J. and RAAB, S. (2001), *Experimental Investigation of the Influence of Differential Stress on Permeability of a Lower Permian (Rotliegend) Sandstone Deformed in the Brittle Field*, Phys. Chem of the Earth A *26(1–2)*, 33–38.
KEANEY, G. M. J., MEREDITH, P. G., and MURRELL, S. A. F. (1998), *Laboratory Study of Permeability Evolution in a 'Tight' Sandstone under Non-hydrostatic Stress Conditions*, SPE paper 47 265.
RUISTEN, H., TEUFEL, L. W., and RHETT, D. (1999), *Influence of Reservoir Deformation and Permeability of Weakly Cemented Sandstone Reservoirs*, SPE Reservoir Eval. and Eng. *2(3)*, 266–272.
STAMMNITZ, T. (1992), *Entwicklung einer Methode zur Untersuchung der Belastungsabhängigkeit von Porosität und Permeabilität sowie deren Beziehungen zur Gesteinsdeformation*, Ph.D. Thesis Techn. Univ. Clausthal.
ZHU, W. and WONG, T. F. (1997), *The Transition from Brittle Faulting to Cataclastic Flow: Permeability Evolution*, J. Geoph. Res. B *102*, 3027–3041.
ZOBACK, M. D. and BYERLEE, J. D. (1975), *The Effect of Microcrack Dilatancy on the Permeability of Westerly Granite*, J. Geoph. Res, *80(5)*, 752–755.

(Received February 2, 2001, revised September 24, 2001, accepted October 1, 2001)

Pure appl. geophys. 160 (2003) 909–927
0033 – 4553/03/060909–19

Pure and Applied Geophysics

Comparison of Measured and Modelled Hydraulic Conductivities of Fractured Sandstone Cores

SALIMA BARAKA-LOKMANE,[1,2] RUDOLF LIEDL[1] and GEORG TEUTSCH[1]

Abstract—A new method for characterising the detailed fracture geometry in sandstone cores is presented. This method is based on the impregnation of samples with coloured resin, without significant disturbance of the fractures. The fractures are made clearly visible by the resin, thus allowing the fracture geometry to be examined digitally. In order to model the bulk hydraulic conductivity, the samples are sectioned serially perpendicular to the flow direction. The hydraulic conductivity of individual sections is estimated by summing the contribution of the matrix and each fracture from the digital data. Finally, the hydraulic conductivity of the bulk sample is estimated by a harmonic average in series along the flow path. Results of this geometrical method are compared with actual physical conductivity values measured from fluid experiments carried out prior to sectioning. The predicted conductivity from the fracture geometry parameters (e.g., fracture aperture, fracture width, fracture length and fracture relative roughness all measured using an optical method) is in good agreement with the independent physical measurements, thereby validating the approach.

Key words: Fractured sandstone, hydraulic conductivity, fracture geometry, visualisation of natural fractures.

1. *Introduction*

The importance of the fracture geometry as a control on flow processes is widely recognised in the disposal of radioactive waste, petroleum exploitation and underground energy storage. The hydrogeological response is governed by the size of the fractures, their connectivity and by the roughness of the fracture surfaces. The key to understanding flow in naturally fractured rocks lies therefore in determining the fracture geometry parameters such as fracture aperture, fracture width, fracture length and fracture relative roughness.

Analysis of fluid flow behaviour in naturally fractured porous media is very difficult to perform because the fracture geometry is generally complex and unknown. Laboratory-scale studies of discrete fractures lead to better understanding

[1] Department of Applied Geology, University of Tübingen, Germany.
[2] Department of Geology and Geophysics, University of Edinburgh, UK.
E-mail: S.Baraka-Lokmane@glg.ed.ac.uk

of the relationship between flow properties and fracture geometry. The determination of the fracture geometry as well as the analysis of the interaction between fracture and matrix is thus necessary to understand flow properties in fractured porous media. In this paper a method is presented which allows a comparison between measured and modelled hydraulic conductivities of fractured sandstone cores, based on a detailed quantitative characterisation of the fracture geometry in three dimensions.

2. *Description of the Samples*

For investigating the complexity of the flow in fractured porous media, it is important to choose an outcrop with high matrix porosity and a relatively dense fracture network. As a minimum, the sample should contain representative and visible fractures. For this study the samples originate from the middle Stubensandstein Unit within the Middle Keuper of the Southwest German Trias (Fig. 1). It is extremely difficult to obtain undisturbed bore core, i.e., without inducing the opening of the natural fractures and therefore some irreversible deformation of the bore core. The manipulation of the bore core drill is then very delicate. Once the cores are obtained, metal rings were placed around the samples prior to further rock cutting to prevent further damage. The sampling was carried out with a special electric bore core drill with a diamond drilling bit (Hilti, DD-160E) (Fig. 2). The bore cores have a diameter of 0.10 m and a length of up to 0.70 m.

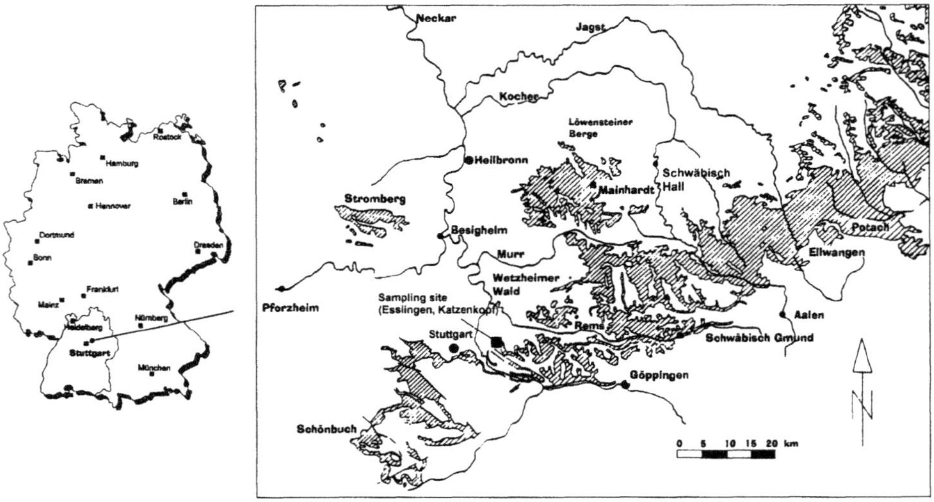

Figure 1

Geographical state of the sampling site (the hatched area represents the location of the sandstones belonging to the Keuper in the middle and eastern part of southwest Germany (ULFRECHT, 1987).

Figure 2
Sampling with the diamond drill.

The type, the nature and the orientation of the fractures have an effect on the permeability of the sample. Poor choice of sample could mean that the permeability of the matrix would dominate the permeability of the whole sample. Initially 41 samples were chosen with a diameter of 0.10 m and a length of up to 0.12 m. Of the 41 samples, 8 were chosen for detailed investigation on the basis of the following criteria:

- Visible and open fractures;
- Unhealed and unfilled fractures;
- Fractures with the same orientation as the bore core axis;
- Fractures cutting the bore core from the top to the bottom.

These criteria preferentially select samples where the fracture is the determining factor for the flow.

It is known that swelling clays can be important in controlling the water permeability. In order to interpret the results of the water permeabilities, a detailed study of the mineralogical nature of the samples and the type of the cement were carried out. The mineralogical study shows that the samples are classified as arkose with well to poorly sorted grain size. Quartz is the dominant framework of the sandstone samples (70% to 80%), feldspars are present with the amount of 10% to 25% and micas are rare. Cements are abundant and the most common are clay minerals (including kaolinite, smectite and illite), carbonate (calcite or dolomite) and quartz. Figure 3 shows an example of the coexistence of clay and carbonate minerals.

The dominant cements of the different samples are listed in Table 1. According to the type of the dominant cement, the samples can be subdivided into three groups:

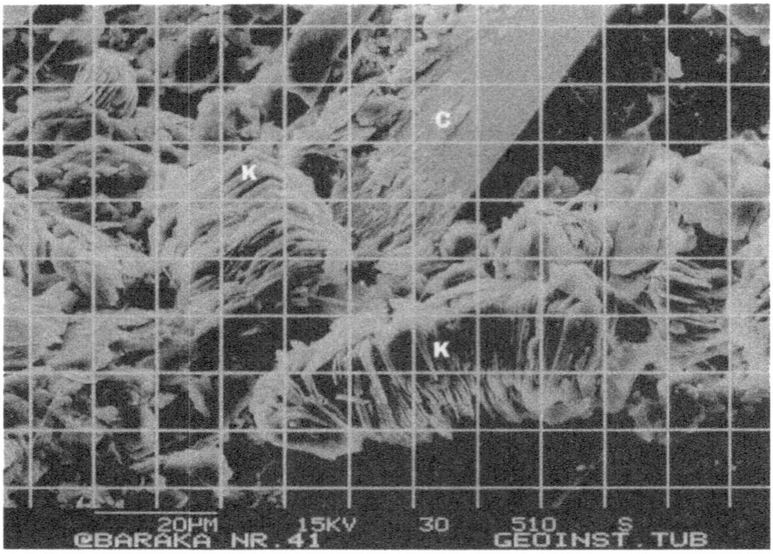

Figure 3
SEM photo of a cement containing kaolinite (K) and carbonate minerals (C).

Table 1

Cement types and the dominant cement for the eight selected samples

Sample Number	Type of Cement	Dominant Cement
2	kaolinite, carbonate, quartz	kaolinite, carbonate
3	kaolinite, carbonate	kaolinite, carbonate
15	carbonate, smectite, illite, kaolinite, quartz	carbonate, smectite
18	carbonate, kaolinite	carbonate
20	carbonate, kaolinite	carbonate
22	carbonate, quartz, smectite	carbonate, smectite
23	carbonate, quartz, smectite	carbonate, smectite
41	kaolinite, quartz, carbonate	kaolinite

- Samples where carbonate is the dominant cement (samples 18 and 20);
- Sample where kaolinite is the dominant cement (sample 41);
- Samples where the dominant cements are clay minerals (kaolinite or smectite) and carbonate (samples 2, 3, 15, 22 and 23).

3. Determination of the Water Permeability

The water permeability test is accomplished with eight chosen samples, which were first water-saturated under vacuum.

3.1. Equipment and Experimental Method

Determination of permeability using water is carried out with the equipment DIN 18130 represented in Figure 4. The sample is sealed in a latex membrane to avoid leakage and a confining pressure of at least 3×10^5 Pa is applied; the equipment allows pressures up to 7×10^5 Pa to be applied. The water present in the storage container is pressed through a saturated sample and with the help of burettes; the volumes of the water flow input and output are measured. The water of the output flows again to a closed storage container where an opposite pneumatic pressure is applied to develop a saturation pressure. To make sure that the samples are completely saturated, the measurements of water permeability are carried out at high saturation pressures ranging from 2×10^5 Pa to 5×10^5 Pa; thereby, the air residue is compressed and the saturation of the sample increases. Measured pressure differences across the sample were in the range 1×10^4 Pa to 7×10^4 Pa and flow rates ranged 1.77×10^{-8} m^3/s to 1.03×10^{-7} m^3/s.

3.2. Calculations and Results

The intrinsic water permeability k_l is calculated using Darcy's law:

$$k_l = \frac{Q \cdot \mu_w \cdot L}{A \cdot \rho_w g \cdot \Delta h}, \tag{1}$$

where Q is the water volumetric rate of flow, μ_w is the dynamic viscosity of water, L is the sample length, A is the sample cross-sectional area, ρ_w is the density of water, g is the acceleration due to gravity and Δh is the difference in hydraulic head in the water between the bottom and the top of the sample, $\Delta h = h_1 - h_2$.

Figure 4
Equipment for the measurement of water permeability ($P_1[Pa]$: input pressure, $P_2[Pa]$: output pressure, $P_z[Pa]$: confining pressure, $Q_1\lfloor m^3/s \rfloor$: water input volumetric rate of flow, $Q_2\lfloor m^3/s \rfloor$: water output volumetric rate of flow).

The water permeability tests were performed under conditions that ensure laminar flow. The intrinsic matrix permeability was measured from small bore cores of a volume of 50 cm^3 (4 cm in length and 4 cm in diameter) containing no visible fracture, taken from each of the eight borehole samples. Intrinsic permeability values k_l measured for the eight samples vary between 12 md and 47 md and intrinsic matrix permeability values k_{lm} vary between 7 md and 26 md. Individual measurements are given in Table 2.

4. *Determination of the Fracture Geometry Parameters*

To date, research work related to the determination of the fracture geometry is in most cases concentrated on low permeability rock (granite) where the hydraulic properties are dominated by fractures (PYRAK-NOLTE *et al.*, 1987; HAKAMI, 1988; GENTIER *et al.*, 1989; COX *et al.*, 1990; DOLLINGER *et al.*, 1994). In contrast, this study focuses on fractured sandstone, a widely used type of hard rock aquifer with matrix porosity and fracture conductivity.

A new method for determining the detailed fracture geometry in sandstone cores is developed. This method is based on the impregnation of samples with resin without disturbing the pore space or the fractures in the samples. For this purpose a resin injection apparatus was developed (Fig. 5). The sample is wrapped in a latex membrane and then fixed into a core holder. The resin is injected through the sample from the top to the bottom using air pressure. At the same time, the apparatus is connected at its base to a vacuum pump.

Several resins were tested with the purpose of seeking the most suitable resin to use for the impregnation. The best resin was found to be the SPURR resin because of its low viscosity (60 cps at 20 °C) and its long pot life (2 days). The technical details and physical properties of the tested resins are displayed in Table 3.

A number of conclusions can be drawn from the resin injection experiments:

- The sample should be oven dried and placed under vacuum before injecting the resin;
- The optimum injection period is around 1 hour;
- The injection pressure varied between 0.05 bar and 0.1 bar; the resin is sequentially injected at 6 pressure levels (0.05 bar, 0.06 bar, 0.07 bar, 0.08 bar, 0.09 bar and 0.1 bar) and the duration of each pressure level is 10 minutes.

Table 2

Hydraulic parameters of the eight selected samples (k_l: intrinsic permeability of the sample, k_{lm}: intrinsic matrix permeability)

Sample Number	2	3	15	18	20	22	23	41
k_l (md)	38.6	27.9	17.3	43.2	46.9	12.3	15.1	25.2
k_{lm} (md)	13.50	26.60	12.45	25.99	26.68	10.42	12.72	7.48

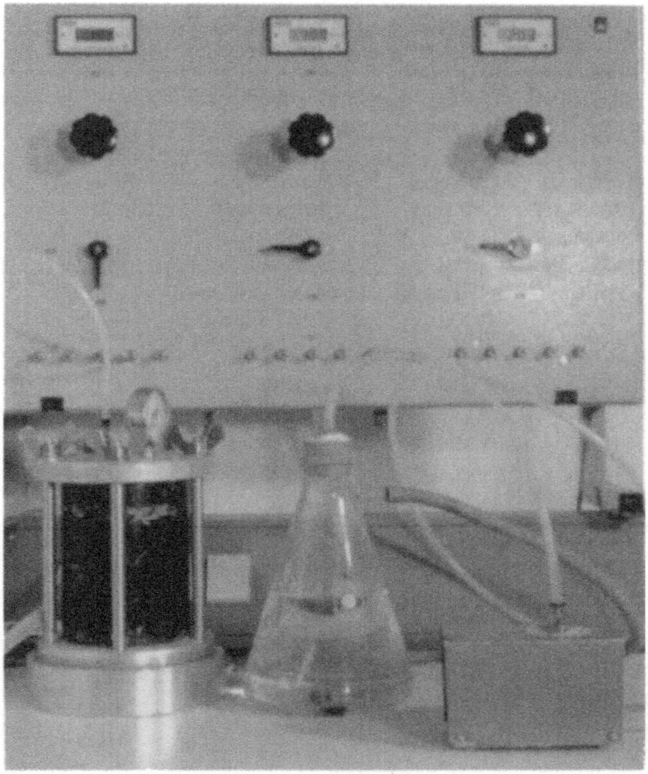

Figure 5
Resin injection equipment.

4.1. Results

Figures 6 and 7 show two examples of the upper and the lower section of the same sample impregnated with the resin injection apparatus using the techniques described above. The results clearly show both the sedimentary layers and the location of the fractures.

For modelling purposes, once the impregnation has been performed, the samples are sectioned serially perpendicular to the flow direction (Fig. 8); the eight samples were cut into core-slices of 5 mm length, resulting in a total of 128 cross-slices exhibiting 196 fracture traces. For each core slice, the parameters of the fracture geometry such as the fracture aperture and the fracture width were calculated using the image analysis program OPTIMAS (Optimas Corporation, Bothell, Washington). Based on this analysis, the rough fracture walls are approximated by a series of rectangles which are numbered by the index i, and the length l_i as well as the aperture e_i of each rectangle are determined (Fig. 9). The accuracy of the measurements was about 0.25–1.25 μm with the image size used.

Table 3

Technical details and physical properties of the tested resins

Method	Resin				
Compound	RTV 141 Resin + Hardener (Rhône Poulenc)	EPOFIX Resin + Hardener (Struers)	Sika Injektion 26 2 Compounds (Sika, Nagra, Switzerland)	EPOXY (Resin + Hardener) (Gbrot, France)	SPURR 4 Compounds (Plano, United Kingdom)
Dosing	70 g Resin + 7 g Hardener	50 g Resin + 6 g Hardener	Compound (A): 1.14 g + Compound (B): 1 g	1.12 g Resin +0.95 g Hardener	VCD: 10 g DER: 6 g NSA: 26 g DMAE: 1 g
Coloration	Coloured in black with araldite colouring paste (Gbrot)				
Viscosity in cps (20 °C)	4000	550	100	150	60
Pot Life	4 Hours	30 Minutes	2½ Hours	50 Minutes	2 Days
Hardening Time	2 Hours at 100 °C	8 Hours at 20 °C	1–2 Days at 20 °C	3 Hours at 60 °C	3 Hours at 70 °C

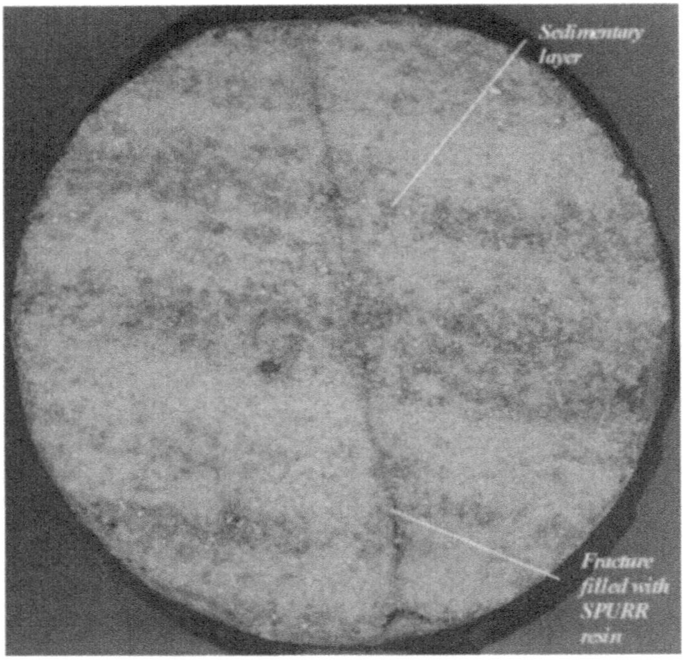

Figure 6
Pixel map of the upper slice of sample 2 impregnated with SPURR resin coloured in black. Note the filled fracture, and the sedimentary layers.

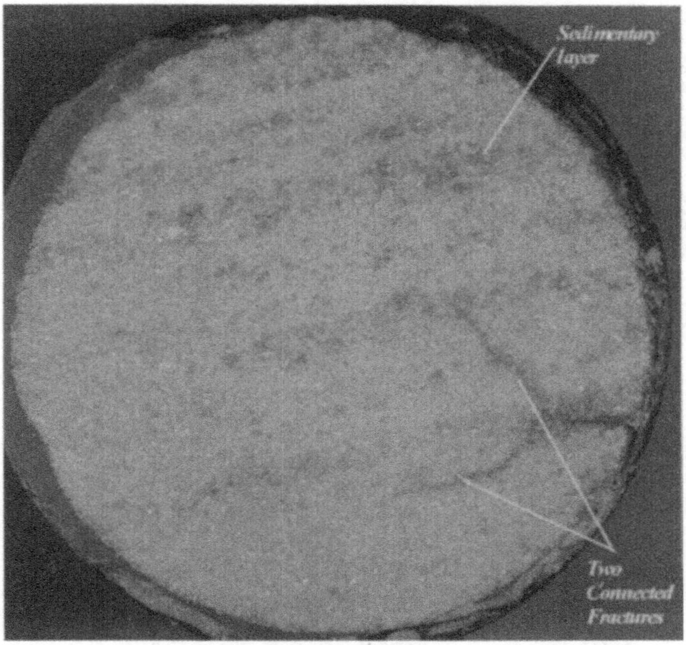

Figure 7
Pixel map of the lower slice of sample 2 impregnated with SPURR resin coloured in black. Note the filled fractures, and the sedimentary layers.

Using series of slice pictures with the fracture traces, various three-dimensional images of the samples were produced. Figures 10 and 11 show three-dimensional visualisations of the fractures of the different samples. The study of the slices of 5 mm thickness and the obtained three-dimensional images of the samples allow a detailed description of the fracture system in terms of the fracture aperture, fracture width, fracture length, and the connection of the fractures.

4.2. Determination of the Fracture Roughness

Laboratory measurements and field observations have shown that fluid flow within natural fractures with high connectivity is controlled by the geometric characteristics of the rough walls (WANG and NARASIMHAN, 1988). Therefore the topography of the fracture walls had to be quantified.

An optical method is used in this study for the determination of the fracture roughness. Various sections of the fracture surface with dimension 1 cm $\times$ 1 cm are photographed using a scanning electron microscope (Fig. 3) and the elevations of the asperities are then calculated.

In the case of rough laminar flow, the effective hydraulic conductivity (K_f) of an equivalent effective fracture is given by DE MARSILY (1981):

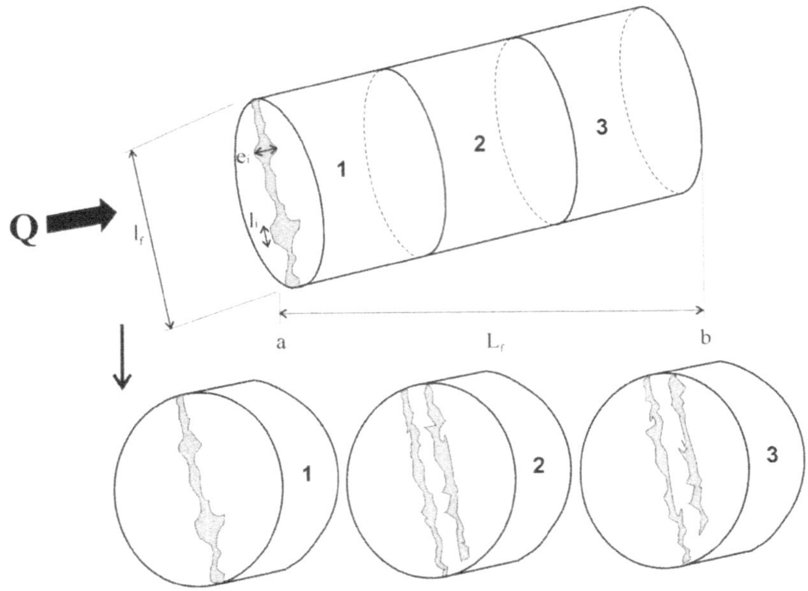

Figure 8
Schematic diagram of a bore core sample in which the fracture aperture is characterised for each slice of the sample as a constant along each streamline but which varies between streamlines ($l_f[m]$: fracture width, $l_i[m]$: fracture width of the section i, e_i $[m]$: effective fracture aperture of the section i, $L_f[m]$: fracture length).

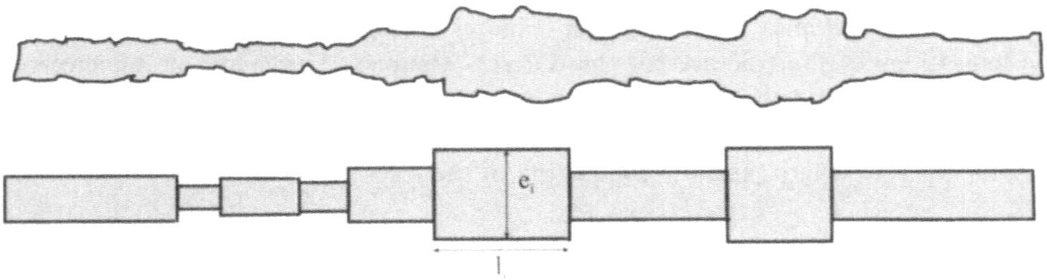

Figure 9
Cross-section of a real fracture and the approximation with rectangles in the digital analysis.

$$K_f = \left[\frac{\rho g}{12\mu(1 + 8.8Rr^{1.5})}\right] e^2 \ , \tag{2}$$

where e is the effective fracture aperture, ρ is the density of fluid, μ is the dynamic viscosity of fluid, g is the acceleration due to gravity, Rr is the fracture relative roughness.

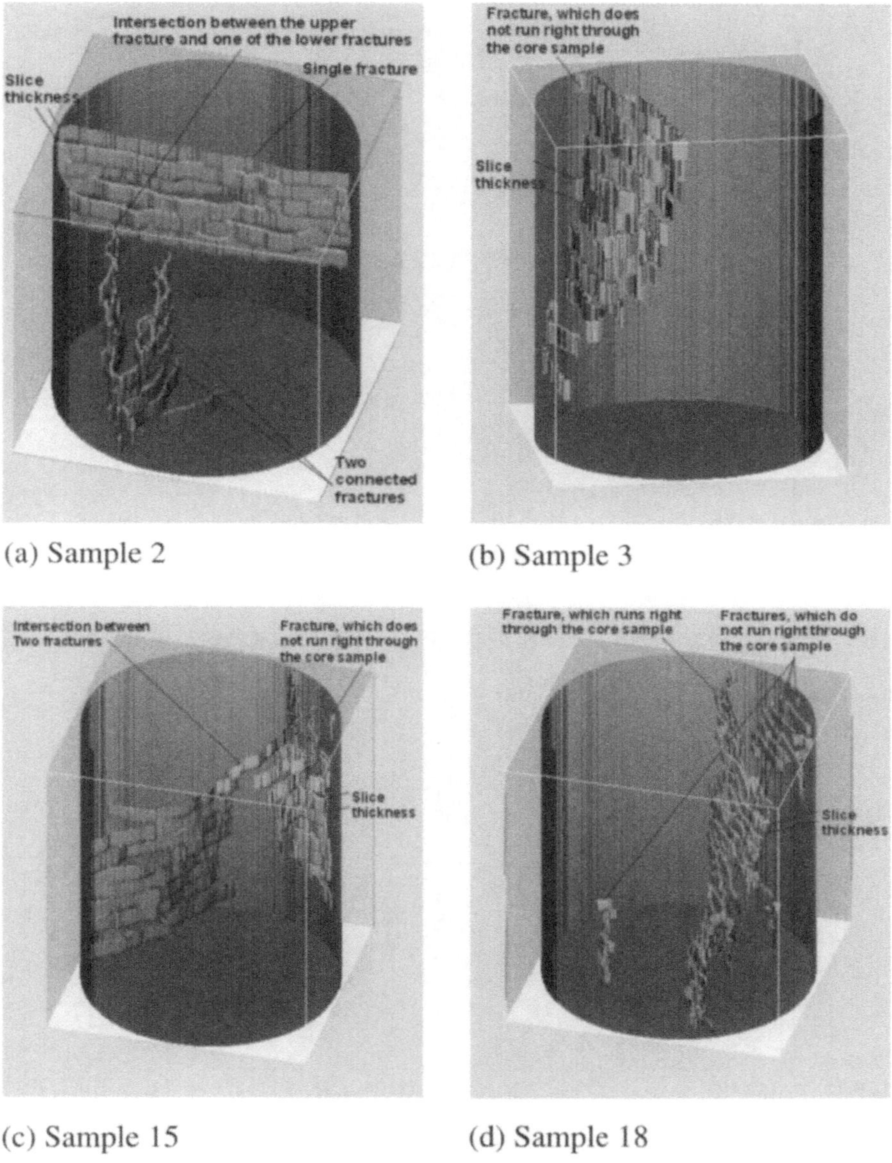

Figure 10
Three-dimensional visualisation of the fractures of samples 2, 3, 15, and 18 (the irregularities and vertical segmentation are artefacts of the interpolation process).

The fracture relative roughness (*Rr*) is the projection of the absolute roughness (ε) onto the equivalent hydraulic diameter (D_h). The fracture relative roughness is defined as

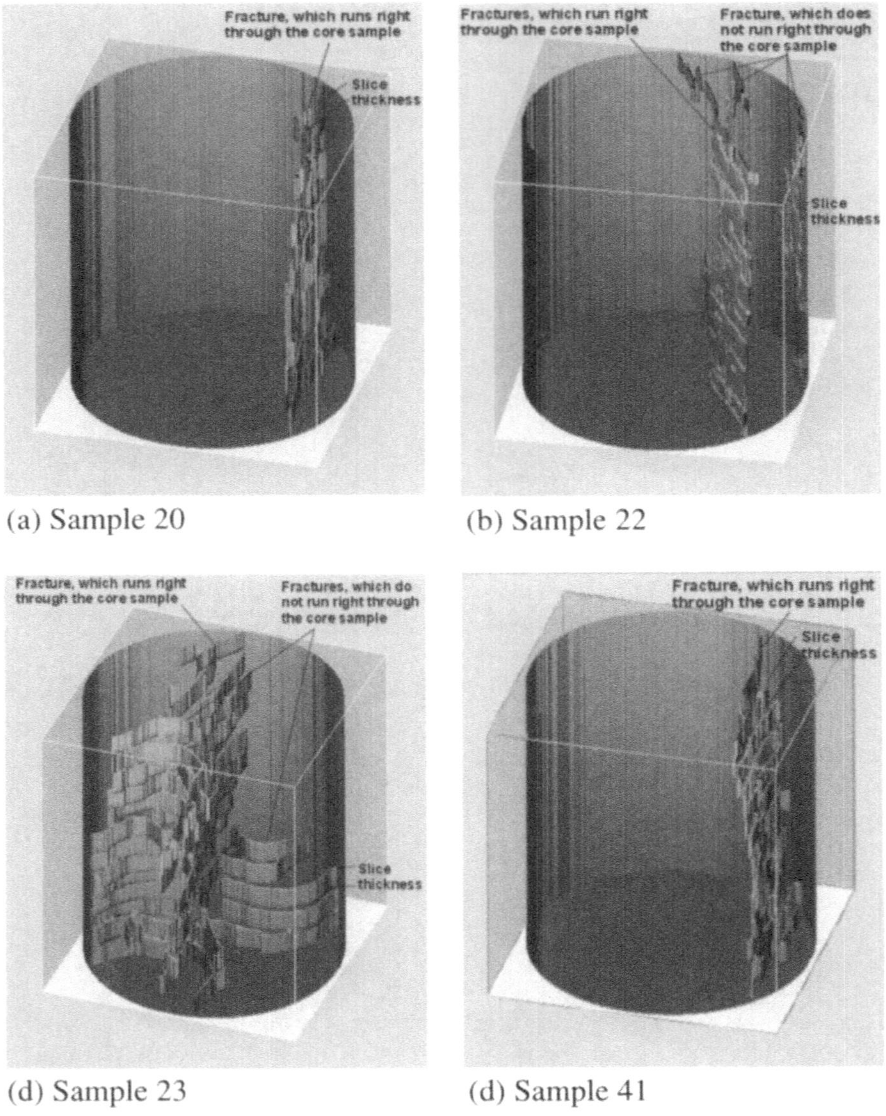

Figure 11
Three-dimensional visualisation of the fractures of samples 20, 22, 23, and 41 (the irregularities and vertical segmentation are artefacts of the interpolation process).

$$Rr = \varepsilon / D_h \ , \tag{3}$$

The absolute roughness (ε) is the mean height of the surface asperities or the irregularities in the fracture, and the equivalent hydraulic diameter (D_h) is defined as

$$D_h = 4A_f/P_f \ , \tag{4}$$

where A_f is the cross-sectional area of flow in the fracture and P_f is the outside perimeter of this cross-sectional area. D_h is estimated to be equal to twice the fracture aperture.

The fracture surface in Figure 9 is viewed at 45 degrees; the correction of each elevation measurement is therefore necessary. For the evaluation of the absolute roughness (ε), ten electron microscope photographs of a fracture surface were taken for each sample. For each electron microscope photograph of a fracture surface, an arithmetic mean of the real elevations was calculated. The absolute fracture roughness (ε) represents the arithmetic mean of the ten different arithmetic means of the real elevations calculated for each electron microscope photograph of a fracture surface for the same sample. The corrected values of the absolute fracture roughness (ε) determined from the photographs of the fracture surfaces are given in Table 4.

5. Calculation of the Parameters of the Fracture Geometry

For each core-slice, the parameters of the fracture geometry such as the effective fracture aperture and the fracture width were calculated using an image analysis program.

Several authors have studied the effect of a variable aperture on the flow in a fracture (e.g., DE MARSILY, 1981; SILLIMAN, 1989; BEAR *et al.*, 1993). The volumetric flux at steady state within a variable fracture aperture can be estimated using the cubic law and an adjusted aperture estimate. For each 5 mm slice, the effective aperture of the fracture n of the slice $z(e_{nz})$ was calculated. It is defined as

$$e_{nz} = \left[\frac{\sum_{i=1}^{m_{nz}} \left(e_i^3 \cdot l_i\right)}{\sum_{i=1}^{m_{nz}} \left(e_i \cdot l_i\right)}\right]^{\frac{1}{2}} \ , \tag{5}$$

where m_{nz} is the number of rectangles representing fracture n in slice z .

Table 4

Values of the absolute fracture roughness

Sample Number	Absolute roughness (ε) 10^{-6} m
2	2.77
15	2.31
18	6.93
20	5.08
22	10.17
23	1.85
41	2.31

The term in square brackets is the third moment of the aperture distribution, normalised to unit cross-sectional area.

5.1. *Effective Hydraulic Conductivity*

Using the parameters of the fracture geometry such as the effective fracture aperture, fracture width and fracture relative roughness, hydraulic conductivities can be calculated.

Following Eq. (2), the hydraulic conductivity for each fracture and for each slice (K_{fnz}) is calculated as

$$K_{fnz} = \left[\frac{\rho g}{12\mu(1 + 8.8Rr^{1.5})}\right] e_{nz}^2 \ , \tag{6}$$

The hydraulic conductivity for each slice (K_{cz}) is estimated by combining the contribution of the matrix and each fracture by a weighted average:

$$K_{cz} = \frac{\sum_{(n,z)=1}^{m} \left[K_{fnz}(e_{nz} \cdot l_{nz})\right]}{\pi R_a^2} + \frac{K_m\left(\pi R_a^2 - \sum_{(n,z)=1}^{m} (e_{nz} \cdot l_{nz})\right)}{\pi R_a^2} \ , \tag{7}$$

where l_{nz} is the width of the fracture n of slice z, K_m is the matrix hydraulic conductivity and R_a is the radius of the sample (BEAR *et al.*, 1993).

Finally, the hydraulic conductivity of the bulk sample (K_c) is estimated by a harmonic average in series along the flow path:

$$K_c = \frac{\sum_{z=1}^{m_z} L_z}{\sum_{z=1}^{m_z} \frac{L_z}{K_{cz}}} \ , \tag{8}$$

where L_z is the slice length and m_z is the number of slices.

5.2. *Results of the Effective Fracture Aperture Calculations*

The effective fracture aperture is calculated for each fracture, in every 5 mm slice and for each sample. The effective fracture aperture (e_n) calculated for the whole sample is estimated by a harmonic average in series along the flow path:

$$e_n = \left[\frac{\sum_{z=1}^{m_z} L_z}{\sum_{z=1}^{m_z} \frac{L_z}{e_{nz}^3}}\right]^{1/3} \ . \tag{9}$$

Table 5 summarises the values of derived effective fracture apertures. Between one and five fractures were observed in each sample (e_1 to e_5). Those marked with an asterisk (*) are related to the dominant fracture system. A dominant fracture is defined as a fracture which traverses the entire sample. Table 5 shows that the values of the effective fracture apertures for the dominant fractures vary between 10 μm and 50 μm.

Table 5

Effective apertures of fractures found in the core samples (values with an asterisk (*) are related to dominant fractures) and contribution of the fractures to the total hydraulic conductivity of the sample

Sample	2	3	15	18	20	22	23	41
Fracture # 1 aperture (e_1) m	4.36×10^{-5}*	5.45×10^{-5}	1.35×10^{-5}*	1.54×10^{-5}*	5.49×10^{-5}	4.11×10^{-5}*	3.18×10^{-5}*	4.90×10^{-5}
Fracture # 2 aperture (e_2) m	4.10×10^{-5}		4.13×10^{-5}	1.45×10^{-4}		3.69×10^{-5}	2.77×10^{-5}	
Fracture # 3 aperture (e_3) m			3.58×10^{-5}*	1.63×10^{-5}		3.53×10^{-5}	2.98×10^{-5}	
Fracture # 4 aperture (e_4) m				2.08×10^{-5}		2.63×10^{-5}		
Fracture #5 aperture (e_5) m				1.15×10^{-4}		3.40×10^{-5}		
K_c m/s	1.41×10^{-6}	1.31×10^{-6}	3.88×10^{-7}	2.14×10^{-6}	1.54×10^{-6}	6.23×10^{-7}	6.31×10^{-7}	9.00×10^{-7}
K_f m/s	1.26×10^{-6}	2.03×10^{-8}	4.42×10^{-8}	1.29×10^{-7}	6.92×10^{-7}	3.74×10^{-7}	4.80×10^{-7}	8.30×10^{-7}
$Q_f/(Q_f + Q_m)$ %	89	1	11	6	45	60	76	92

6. Results of the Model

Results of this geometric method are compared with actual physical conductivity values that were measured directly from fluid experiments conducted with water as a test fluid prior to resin injection. The flow model provides good results with a correlation coefficient R^2 of 0.81 (Fig. 12). However, the calculated values of hydraulic conductivity are systematically about four times higher than the hydraulic conductivity values obtained from the experiments.

The results of the individual contribution of the fracture to the total hydraulic conductivity of the sample are represented in Table 2. K_c represents the total hydraulic conductivity of the sample and K_f represents the hydraulic conductivity of the fracture in the case where the matrix is not conductive. $Q_f/(Q_f + Q_m)$ represents the contribution of the water flow in the fracture compared to the total water flow in the sample.

7. Discussion of Results

The results show that in spite of the high aperture (5.45×10^{-5} m); the fracture in sample 3 transports only 1% of the flow. The fracture does not dominate the flow because of its limited vertical extent (Fig. 10b). The partial blocking of the fracture aperture by clay minerals in the case of samples 20 and 22 appears to be the reason for the reduction of the hydraulic conductivity within the fracture. The fractures in samples 20 (Fig. 11a) and 22 (Fig. 11b) transport only 45% and 60% of the flow, respectively, in spite of the high apertures (more than 4.0×10^{-5} m). The partial

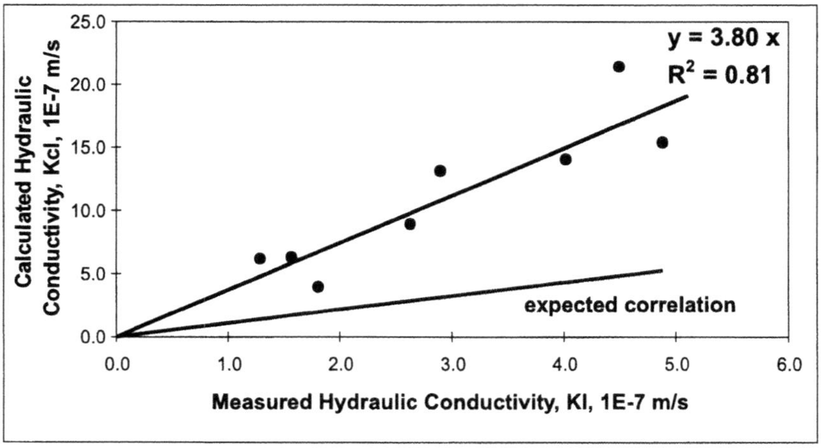

Figure 12
Correlation between measured and calculated hydraulic conductivities.

sealing nature of the fractures by clay minerals was verified by two independent methods. The Magnetic Resonance Imaging (MRI) technique was used to determine the geometry of a flood front during a water imbibition experiment for quantifying the distribution of water saturation as the flooding progresses. MRI measurements carried out on sample 20 indicated that the fracture shows up as a lighter tone, implying lower water saturation. Some residual water remains in the fracture, although at a reduced intensity (BARAKA-LOKMANE, 1999). The partial sealing of the fractures in the case of samples 20 and 22 was also verified by resin casting, showing the clay fill directly after the test as well as the results of the flow measurements (BARAKA-LOKMANE, 1999).

Two factors may cause the observed systematic difference between the measured and the calculated hydraulic conductivity. These factors are:

1. The reduction of the water permeability during the experiments, which is governed by three phenomena:

a) Chemically, by the migration of the swelling clay particles which clog the pores (SAMPATH and KEIGHIN, 1982; BLOOMFIELD and WILLIAMS, 1995);

b) Mechanically, by the breakdown of original fabrics caused by the passage of wetting fronts across relatively delicate clay mineral complexes (LOVELOCK, 1977; BLOOMFIELD and WILLIAMS, 1995);

c) Experimentally, by the undersaturation of samples during water permeability measurements (LOVELOCK, 1977).

2. A slight overestimation of the fracture aperture. For the samples investigated here, the systematic difference between measured and calculated hydraulic conductivity is most likely to be caused by the interaction between water and clay minerals inducing the reduction of the hydraulic conductivities during the water flow experiments. Indeed, clay particles tend to be mobile under flow-through conditions. They absorb water, inducing swelling. In addition, the undersaturation of samples during water permeability measurement was responsible for the low value of water permeability; this is the case for sample 41 (Fig. 11d), where the saturation degree is lower than 80%. This phenomenon occurs for finer-grained sandstones, as is the case for sample 41 (BARAKA-LOKMANE, 1999).

The second possibility is significant because the fracture hydraulic conductivity is proportional to the square of fracture aperture (eq. 6); therefore the calculated hydraulic conductivity of each slice (K_{cz}) (eq. 7) is proportional, as a first-order estimation, to the cube of the effective fracture aperture. In the case where the calculated hydraulic conductivities are four times higher than the measured hydraulic conductivities and if we consider only the overestimation of the fracture aperture as a factor causing the difference between these two parameters, the effective fracture aperture determined from the resin casting method would only have to be about 1.6 times higher than the effective fracture aperture inferred from measuring the hydraulic conductivity. That is, a small systematic error on the effective fracture aperture can propagate into a considerably larger systematic error in the implied hydraulic conductivity. However, according to the accuracy of the fracture aperture measurements, which was about 0.25–1.25 μm, the overestimation of the fracture aperture cannot alone explain the systematic difference between the measured and the calculated hydraulic conductivities.

8. Summary and Conclusions

A petrographical study of the samples, as well as permeability measurements, was conducted in order to characterise the samples and to establish the type of relationship between the petrographical and the flow properties. A new method for determining the detailed fracture geometry in sandstone cores was developed. The objective was to impregnate the sample with resin without significantly disturbing the fractures in the sample. The results using the injection of coloured resin (SPURR) into samples are very satisfactory. The resin makes the fracture space visible, allowing the fracture geometry to be examined digitally. Once the impregnation has been performed, the eight samples were cut into core sections of 5 mm length, resulting in a total of 128 cross sections and 196 fracture traces. For each core section, the parameters of the fracture geometry such as the fracture aperture and the fracture width were calculated using an image analysis program. For each fracture, the effective fracture aperture was calculated. Also, a reconstruction of the three-dimensional fracture system was performed. An optical method was subsequently used for determining the fracture roughness of the sandstone cores. Ten electron microscope photographs of a fracture surface were taken for each sample, and the elevations of the asperities were calculated. The fracture relative roughness was calculated as the ratio between the absolute fracture roughness and twice the fracture aperture. Using the parameters of the fracture geometry such as the effective fracture aperture, fracture width and fracture relative roughness, hydraulic conductivities were calculated.

The correlation between the measured and the calculated hydraulic conductivity values give good results with a correlation coefficient R^2 of 0.81. However, the calculated values of hydraulic conductivity are systematically about four times higher than the hydraulic conductivity values obtained from water experiments.

This study demonstrates that the determination of the conductivity from fracture geometry parameters such as the effective fracture aperture, fracture width, fracture length and relative roughness measured using an optical method can be valid. This is consistent with the strong geometric constraints on fluid flow in network models for macroscopic fluid flow (ZHU and WONG, 1996).

Acknowledgements

This work was carried out as part of the project "Festgesteins-Aquiferanalog: Experimente und Modellierung," funded by Deutsche Forschungsgemeinschaft

(DFG), Germany. S. Baraka-Lokmane was supported by Engineering and Physical Sciences Research Council, grant GR/M6 2150 during the writing of the paper. The authors wish to thank Prof. I.G. Main for his constructive and detailed review of the final version of the manuscript. Dr. G. Simpson and Dr. J. Heiland are thanked for their constructive critical reviews of the original manuscript.

References

Baraka-Lokmane, S. (1999), *Determination of Hydraulic Conductivities from Discrete Geometrical Characterisation of Fractured Sandstone Cores* (Dissertation. Applied Geology, University of Tübingen, Germany).

Bear, J., Tsang, C. F., and de Marsily, G., *Flow and Contaminant Transport in Fractured Rock* (Academic Press, Inc, San Diego, California, USA 1993).

Bloomfield, J. P. and Williams, A. T. (1995), *An Empirical Liquid Permeability–Gas Permeability Correlation for Use in Aquifer Properties Studies*, Quarterly J. Engin. Geology *28*, 143–150.

Cox, B. L., Pruess, K., and Persoff, P. (1990), *A casting and Imaging Technique for Determining Void Geometry and Relative Permeability Behaviour of a Single Fracture Specimen* (Earth Sciences Division Lawrence Berkeley Laboratory University of California Berkeley, California 94720), pp. 1–6.

de Marsily, G., *Hydrogéologie Quantitative* (Edition Masson 1981).

Dollinger, H., Bühler, C., and Bossart, P. (1994), *Exkavationsprojekt: Epoxidharz-Imprägnations-Experimente*, Interner Bericht 94–24, Nagra, National Genossenschaft für die Lagerung radioaktiver Abfälle.

Gentier, S., Billaux, D. and van Vliet, L. (1989), *Laboratory Testing of the Voids of a Fracture*, Rock Mechanics and Rock Engin. *22*, 149–157.

Hakami, E. (1988), *Water Flow in Single Rock Joints* (Licentiale Thesis, Lulea University of Technology, Lulea Sweden, November).

Lovelock, P. E. R. (1977), *Aquifer Properties of Permo-Triassic Sandstones in the United Kingdom*, Bull. Geological Survey of Great Britain. *56*, 1–40.

Pyrak-Nolte, L. J., Myer, L. R. O., Cook, N. G. W., and Witherspoon, P. A. (1987), *Hydraulic and Mechanical Properties of Natural Fractures in Low Permeability Rock*, Paper presented at sixth international Congress on Rock Mechanics, Montreal, Canada.

Sampath, K. and Keighin, C. W. (1982), *Factors Affecting Gas Slippage in Tight Sandstones of Cretaceous Age in the Uinta Basin*, J. Petroleum Technology, November, 2715–2720.

Silliman, S. E. (1989), *An Interpretation of Difference between Aperture Estimates derived from Hydraulic and Tracer Tests in a Single Fracture*, Water Resour. Res. *25*, *10*, 2275–2283.

Ulfrecht, W (1987), *Zur Hydrogeologie und Hydrochemie des Sandsteinkeupers in Mittel- und Ostwürttemberg*, Arbeiten aus dem Institut für Geologie und Paläontologie an der Universität Stuttgart *83*, 1–136.

Wang, J. S. Y. and Narasimhan, T. N. (1988), *Fractal and Statistical Characterization of Rough Fractures*, Second Berkeley Symposium on Topics in Petroleum Engineering, 33–37.

Zhu, W. and Wong, T. (1996), *Permeability Reduction in a Dilating Rock; Network Modelling of Damage and Tortuosity*, Geophys. Res. Lett. *23*, 22, 3099–3102.

(Received January 3, 2001, revised/accepted October 4, 2001)

Pure appl. geophys. 160 (2003) 929–936
0033–4553/03/060929–8

Pure and Applied Geophysics

An Integrated Study on Physical Properties of a KTB Gneiss Sample and Marble from Portugal: Pressure Dependence of the Permeability and Frequency Dependence of the Complex Electrical Impedance

Stephanie Heikamp[1] and Georg Nover[1]

Abstract—Pressure-induced variations in pore geometry were studied on dry- and fluid- saturated samples by means of electrical impedance spectroscopy and permeability measurements. Hydrostatic pressures (up to 120 MPa) and uniaxial pressures (up to failure) were applied. Hydrostatic pressures reduce the aspect ratio of cracks and thus cause a decrease of permeability and electrical bulk conductivity. The opposite was observed in uniaxial pressure experiments where new cracks were formed and consequently permeability and electrical conductivity were increased. More specific informations of these generated observations were derived from the interpretation of the frequency dispersion of the complex electrical conductivity. This least-squares-refinement considers electrochemical interactions between the fluid pore electrolyte and the inner surface of the sample, thus providing informations on the pore geometry and pressure-induced variations. Consequently changes in aspect ratio, size and geometry of the pore system can be detected by means of impedance spectroscopy.

Key words: Pressure, permeability, electrical conductivity, pore geometry.

1. *Introduction*

The Archie equation relates in a simple way the electrical conductivity of fluid-saturated rocks with petrophysical properties like porosity, permeability, degree of pore filling and conductivity of the pore fluid. It cannot be used to propagate the variation of electrical rock conductivities being caused by external pressures applied. Such information can be obtained from frequency-dependent electrical conductivity measurements (Glover and Vine, 1992; Duba *et al.*, 1994; Nover *et al.*, 1998). This method is sensitive to electrochemical interactions between ions and dipoles in the fluid pore electrolyte and the charged mineral surfaces. The electrochemical reactions cause the formation of "clouds" of charged particles (adsorbed on the inner surface of the pore system). These oscillate in the stimulating electrical field applied. If the frequency is varied, e.g., from Hz up to MHz, the relaxation times of these polarizations can be

[1] Mineralogical Institute, University of Bonn, Poppelsdorfer Schloss, 53115 Bonn, Germany.
E-mail: heikamp@uni-bonn.de

measured by means of impedance spectroscopy. Hydrostatic and triaxial pressures induce variations in pore geometry and thereby change both, the overall electrical (bulk)-conductivity by closing and opening of conduction paths, and the frequency dispersion by variations in pore size that cause e.g. the overlapping of such "clouds" on both sides of a fracture thus changing the time constant of the relaxation process (NOVER *et al.*, 1999).

2. *Experimental*

Carbonate rock samples from a surface outcrop (Portugal) and amphibolites and gneisses from the German Continental Deep Drilling – program (KTB) were chosen for this study. The petrophysical parameters porosity, permeability, inner BET surface and complex electrical conductivity were measured. The pressure dependence of the permeability was determined on cylindrical plugs in an autoclave reaching confining gas-pressures of 125 MPa, using a pressure transient technique (details: NOVER *et al.*, 1995, 1998). The anisotropy of the permeability and electrical conductivity were considered by studying plugs being drilled parallel and perpendicular to the bore hole axis.

The electrical conductivity measurements were performed in an autoclave that allowed the independent adjustment of confining and uniaxial pressure. Confining pressures are limited to a maximum pressure of 350 MPa, the uniaxial load could be increased up to 800 MPa. Hydrostatic pressures were increased stepwise up to 120 MPa, while the uniaxial pressures were increased stepwise (4 MPa for each step) until failure that occurred in the pressure range 80 to 120 MPa. The confining pressure was fixed at 2 MPa. Figure 1 displays a principle sketch of the hydrostatic (left) and stress loading equipment (right). The cylindrical plugs were saturated with a NaCl solution (0.1 M) and were subsequently enclosed in a shrink tubing to prevent contamination of the sample with the pressure medium (glycerine). The complex electrical conductivity was measured in the frequency range 1 kHz up to 1 MHz using the two electrode technique and 15 frequencies per decade.

The electrical charge transport in fluid-saturated rock samples consists of both, ohmic conductivity and diffusion controlled electrochemical processes, thus causing the conductivity to be a complex quantity: $\sigma^* = \sigma' + i\sigma''$. Measurements of the real σ' and imaginary part σ'' of the conductivity σ therefore includes information on pure electrolytic conduction and surface-related polarizations that take place between the ions and water dipoles in the pore saturand and the charged mineral surfaces. Consequently, pressure-induced variations in pore geometry can be interpreted more detailed by fitting model data of an electrochemical equivalent circuit model to the measured impedance data (experimental details in NOVER *et al.*, 1995, 1998 and 1999). Two parallel RC elements in series were used to consider pure electrolytic conduction and surface-related conduction (Fig. 2). The time constants

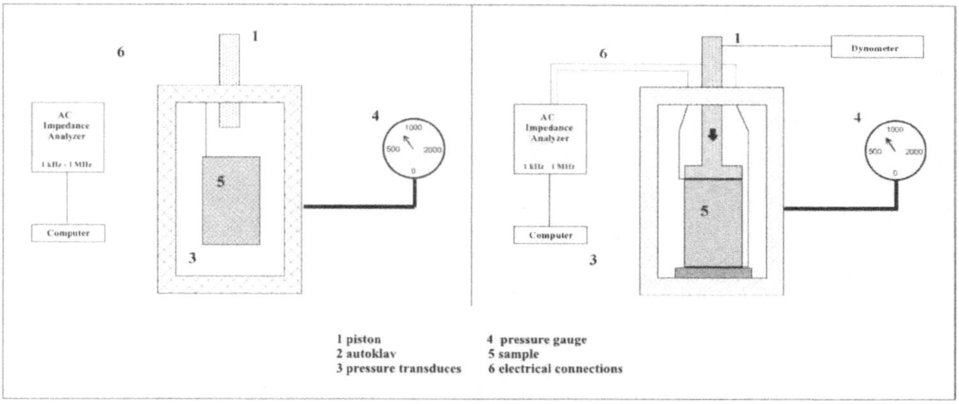

Figure 1
Sketch of the experimental setup used for the electrical conductivity measurements. Left: hydrostatic pressures, right: uniaxial loading with fixed confining pressure.

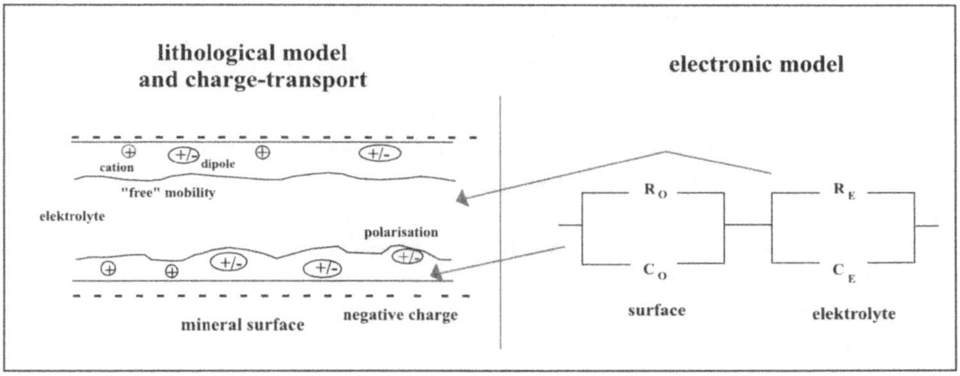

Figure 2
Equivalent circuit model used in the least-squares-refinement and its correlation to petrophysical phenomena.

of the polarizations are given by their relaxation times $\tau = R*C$. Using a nonlinear least-squares-refinement resulted in four model parameters (R1,C1,R2,C2), thus describing the electrical charge transport of bulk conduction and surface conduction.

3. Results and Discussion

3.1 Petrophysical Properties

The three samples (marble, amphibolite, gneiss) selected for this study were nearly identical in grain size and porosity. The textured amphibolite and the less

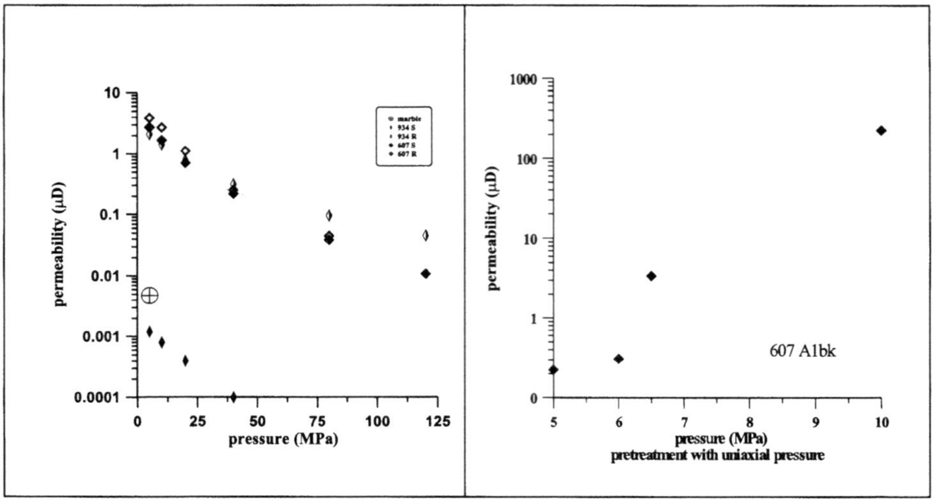

Figure 3
Pressure dependence of the permeability of the marble and the KTB rock samples. The permeability was measured using a pressure transient technique with a pressure gradient of 5 MPa across the sample. The transport medium was Argon-gas. The confining (hydrostatic) pressure could be increased up to 300 MPa.

textured gneiss (KTB) were fine grained and dense with only minor occurrence of open fissures which were mainly caused by pressure and temperature release when the sample was brought up to the surface. Porosities are less than 3% vol. and permeabilities range from 1 μD to less than 10 nano-Darcy (Fig. 3) depending on the orientation of the sample, thus revealing a significant anisotropy in crack orientation and permeability. Permeability in radial orientation was about three orders in magnitude higher (amphibolite) than in the direction of the bore hole axis. The progressive closure of microcracks results in a pronounced permeability decrease at low confining pressures extending to about 80 MPa. At pressures above 100 MPa the pressure permeability relation (p/k) tends to be more linear, reflecting the closure of texture related fractures (FREUND and NOVER, 1995; NOVER *et al.*, 1998).

3.2 Complex Electrical Conductivity in Hydrostatic and Triaxial Pressure Experiments

The bulk conductivity was found to react very sensitive on either hydrostatic or triaxial pressures applied (Fig. 4). Hydrostatic pressures decreased the bulk conductivity by a progressive closure of existing fractures. First fractures with a high aspect ratio were closed, then, at higher pressures, the intrinsic properties control the pore volume reduction (NOVER *et al.*, 1998, 1999, 2000). The reverse observation was made in triaxial pressure experiments, where the axial stress was increased gradually until failure of the sample (strain: 6×10^{-4}, strain-rate: 10^{-7} s^{-1}). Here we could measure an increase of the bulk conductivity being caused by the

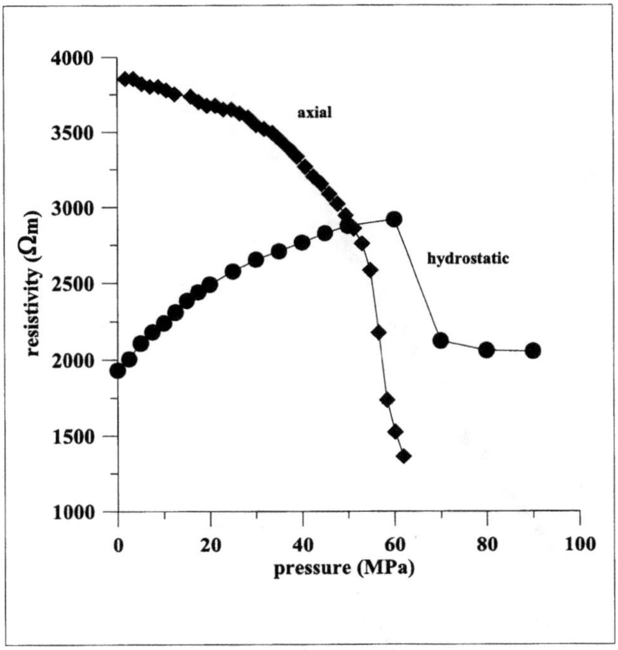

Figure 4
Pressure dependence of the volume (bulk) resistivity of samples that were used in a hydrostatic (resistivity increase) and a triaxial pressure experiment (resistivity decrease).

formation of new fractures. Figure 5 shows as an example (KTB sample VB934C1e) the displacement and axial stress.

Figures 6a,b exhibit the results of hydrostatic and axial stress experiments performed on KTB and marble samples. The Fig. 6 displays the bulk-resistivity (resistivity = reziprocal of conductivity) as a function of a hydrostatic (left) and a uniaxial pressure increase (right). The insert displays the complex response in the frequency range kHz up to MHz. Different curves display the response of the sample for each step in pressure. The uniaxial loading experiments were performed with a stepwise increase of the load until failure of the sample. Complex AC-impedance data are presented as Cole-Cole diagrams where the real part of the impedance (Z') is plotted versus the imaginary part ($-Z''$). The shape of the complex response is a "semicircle," where the frequency develops from right to left from 700 Hz up to 1 MHz. The step rate was 3 MPa in the case of the KTB amphibolite and gneiss (Figs. 6a) and 2 MPa for the marble (Fig. 6b). The amphibolite resists axial stresses up to 91 MPa, but at higher pressures the shape of the semicircles was varied significantly, thus indicating that the pore geometry was changed by being close to the limit of the uniaxial strength. Under these conditions new fractures were generated and existing fractures form clusters with the consequence that the

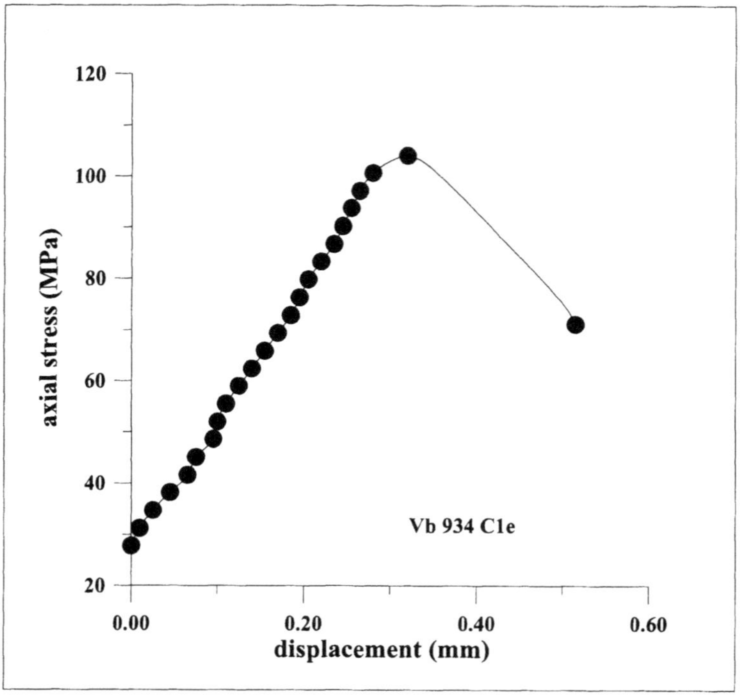

Figure 5
Complete load displacement of a KTB amphibolite. The load was increased stepwise, because we measured the complex electrical response for each step in pressure and 15 frequencies per decade in frequency, which took about 15 minutes to complete a spectrum from kHz up to MHz.

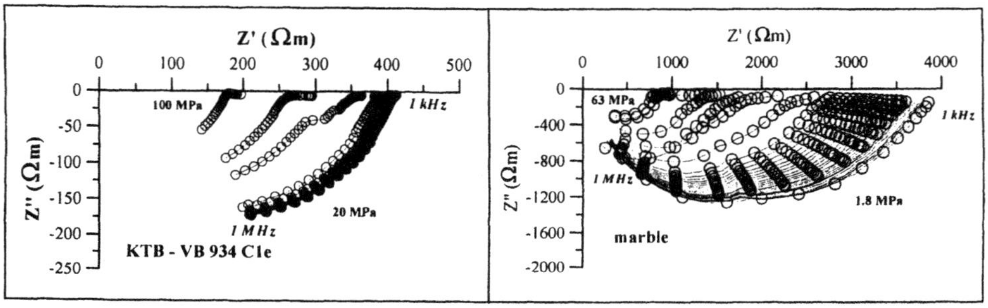

Figure 6
Complex response of a carbonate rock sample and an amphibolite from the KTB used in triaxial pressure experiments. The data are displayed as Cole-Cole-diagrams where the real part of the impedance is plotted versus the imaginary part. The intersection of the semicircle with the real axis gives the volume or bulk conductivity, which was usually measured at frequencies of about 1–5 kHz. Frequency develops from right to left, the left end of the "semicircles" corresponds to the 1 MHz data point.

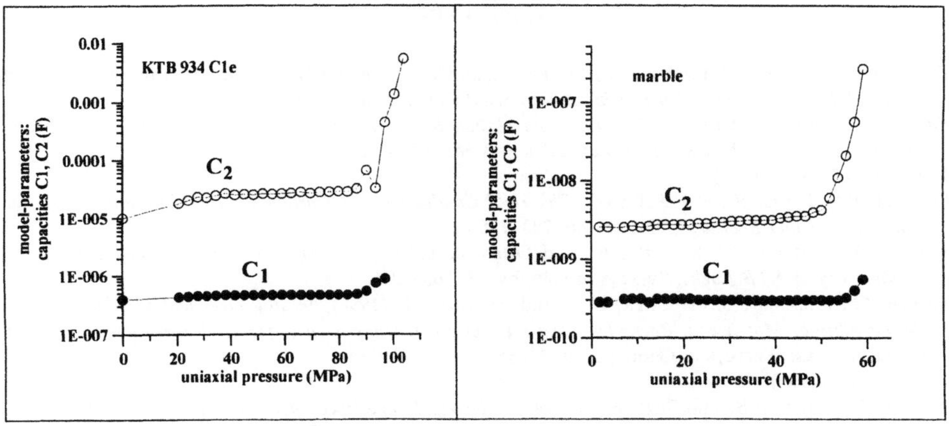

Figure 7
Variation of the refined model parameters C1 and C2 as derived from the equivalent circuit model. Capacity C1 is the electrical parameter that correlates with the inner surface of the sample.

mechanic stability was reduced and the pore volume was increased. The point of intersection of the extrapolated semicircle with the real axis (Z') is defined as the bulk resistivity. It is decreased as a function of increasing load (Fig. 4) thus indicating the forming and clustering of new cracks. Failure occurred at axial stress above 100 MPa. A slightly different result characterizes the marble and the gneiss sample. Well developed Cole-Cole semicircles could be measured, and in contrast to the amphibolite, the shape of these semicircles was changed in each step in pressure (Figs. 4b (left), 6b). Consequently these samples already deform at low axial-stresses (1.8 MPa). But the final steps of the loading process (above 50 MPa) result in a significant variation of the complex response, thus indicating clustering of fractures and a dramatic increase in inner surface, porosity and permeability.

Figure 7 finally displays the variations of two of the model parameters, C1 and C2. Both capacitors describe polarizations. C2 considers bulk and C1 surface-related polarizations. C2 does not change significantly as a function of pressure, while C1 increases continously and most remarkably results in a sharp increase before the sample fails (see last 5 data points in Fig. 4). Thus we can clue that impedance spectroscopy is a sensitive tool to study pressure induced variations in pore geometry under *in situ* pressure conditions. However, the precursor signal (increase of C2 before failure) requires further confirmation in that it is an unexpected result in this study.

Acknowledgement

This work was financed by the Deutsche Forschungsgemeinschaft (DFG) under grant No. 294/8-1 which is gratefully acknowledged.

References

Duba, A., Heikamp, S., Meurer, H.J., Nover, G., and Will, G. (1994), *Evidence from Borehole Samples for the Role of Accessory Minerals in Lower-crustal Conductivity*, Nature *367*, 59–61.

Dürrast, H. and Siegesmund, S. (1999), *Correlation Between Rock Fabrics and Physical Properties of Carbonate Reservoir Rocks*, Int. J. Earth Sciences *88*, 392–408.

Freud and Nover (1995)

Glover, P.W.J. and Vine, F.J. (1992), *Electrical Conductivity of Carbon Bearing Granulite at Raised Temperatures and Pressures*, Nature *360*, 723–725.

Nover, G., Heikamp, S., Kontny, A., and Duba, A. (1995), *The Effect of Pressure on the Electrical Conductivity of KTB Rocks*, Surveys in Geophys. *16*, 63—81.

Nover, G., Heikamp, S., Meurer, H.J., and Freund, D. (1998), *In situ Electrical Conductivity and Permeability of Mid-crustal Rocks from the KTB Drilling: Consequences for High Conductive Layers in the Earth Crust*, Surveys in Geophys. *19*, 73–85.

Nover *et al.* (1999)

Nover, G., Heikamp, S., and Freund, D. (2000), *Electrical Impedance Spectroscopy Used as a Tool for the Detection of Fractures in Rock Samples Exposed to a Stress Field*, Natural Hazards *21*, 317–330.

Nover, G. and Heikamp, S. (2001), *The Electrical Signature of Rock Samples Exposed to Hydrostatic and Triaxial Pressures*, Annali di Geofisika, in press.

Pape, H., Clauser, C., and Iffland, J. (1999), *Variation of Permeability with Porosity in Sandstone Diagenesis Interpreted with Fractal Pore Space Model*, Pure Appl. Geophy. *157*, 603–619.

(Received January 29, revised September 12, 2001, accepted October 1, 2001)

Pure appl. geophys. 160 (2003) 937–960
0033 – 4553/03/060937–24

Pure and Applied Geophysics

Permeability-porosity Relationships in Rocks Subjected to Various Evolution Processes

Y. Bernabé,[1] U. Mok,[2] and B. Evans[2]

Abstract—It is well known that there is no "universal" permeability-porosity relationship valid in all porous media. However, the evolution of permeability and porosity in rocks can be constrained provided that the processes changing the pore space are known. In this paper, we review observations of the relationship between permeability and porosity during rock evolution and interpret them in terms of creation/destruction of effectively and non-effectively conducting pore space. We focus on laboratory processes, namely, plastic compaction of aggregates, elastic-brittle deformation of granular rocks, dilatant and thermal microcracking of dense rocks, chemically driven processes, as a way to approach naturally occurring geological processes. In particular, the chemically driven processes and their corresponding evolution permeability-porosity relationships are discussed in relation to sedimentary rocks diagenesis.

Key words: Permeability, porosity.

1. Introduction

It is well known that there is no one-to-one relationship between porosity and permeability applicable to all porous media. One reason is that porosity is invariant under a homothetic transformation (e.g., uniform, isotropic stretching) of the pore space whereas permeability is not. A second reason is that, in a given material, not all pores are equally effective in conducting fluid flow. Two media with the same porosity but different proportions of effective and non-effective pore space must therefore have different permeabilities.

In the earth, rocks and their pore space evolve according to various geological processes (e.g., compaction during burial, depressurization and cooling during uplift, diagenesis and metamorphic reactions, deformation under tectonic stresses). Some of these processes produce pores and others destroy them; all change permeability. Thus, each process defines a specific evolution permeability-porosity relationship (EPPR). For a given a process, it may be possible to construct the associated EPPR

[1] Institut de Physique du Globe, CNRS – Université Louis Pasteur, 5 rue René Descartes, 67084, Strasbourg, France. E-mail: ybernabe@eost.u-strasbg.fr

[2] Department of Earth, Atmospheric and Planetary Sciences, Massachusetts Institute of Technology, Cambridge, MA 02139, U.S.A.

and thus predict the transport properties of rock at depth. Conversely, knowledge and understanding of a particular EPPR may help to characterize its underlying process. Natural processes are complex and can often be described as the superposition of several coupled "simple" processes. Our goal here is to review a number of "simple" laboratory processes and their corresponding permeability-porosity relationships.

In Section 2 we will show how EPPRs can be interpreted in terms of creation/destruction of effective and non-effective porosity. In Sections 3 to 6 we will discuss a variety of materials and processes, beginning with synthetic rock analogues undergoing plastic compaction. In Sections 4 and 5 we will consider elastic and brittle compaction of granular rocks such as sandstones, microcracking of dense rocks, and finally chemically driven processes in Section 6. Studying this last group of processes is a way to approach the important problem of diagenesis of natural sedimentary rocks. We close in Section 7 with general conclusions.

2. Evolution Permeability-porosity Relationships: Creation/Destruction of Effective and Non-effective Porosity

In Kozeny-Carman's and other classic models (e.g., WALSH and BRACE, 1984), permeability is described as proportional to simple integer powers of the relevant pore geometry parameters, i.e., porosity, hydraulic radius, tortuosity and/or specific surface area. For a given process, these parameters are usually assumed to be related to each other through power-law relationships, therefore leading to a power-law dependence of permeability on porosity, possibly with a non-integer exponent (e.g., DAVID *et al.*, 1994). Experimental evidence however indicates that a single power-law exponent does not always hold as porosity changes. One popular approach is to keep the power-law representation but with a variable exponent. Accordingly, we depict the EPPRs as oriented lines in log(porosity)-log(permeability) space. The slope α of the tangent at any point on such a curve can be understood as the exponent of a local power law $k \propto \phi^{\alpha}$, where k is the permeability and ϕ the porosity. Our goal here is to show that α is related to changes in the ratio $\xi = \phi_e/\phi_{ne}$ of effective (ϕ_e) over non-effective porosity (ϕ_{ne}).

In order to devise a precise definition of ϕ_e and ϕ_{ne} we begin by considering steady-state flow in a single capillary. We define the non-effective pore space as the union of all points where the fluid velocity is less than a small fraction (e.g., 1%) of the mean velocity. It is easy to see that the non-effective pore space in a single capillary is a thin boundary layer along the capillary walls. From Poiseuille velocity formula, $u(r) = 2\langle u\rangle\,(1 - r^2/r_0^2)$, where $u(r)$ is the local velocity, $\langle u\rangle$ the mean velocity and r_0 the radius of the capillary, we calculate $\xi = 199$ for a threshold value of 1%. Clearly, the definition above can be applied to any porous media although the thickness of the non-effective boundary layer is not necessarily constant over the whole pore space. In fact, entire pores may be counted as non-effective. However a new problem arises. For example, we imagine that a steady-state flow is generated in a homogeneous, simple cubic network by applying a pressure gradient along one of the principal directions. All the capillaries perpendicular to this direction are stagnant and thus become improperly included into the non-effective pore space. One way to overcome this difficulty is to define the non-effective pore space as the intersection of the apparent non-effective pore space obtained in three mutually perpendicular flow experiments. Note that a non-effective, boundary layer coating the entire pore-solid interface must always exist. Consequently ξ cannot become infinite.

What are other sources of non-effective porosity? Obviously, roughness of the pore walls can increase the size of the nearly stagnant boundary layer that coats all pores. In addition, connectivity loss and pore-scale heterogeneity can generate non-effective pore space. To investigate connectivity loss, we calculated ϕ_e and ϕ_{ne} in homogeneous, three-dimensional, simple cubic networks in which bonds were randomly removed according to a probability q (note that percolation theory uses the occupation probability $p = 1 - q$). According to percolation theory (STAUFFER and AHARONY, 1992), simple cubic networks become disconnected when q reaches a critical value $q_c = 0.75$ (or $p_c = 0.25$). Figure 1 shows the variations of ξ as a function of $2\ q/q_c$. This parameter varies from zero for fully connected networks to

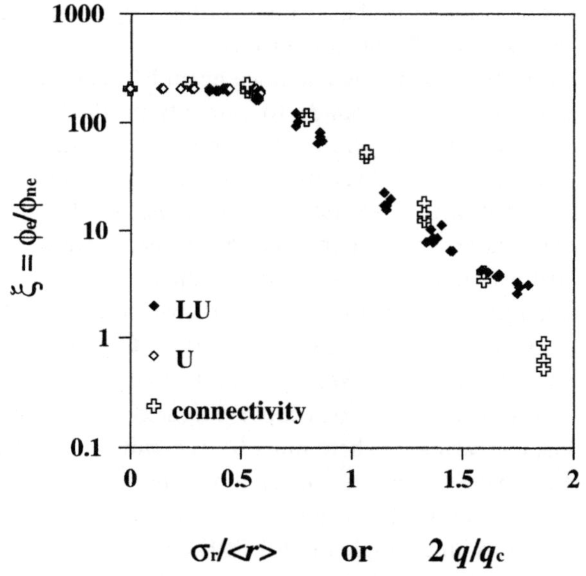

Figure 1
A semi-log plot of ξ, the ratio of effective to non-effective porosity, as a function of heterogeneity level $\sigma_r/\langle r\rangle$ (open diamonds refer to uniform and solid diamonds to log-uniform distributions) and connectivity $2\ q/q_c$ (crosses).

two for disconnected ones. We can see that ξ remains near 200 (values slightly greater than 199 can be obtained) until $2\ q/q_c = 0.5$, then drops dramatically reaching values lower than unity near the percolation threshold (clearly, ξ must vanish when q reaches q_c).

We also calculated ϕ_e and ϕ_{ne} in fully connected networks in which the radii r_i of the capillaries were randomly assigned according to variable uniform and log-uniform distributions. In Figure 1 we plotted ξ as a function of the normalized standard deviation $\sigma_r/\langle r\rangle$. This parameter is equal to zero in homogeneous networks and increases with increasing heterogeneity. Hydrology studies report values of the variance of the distribution of $Y = \ln(k)$, σ_Y^2, up to 1 for loose sands and soils and 5 for sandstones (GELHAR, 1993). Using the properties of the log-normal distribution and the fact that k is proportional to r^4 we can relate $\sigma_r/\langle r\rangle$ to σ_Y^2 by $\sigma_r/\langle r\rangle = [\exp(\sigma_Y^2/16) - 1]^{1/2}$. Then typical values of $\sigma_r/\langle r\rangle$ should not be greater than 0.3 for loose sand and 0.6 for sandstones. Note that $\sigma_r/\langle r\rangle$ is theoretically unbounded in the case of the log-uniform distribution. Nonetheless we were unable to generate values greater than 1.8 (in the case of the uniform distribution the maximum value is 0.58). To our surprise, we discovered that the dependence of ξ on $\sigma_r/\langle r\rangle$ is quite similar to that of ξ on $2\ q/q_c$. However this coincidence is only approximate and is visually enhanced by the use of semi-log scale, and, it must breakdown for $\sigma_r/\langle r\rangle$ greater than 2, since $2\ q/q_c$ is no longer defined. Nevertheless, this unexpected observation suggests a strong link between these two apparently unrelated problems. Finally, note that it takes rather large heterogeneity levels (i.e., $\sigma_r/\langle r\rangle > 0.6$) to generate significant non-effective porosity.

We can now return to interpreting α along a given EPPR in terms of variations of ξ. In Figure 2, quadrant I corresponds to porosity-producing and permeability-enhancing processes whereas quadrant III contains the porosity-destroying and permeability-reducing processes. Processes belonging to quadrants II and IV are possible (e.g., ZHU and WONG, 1996, 1997; DAROT *et al.*, 1992), but they are rare and will not be considered hereafter. For the sake of precision, we will use the following terminology: "transformation" has its general mathematical meaning and "process" refers to an actual, physical transformation. In quadrant I, $\alpha = 0$ corresponds to transformations that increase ϕ but leave k constant. Obviously only non-effective porosity, is created and ξ decreases. We can also see that $\alpha = \infty$ in quadrant I should normally correspond to transformations converting non-effective pores into effective pore space and thus increasing ξ. There must be one finite value of α associated with transformations that leave ξ constant, α_0, as illustrated in Figure 2a.

Figure 2b shows a family of constant-ξ transformations (indicated by parallel straight lines, $\xi = \xi_i$) and an example of an EPPR curve cutting through them. Note that $\xi_{i+1} < \xi_i$ in Figure 2b and that the ratio k/ϕ^{α_0} must be constant along each constant-ξ curve. Obviously, the greater the slope of the EPPR the greater the rate of change of non-effective porosity. We can thus interpret the underlying EPPR's process in terms of creation/destruction of effective and non-effective porosity.

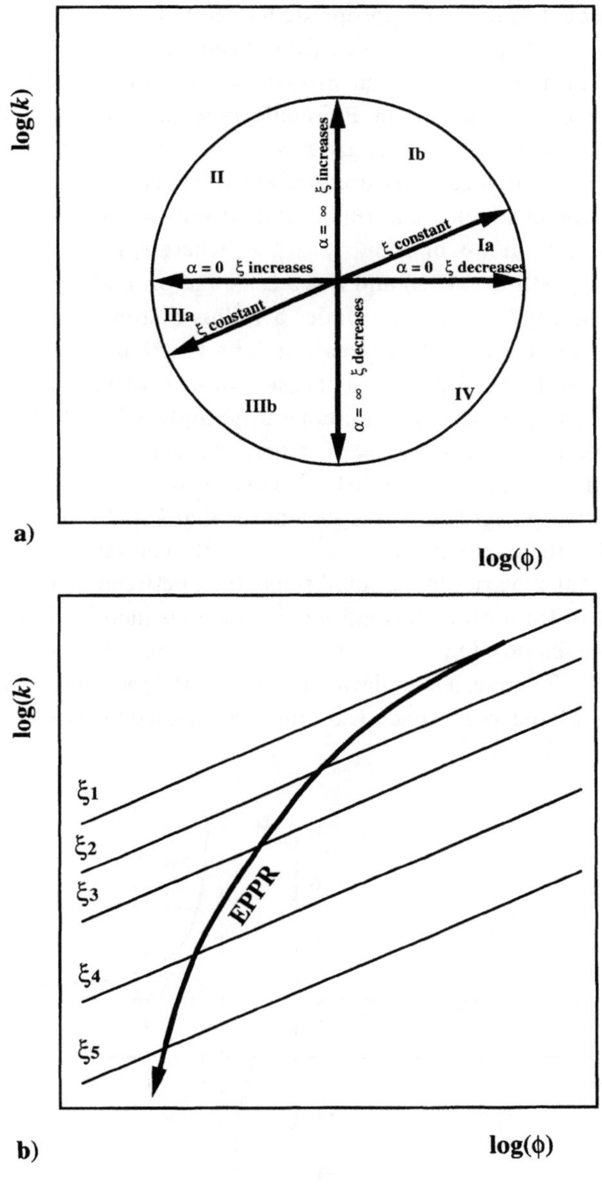

Figure 2

a) All the possible directions of the oriented tangent at an arbitrary point on an EPPR curve in $\log(k) - \log(\phi)$ space and their interpretation in terms of ξ evolution. b) An example of a quadrant III EPPR curve cutting through a family of constant-ξ lines, illustrating how the shape of the EPPR curve can be related to variations of ξ (the non-effective porosity fraction increases along this particular EPPR, i.e., $\xi_5 < \xi_4 < \xi_3 < \xi_2 < \xi_1$).

Now we must determine the appropriate value(s) of α_0. Initially, note that we are only interested in deterministic transformations although we are aware that stochastic transformations can often provide a good description of some natural processes. We will focus again on Poiseuille flow in capillary networks (in each capillary i the fluid volume flux q_i is given by $q_i = -\pi r_i^4/8\eta \, \nabla P_i$, where η is the fluid viscosity and ∇P_i is the local pressure gradient). We can identify two independent transformations under which the normalized fluid velocity field in the network, $\boldsymbol{u}/\langle \boldsymbol{u} \rangle$, remains invariant: 1) radius dilation, $r_i \Rightarrow \beta \, r_i$, where r_i is the radius of capillary i (with i taking all possible values) and β is a scale factor, and, 2) length dilation, $l \Rightarrow \beta \, l$, where l is the capillary length. Under a radius dilation we have $k \Rightarrow k \, \beta^4$ and $\phi \Rightarrow \phi \, \beta^2$, and therefore $k \propto \phi^2$, whereas, under a length dilation we have $k \Rightarrow k \, \beta^{-2}$ and $\phi \Rightarrow \phi \, \beta^{-2}$, and therefore $k \propto \phi$. These two transformations can of course be combined. For example, we can simultaneously apply a length dilation of factor β and a radius dilation of factor $\beta' = \beta^\gamma$, where the exponent γ can take any value between $-\infty$ and $+\infty$ (other relationships between β and β' can be considered but they do not result in power laws). It is easy to see that k and ϕ are related by a power law $k \propto \phi^{\alpha_0}$ with the exponent $\alpha_0 = (4\gamma-2)/(2\gamma-2)$. The curve α_0 vs. γ is shown in Figure 3. Note that α_0 varies in a limited range (i.e., between 1 and 2) for all negative values of γ, that is, for all transformations in which l is increased (decreased) while r_i are decreased (increased). On the other hand, α_0 can take any value between $-\infty$ and $+\infty$ for positive γ. There is a singularity at $\gamma = 1$, corresponding to pure homothetic transformations of the pore space (i.e., uniform, isotropic stretching). Porosity is

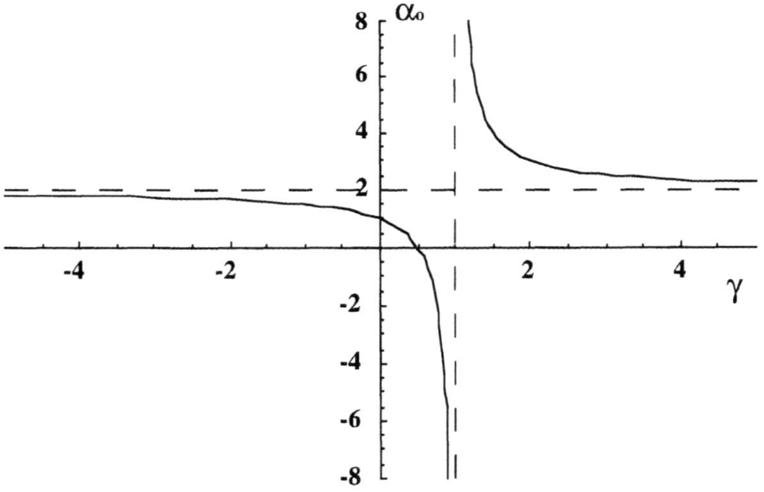

Figure 3
The exponent α_0 for a constant-ξ transformation combining length dilation and radius dilation with scaling factors β and β^γ, respectively, as a function of γ.

invariant under such a transformation whereas permeability is not, thus explaining the singularity. It is easy to see that, in this case, k should be proportional to the square of the scaling factor $\beta = \beta'$, as was experimentally verified by ALMOSSAWI (1988). This proportionality is the basis for the widely used linear relation between k and the square of the grain size. We propose that the network-based concepts of length dilation and radius dilation can be extended to all rocks in which a medial axis (LINDQUIST *et al.*, 1996) can be determined (i.e., a set of connected lines topologically equivalent to the pore network, obtained by repeated application of an erosion operator). Length dilation is then simply an affine transformation of the medial axis without changing the local values of pore cross-section dimensions along the medial axis. Under radius dilation the medial axis is unchanged while the pore cross sections are dilated by a constant scale factor. ALMOSSAWI's (1988) study shows that it is indeed reasonable to expect that general results obtained for networks may be verified in real porous media as well.

Now we will argue that the processes considered in the following sections are more appropriately associated with negative rather than positive γ. For example, during a dissolution-precipitation process, grain size and, therefore, pore length remains unchanged. In this case, the appropriate ξ constant transformation should be a pure radius dilation with $\alpha_0 = 2$. The same conclusion can be drawn for elastic deformation because grains are essentially incompressible compared to pores. During dilatant microcracking l tends to decrease (note that l is not the crack length but the distance between two crack intersections) while the aperture tends to increase, thus corresponding to a negative γ. During plastic compaction processes r_i decrease while l is more likely either to remain constant or to increase if grain growth occurs. Cataclastic compaction of granular rocks with grain crushing (i.e., decrease of l) and pore collapse (i.e., decrease of r_i) apparently corresponds to a positive γ. However, the important pores for hydraulic conduction are the pore throats, whose apertures are probably much less affected during cataclastic compaction than the grain size. Hence the appropriate α_0 may be close to 1. In conclusion, we will assume that α_0 lies between 1 and 2 for all processes considered hereafter.

3. Plastic Compaction

Several recent studies (e.g., BERNABÉ *et al.*, 1982; ZHANG *et al.*, 1994a; WARK and WATSON, 1998) document the relationship between k and ϕ during plastic compaction in synthetic rock analogues, i.e., mono- or polymineralic aggregates compacted under high pressure (0.1–1 GPa) and temperature (500–800°C). Synthetic rock analogues are very useful materials. They allow good reproducibility of the starting material, good control of the process and precise characterization of the material during evolution. Here we will focus on monomineralic aggregates compacted during a process known as hot isostatic pressing (HIP). HIP can be performed dry or with water as a pore fluid. In this last case, diffusion creep, pressure solution, and/or grain growth may occur. HIP is a quadrant III process (see Fig. 2a) and is capable of spanning the entire porosity range. In Figure 4a we show the EPPRs measured for HIP calcite in dry condition (BERNABÉ *et al.*, 1982; ZHANG *et al.*,

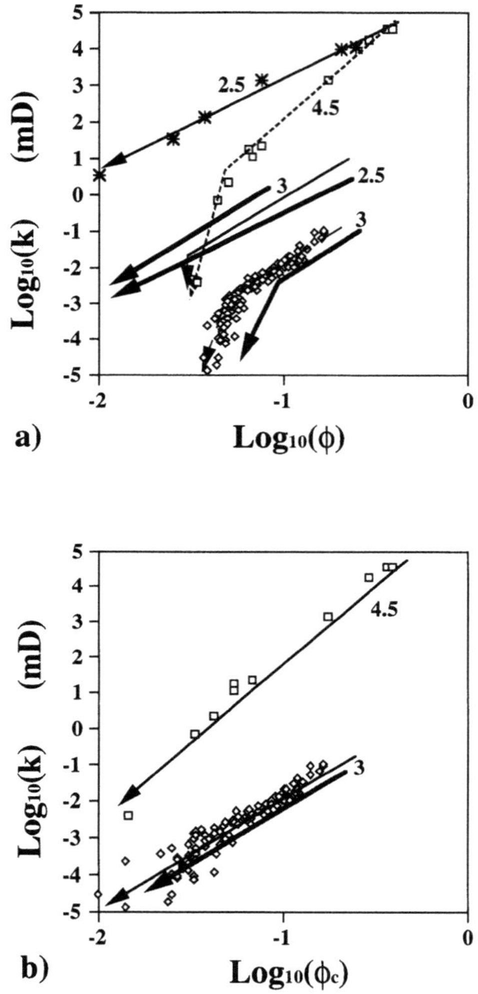

Figure 4
a) EPPR curves for HIP calcite (open diamonds, ZHANG *et al.*, 1994a; thin solid arrow, WARK and WATSON, 1998; and thick patterned arrow, BERNABÉ *et al.*, 1982), HIP quartz (thick solid arrows, WARK and WATSON, 1998), sintered glass (open squares, BLAIR *et al.* (1996), MOK *et al.* (2001)), HIP glass (stars, MAALØE and SCHEIE, 1982). Arrows are visual best fits through the original data. They are used to avoid cluttering the diagram. The slopes of the various EPPRs are indicated. b) Same as above except the log(k) is plotted as a function of log(ϕ_c).

1994a) and in the presence of water (WARK and WATSON, 1998) (in this last case, we recalculated the measured permeabilities using the reported values of the grain size). The behavior is essentially the same in both cases. Namely, for ϕ greater than about 0.07, the $\log(k)$ vs. $\log(\phi)$ curve is approximately a straight line with a slope α very close to 3. For $\phi < 0.07$, the $\log(k)$ vs. $\log(\phi)$ curve becomes increasingly steeper with decreasing ϕ, indicating an accelerating decrease of ξ during the process. Furthermore it was experimentally verified that k becomes effectively zero for ϕ lower than a finite value between 0.02 and 0.04, implying that the pore space is completely disconnected for these small values of ϕ.

ZHU *et al.* (1999) showed that HIP of dry calcite involves two mechanisms: namely, pore space shrinking by plastic deformation of the grains (power-law creep in the case of calcite, ZHANG *et al.*, 1994a) and pinching off of pores along three-grain edges by tube ovulation. This last mechanism eventually leads to a complete disconnection of the pore space. Porosity can be decomposed into connected and unconnected porosity, $\phi = \phi_c + \phi_u$. Plotting $\log(k)$ as a function of $\log(\phi_c)$ is a way to eliminate the disconnection effect and isolate the effect of plastic pore shrinking. In Figure 4b we see that the measured $\log(k)$ vs. $\log(\phi_c)$ curves (BERNABÉ *et al.*, 1982; ZHANG *et al.*, 1994a) are straight lines with a slope of 3. This small value indicates that plastic deformation only produces a moderate decrease of ξ. However, ϕ_u should not be identified with ϕ_{ne} since a slope of 3 is greater than the neutral slope α_0, which should be between 1 and 2 as discussed in Section 2 (note that the model of ZHU *et al.*, (1999) leads to $\alpha_0 = 2$ exactly).

The situation is quite different for quartz aggregates hot-pressed at higher pressures, in the presence of distilled water or brine. Permeability does not vanish for finite values of ϕ and the slope of the $\log(k)$ vs. $\log(\phi)$ curve remains constant near 2.5 for distilled water and 3 for brine (WARK and WATSON, 1998), suggesting that disconnection does not occur. The tube ovulation mechanism observed by ZHU *et al.* (1999) may be inactive in water-saturated quartz or may only occur in extremely narrow conduits, i.e., at practically zero porosity.

Another interesting related process is sintering, i.e., submitting the aggregates to high temperatures alone. During sintering, densification is driven by interfacial energy reduction, a very small force. It is therefore a rather ineffective densification process except at temperatures near the melting point. In Figure 4a we see the EPPR curve for 120 μm glass beads sintered at around 600°C (BLAIR *et al.*, 1996; MOK *et al.*, 2001). The behavior is similar to that of HIP calcite except that the high-ϕ slope is 4.5 instead of 3. In Figure 4b we plotted the $\log(k)$ vs. $\log(\phi_c)$ curve in the case of sintered glass and found a straight line with again a slope of 4.5, suggesting that sintering transforms significant amounts of connected porosity into non-effective porosity. This is consistent with the fact that pore space reduction during sintering results from the growth of necks at grain contacts while large pores at 4-grain vertices (i.e., "nodal pores" using medial axis terminology) remain largely unchanged. Indeed, owing to mass conservation and their comparatively big size, the nodal pores have a lower mean fluid velocity than the neck constrictions and therefore contain a higher proportion of non-effective pore space. Finally, it is interesting to compare the sintered glass EPPR just discussed to that of hot-pressed glass. Compression experiments were performed by MAALØE and SCHEIE (1982) on packs of 3 mm glass beads at 500°C, a value too small to induce significant sintering. Their results are plotted in Figure 4a showing that no disconnection occurs and the log(k) vs. log(ϕ_c) curve is a straight line with a slope of 2.5 as in the case of HIP quartz.

As a transition to the next section, we now consider room temperature compaction of salt aggregates, in which both plastic and brittle deformation occur simultaneously (e.g., SPANGENBERG *et al.*, 1998; POPP *et al.*, 2001). Experimental results from SPANGENBERG *et al.* (1998) and MÜLLER-LYDA *et al.* (1999) are represented in Figure 5. Apparently disconnection did not occur in compacted salt, but the EPPR curves are much steeper (i.e., $\alpha = 5$ to 7) than for HIP quartz. It is possible that brittle compaction is more efficient in diminishing the effective porosity fraction than plastic compaction. MÜLLER-LYDA *et al.* (1999) reviewed several data sets and found that the salt EPPR's were not statistically sensitive to moisture conditions but observed a systematic influence of grain size (Fig. 5). One of their data sets also revealed a surprising leveling off of the EPPR's for $\phi < 0.01$ (Fig. 5). However, we suspect that this trend may be an experimental artifact, as suggested by the dramatic increase of the scatter in the data for $\phi < 0.01$.

Finally, it would have been very interesting to compare the EPPR curves for HIP and sintering to those associated with pressure solution. Although theoretical and numerical models abound (e.g., ANGEVINE and TURCOTTE, 1983; DEWERS and

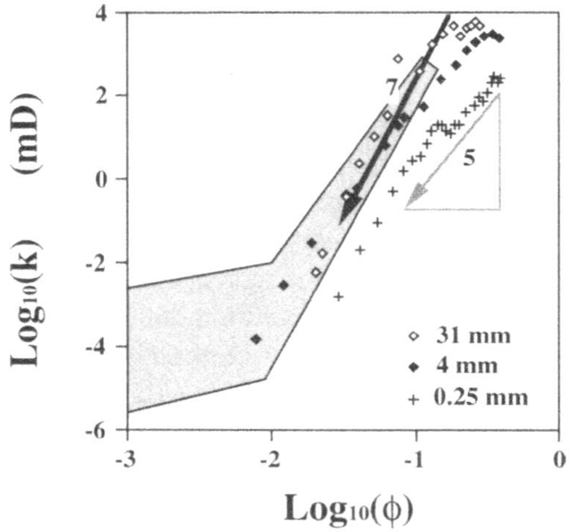

Figure 5
EPPR curves for compacted salt aggregates (thick solid arrow, SPANGENBERG *et al.*, 1998; symbols indicating different grain sizes and gray area, MÜLLER-LYDA *et al.*, 1999). The patterned arrow is not an EPPR but just visualizes a slope of 5.

ORTOLEVA, 1990; GAVRILENKO and GUÉGUEN, 1993), we are not aware of any published data relating k to ϕ.

4. Elastic and Brittle Compaction

We now consider granular rocks such as sandstones, which deform in an elastic-brittle manner in dry, moderate-temperature, high-pressure conditions. Figure 6a shows EPPR's for elastic (i.e., reversible) hydrostatic compression of a variety of sandstones compiled by BERNABÉ (1991) (see the original references therein), FREDRICH *et al.* (1993), and DAVID *et al.* (1994). Firstly, note that each EPPR covers a very small porosity range. The pore space of most sandstones is indeed dominated by pressure-insensitive pores with aspect ratios close to 1 (e.g., EHRLICH *et al.*, 1991; FREDRICH *et al.*, 1993; FREDRICH, 1999). Furthermore, sandstones have a rather limited elastic domain (e.g., DAVID *et al.*, 1994). Another striking result is the large domain covered by the exponent α for this set of rocks, namely from 1.2 to 21. Among the 33 rocks compiled here, 4 had α between 1 and 2, 16 between 2 and 4, 8 between 4 and 10, and 5 above 10. Ordinary occurrence of very high α was confirmed by DAVID *et al.* (1994) and ZHU and WONG (1997) (a selection of data from this last paper is included in Figure 6a). Although there is no obvious visible trend linking α and k or ϕ, we can see that the highest α occur to the right and bottom of the cloud of data points in Figure 6a, i.e., for high ϕ – low k combinations. This correlation suggests that high-α rocks contain large, pressure-insensitive nodal pores connected to each other by narrow, pressure-sensitive throats. As explained earlier, the throats belong to the effective pore space except for the usual boundary layer. Consequently, closing the throats greatly decreases ξ. It should be also noted that α is not necessarily constant during elastic compression. The detailed data for Berea and Boise sandstones in Figure 6a show that α increases with increasing pressure (ZHU and WONG, 1997). Therefore some of the low α values reported in BERNABÉ (1991) may in fact correspond to low pressure experiments that did not cover the complete elastic domain of the rocks studied.

During hydrostatic compression tests the elastic domain ends at the critical pressure P^* (DAVID *et al.*, 1994), above which irreversible compaction occurs. This domain is characterized by cataclastic deformation, i.e., grain crushing and pore collapse. Examples of hydrostatic compression EPPRs in the cataclastic domain are shown in Figure 6b (ZHU and WONG, 1997). The behavior depends on the initial porosity. In high-porosity ($\phi = 0.35$) Boise sandstone the EPPR becomes steeper ($\alpha = 19.5$) after entering the cataclastic domain. In contrast, low-porosity ($\phi = 0.14$) Darley Dale sandstone shows a decrease of α from 19.5 to 11.3, however for medium-porosity ($\phi = 0.21$) Berea sandstone there is no sharp change in α. If we consider the union of the two domains, the Berea sandstone EPPR curve appears like a smooth S-shaped curve with low α at high ϕ (i.e., below P^*), increasing to a maximum of about 25 at P^* and decreasing at low ϕ (i.e., above P^*).

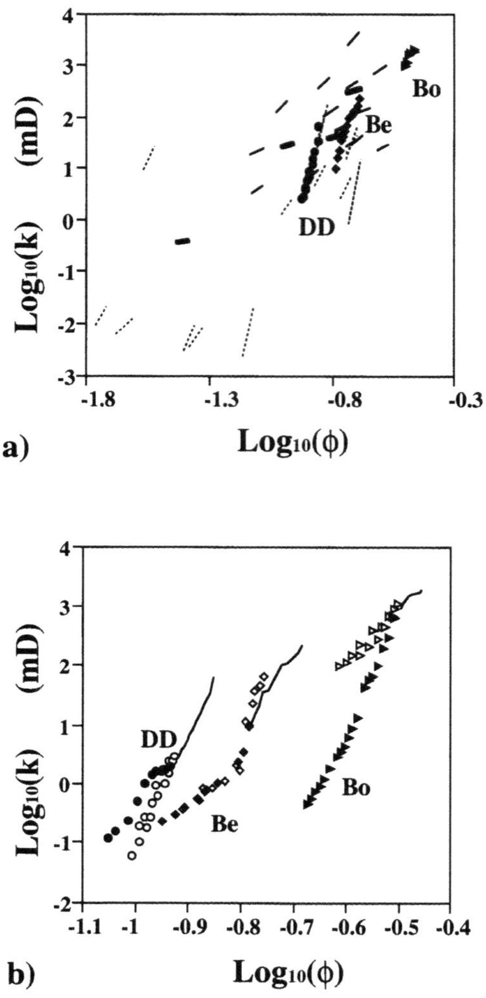

Figure 6

a) EPPR curves for elastic hydrostatic compression of sandstones. We used the following convention: rocks from BERNABÉ 1991; FREDRICH *et al.* 1993 and DAVID *et al.*, 1994, with $1 < \alpha < 2$ (thick segments), $\alpha < 5$ (thin segments) and $\alpha > 5$ (dashed segments), Darley Dale sandstone (solid circles), Berea sandstone (solid diamonds) and Boise sandstone (solid triangles) (ZHU and WONG , 1997). Note that in elastic EPPR's both orientations are possible. b) EPPR curves for cataclastic compression of sandstones (solid symbols refer to hydrostatic compression and open ones to triaxial compression, ZHU and WONG, 1997).

A possible explanation could be that in low-porosity sandstones the "pore-throat" microstructure is predominant (EHRLICH *et al.*, 1991; MCCREESH *et al.*, 1991; FREDRICH *et al.*, 1993; FREDRICH, 1999; LINDQUIST and VENKATURAGAN, 1999). Because the nodal pores are relatively large and connected only through the much narrower throats, they experience comparatively small fluid velocities. Their non-effective boundary layer must therefore be larger than in the throats. Consequently, nodal pores contribute primarily to non-effective pore space. The main cataclastic compaction mechanism is grain crushing and pore collapse (e.g., MENÈNDEZ *et al.*, 1996). Although throat collapse must also take place, it is clear that nodal pore collapse is the major porosity reduction mechanism. Because nodal pore collapse destroys non-effective porosity, it should result in a lowering of α, as is observed. To the contrary, in high-porosity sandstones the grains are loosely packed and weakly cemented. The pore geometry is not dominated by the "pore-throat" microstructure in the sense that the throat size is comparable to pore size (EHRLICH *et al.*, 1991; MCCREESH *et al.*, 1991). In these conditions cataclastic deformation is probably accommodated by grain sliding and rotation rather than by grain crushing. Hence, we expect that porosity reduction should be dominated by throat collapse, resulting in strongly decreasing ξ and a comparatively high α as observed.

ZHU and WONG (1997) also investigated the changes in porosity and permeability of sandstone samples triaxially deformed. At low confining pressures, dilatancy occurs in the bulk of the rock while shear deformation localizes on highly compacted shear bands, leading to a decrease of permeability despite a net porosity increase (i.e., quadrant IV in Fig. 2a; see also ZHU and WONG, 1996). Our analysis assumes that identical mechanisms are activated in all points of the material and, consequently, cannot be applied to these cases. We will therefore restrict our study to high-confining pressure, triaxial-compression tests in which cataclastic deformation is uniformly distributed. Results for Boise, Berea and Darley Dale sandstones (ZHU and WONG, 1997) are superimposed on the hydrostatic EPPR curves in Figure 6b. In Boise sandstone the triaxial EPPR curves have a lower α than the hydrostatic one, while the opposite is observed in Darley Dale. In Berea sandstone, both types of EPPR curves appear to coincide. Although it is not easy to figure out why it is so, shear deformation counters the hydrostatic compaction behavior described above in both high- and low-porosity sandstones. In addition, ZHU *et al.* (1997) demonstrated that permeability of sandstones became significantly anisotropic during triaxial compression tests (see also BRUNO, 1994); k along the major principal stress direction is higher than in a direction perpendicular to it.

5. *Microcracking of Dense Rocks*

We now examine the case of dense rocks. At low to moderate temperature conditions and under moderate to high non-hydrostatic stresses, intact dense rocks deform by dilatant microcracking (BRACE *et al.*, 1966), a quadrant I process (see Fig. 2a). In Figure 7 we show the EPPR curves of a variety of rocks undergoing dilatancy: Westerly granite (ZOBACK and BYERLEE, 1975), Carrara marble (FISCHER and PATERSON, 1992; ZHANG *et al.*, 1994b), and rock salt (STORMONT and DAEMEN,

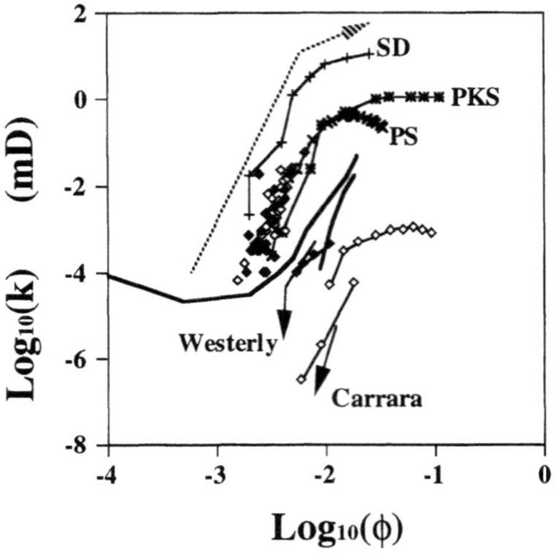

Figure 7

EPPR curves for dilatant microcracking of Westerly granite (joined solid diamonds, ZOBACK and BYERLEE, 1975), Carrara marble (joined open diamonds, FISCHER and PATERSON, (1992) and ZHANG *et al.*, 1994a), rock salt (joined stars, x and + referring to POPP *et al.*, 2001; PEACH and SPIERS, 1996 and STORMONT and DAEMEN, 1992, respectively), for thermal microcracking of La Peyratte granite (curvy, thick line, DAROT *et al.*, 1992) and Bresse mylonite (fairly straight, thick line, LE RAVALEC *et al.*, 1996), and for two natural granite formations (solid and open diamonds, KATSUBE and WALSH, 1987). The two descending, solid arrows correspond to hydrostatic compression of Westerly granite (BRACE *et al.*, 1968) and Carrara marble (FISCHER and PATERSON, 1992). The patterned, ascending arrow is not an EPPR. It is used to remind the reader that the microcracking processes belong to quadrant I (see Fig. 2a).

1992; PEACH and SPIERS, 1996; POPP *et al.*, 2001). For comparison, we also plotted the room-temperature hydrostatic compression curves of Westerly granite (BRACE *et al.*, 1968) and Carrara marble (FISCHER and PATERSON, 1992), both characterized by α values around 7–8. The dilatancy EPPR curves all present the same features more or less clearly (e.g., intact Westerly granite is already cracked so the early increase of k is more gradual than in other rocks). At the onset of dilatancy, k increases dramatically and porosity only slightly (i.e., large α). After a certain amount of dilation, the behavior changes sharply. The slope α decreases greatly and stabilizes at a value between 1 and 2, consistent with our anticipation that α_0 should be lower than 2 (Section 2).

Values of α lower than 1 are observed in the last stage of dilatancy in rock salt and Carrara marble ($\alpha = 0$, STORMONT and DAEMEN (1992), α negative, PEACH and SPIERS (1996) and ZHANG *et al.*, (1994b)), but these values probably do not correspond to purely brittle dilatancy. Indeed, in addition to brittle microcracking, plastic deformation mechanisms are active in these rocks, although their contribution to the overall deformation remains insignificant until the last stage of dilatancy. When the plastic deformation becomes important, mechanical behavior changes (i.e., onset of work hardening) as does the EPPR (i.e., $\alpha < 1$).

If we restrict our analysis to purely brittle dilatancy, the EPPR curves can be interpreted in terms of percolation theory (e.g., ZHU and WONG, 1999). The early increases of permeability result from the increasing connectivity of the microcrack network as dilatant cracking takes place. When a critical crack density is reached (i.e., the cross-over porosity of ZHU and WONG, 1999), the microcrack network is thought to have reached full connectivity (i.e., $=0$). This transition defines a second stage during which permeability is only affected by aperture increase, corresponding to a decrease of α. However, based on Figure 1, we can see that a small value of α should be produced as soon as $2\ q/q_c$ reaches 0.5, that is, long before full connectivity is achieved.

During dilatant failure of rocks most cracks are oriented sub-parallel to the major principal stress direction, therefore producing a highly anisotropic crack distribution (DAVID *et al.*, 1999). To appreciate the importance of anisotropy we compare the dilatant EPPR to that of an isotropic microcracking process, i.e., thermal cracking (in La Peyratte granite, DAROT *et al.*, 1992, and Bresse mylonite, LE RAVALEC *et al.*, 1996). The thermal cracking EPPR curves do not level off at high ϕ as dilatancy ones do, indicating that the condition $2\ q/q_c = 0.5$ was not reached in these experiments. It may be that thermal cracks are not as connected as dilatant cracks, perhaps because they are distributed isotropically. Alternatively, it may be that dilatant cracks, unlike thermal cracks, tend to nucleate and/or arrest at intersections with other cracks (T. Madden, personal communication). Finally, note the surprising start of the EPPR curve in La Peyratte granite. During the initial stage of thermal microcracking (i.e., at temperatures below 100°C), α takes very small, even slightly negative, values, possibly a consequence of a significant pre-existing crack density in "intact" La Peyratte granite.

For comparison, we also included two data sets collected from well-bores in granite formations (KATSUBE and WALSH, 1987). Although these data sets cannot rigorously be considered as EPPRs, it can be argued that they each come from a single formation and thus, at least partially, represent the evolution of the rock. Indeed, the data point clouds in Figure 7 are fairly linear with $\alpha \approx 5$, a value comparable to the values found for thermal cracking as well as hydrostatic compression. These data are thus consistent with uplifting, i.e., cooling off and depressurization.

6. Chemical Alteration and Diagenesis

Thus for we have only considered mechanical processes. However, we know that chemical reactions between the saturating fluids and rock forming minerals can also produce dramatic changes in the pore geometry. Chemically driven processes are an essential part of the diagenesis of sedimentary rocks. There have been numerous attempts to study chemical diagenesis experimentally. For example, MORROW *et al.*

(1981), MOORE *et al.* (1983), and VAUGHAN *et al.* (1986) circulated water down a temperature gradient in crystalline rock samples and measured the associated permeability changes (i.e., a 96% permeability drop corresponding to a calculated 10% porosity decrease). SCHOLZ *et al.* (1995), TENTHOREY *et al.* (1998), and AHARONOV *et al.* (1998) induced chemical reactions in quartz-feldspar aggregates under constant confining pressure and measured permeability and volumetric strain as a function of reaction time. The volumetric strain was very small, suggesting that porosity remained nearly constant while permeability decreased dramatically. REIS and ACOCK (1994) and TODD and YUAN (1992) flowed super-saturated fluids in Berea sandstone samples and measured porosity and permeability decrease during precipitation. MCCUNE *et al.* (1979) circulated HF-HCl mixtures in Phacoides sandstone and Spergen limestone, and measured k and ϕ during dissolution of the rock matrix.

Recently, MOK *et al.* (2001) induced chemical alteration of sintered glass while continuously measuring porosity and permeability. Examples of the EPPR curves by REIS and ACOCK (1994), TODD and YUAN (1992), and MOK *et al.* (2001) are represented in Figure 8 (the first two studies reported normalized k and ϕ; we recalculated these quantities based on typical Berea sandstone values). We observe an initial, very sharp decrease of permeability (i.e., large values of α) followed by a stage of moderate k decrease (i.e., $\alpha \approx 2$). From detailed microstructural observations, MOK *et al.* (2001) concluded that this behavior was primarily caused by a dramatic

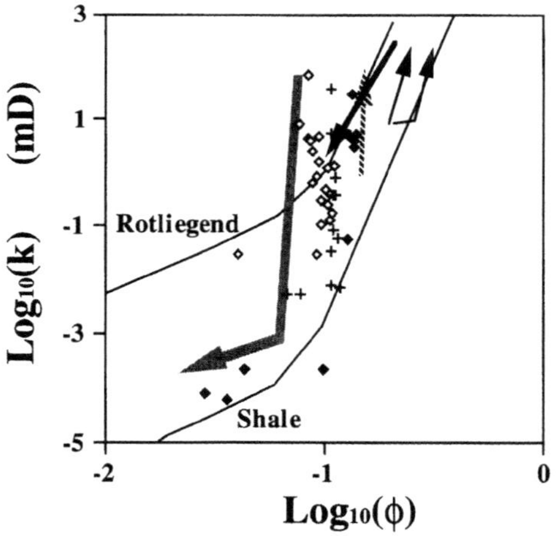

Figure 8
EPPR curves for chemical processes (+, open and solid diamonds, altered silica glass (MOK *et al.*, 2000); thick solid arrow, precipitation in Berea sandstone (TODD and YUAN, 1992; REIS and ACOCK, 1994); thin solid arrows, acidization of Phacoides sandstone; thick patterned arrow, acidization of Spergen limestone MCCUNE *et al.*, 1979). The wide gray arrow was added to help visualize the altered glass EPPR. In order to allow easy comparison with the data represented in Figure 9, the curves for Rotliegend sandstone and shale PAPE *et al.*, 1999, 2000) are indicated by the thinly dashed lines.

change of the pore-solid interface, from very smooth (i.e., intact sintered glass) to very rough (note that REIS and ACOCK (1994), and TODD and YUAN (1992) started with an already rough natural rock and therefore observed a less extreme behavior than in perfectly smooth sintered glass). During this first stage, large volumes of previously effective pore space were converted into non-effective pore space by roughness growth. Indeed, small wavelength irregularities in the pore wall topography delimit stagnant side pockets, thus increasing the size of the non-effective boundary layer (e.g., BERNABÉ and OLSON, 2000). The size of these pockets cannot increase indefinitely however and, at the onset of the second stage, a maximum was reached, followed by a nearly equal reduction of effective and non-effective porosities, thus explaining that α ended nearly equal to 2.

Figure 8 also shows the EPPR for acidization (i.e., dissolution) of sandstone and carbonate rocks (MCCUNE *et al.*, 1979). Here again, normalized k and ϕ were reported. In Figure 8, we used the reported values for Spergen limestone, $\phi = 0.15$ and $k = 1$ mD, and, since no values were indicated for the Phacoides sandstone, we arbitrarily chose $\phi = 0.2$ and $k = 10$ mD. In Spergen limestone, k increased by nearly two orders of magnitude while porosity remained nearly constant (i.e., nearly infinite α). In Phacoides sandstone, the behavior was not as extreme (i.e., $\alpha = 8$–10). There is even one experiment that started with a very slow permeability increase (i.e., α near zero). The difference in behavior can be explained by the mineralogy and the mineral properties. Spergen limestone is 96–98% $CaCO_3$, dissolution of which is transport-limited. Local dissolution rates are therefore proportional to local flow velocities, leading to selective enlargement of the well-conducting pores and therefore to a dramatic increase of ξ. Phacoides sandstone contains quartz, microcline, albite, dolomite and illite. After very rapid dissolution of dolomite, the process is dominated by the reaction-limited dissolution of the aluminosilicate feldspars and clays. Clearly, this is less efficient in creating effective porosity as in the case of Spergen limestone.

One goal is, whenever possible, to compare the laboratory processes to their natural counterparts. We thus selected several k–ϕ data sets thought to characterize the diagenetic evolution of various natural sedimentary rocks. Our selection is only a minute portion of the enormous literature on diagenesis. We readily admit overlooking many interesting and important studies. In Figure 9 we plotted the EPPR curves for Fontainebleau sandstone (BOURBIÉ and ZINSZNER, 1985), Rotliegend sandstone, shale, and other shaly sand formations represented by an "average" curve (PAPE *et al.*, 1999, 2000), sandstones from the South China Sea (BLOCH, 1991; WORDEN *et al.*, 2000) and from the North Sea (EHRENBERG, 1997), fine-grain and silty sandstones (REVIL and CATHLES, 1999), and dolomite formations from North America (MOWERS and BUDD, 1996). Most of these rocks exhibit behavior similar to that of altered silica glass (see Fig. 9a) at high porosity, namely, large values of α (i.e., around 10). Furthermore, the rocks with the broadest porosity range show a similar sharp decrease of α at low ϕ. What causes this behavior?

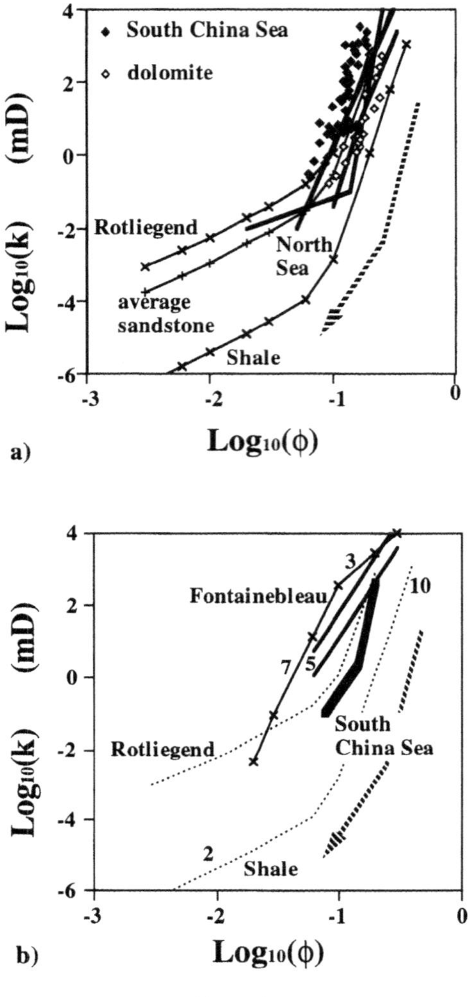

Figure 9

k-ϕ data sets for natural diagenesis. a) Various sandstones and shales (joined x for Rotliegend sandstone and shale, joined + for "average" sandstone, PAPE *et al.*, 1999, 2000), South China Sea sandstones (solid diamonds, BLOCH, 1991), dolomites (open diamonds, MOWERS and BUDD, 1996), and North Sea sandstones (solid lines, EHRENBERG, 1997). b) Fontainebleau sandstone (joined x, BOURBIÉ and ZINSZNER, 1985), fine-grain and silty sandstones (thin solid lines, REVIL and CATHLES, 1999) and South China Sea sandstones (thick solid line, WORDEN *et al.*, 2000). The curves for Rotliegend sandstone and shale (thinly dashed lines) are drawn again here to make comparison easier. The thick patterned arrows do not represent EPPR curves. They are just meant to remind the reader that diagenesis is assumed to be a quadrant III process (see Fig. 2a).

As shown in Section 4, large values of α at high ϕ can be generated by brittle compaction as well as by chemical alteration. But, from the observed omnipresence of rough pore-solid interfaces in most sandstones (KROHN, 1988a,b) we surmise that chemically driven mechanisms played the essential role in the diagenesis of these natural rocks. However, not all sedimentary rocks present this behavior. Some exceptions are shown in Figure 9b. The fine-grain and silty sandstones from REVIL and CATHLES (1999) have α near 5, and, more dramatically, Fontainebleau sandstone (BOURBIÉ and ZINSZNER, 1985) shows an EPPR curve similar to that of HIP calcite (see Fig. 4a). Despite its popularity as a laboratory rock, Fontainebleau sandstone is quite exceptional and must have undergone a rather unusual diagenesis. It is well sorted, has a 100% quartz composition, but is extremely variable in porosity (BOURBIÉ and ZINSZNER, 1985). It is also notable for the smoothness of the pore-solid interface.

Chemical processes are known to produce structural variations and/or irregularities at all scales in sedimentary rocks (e.g., ORTOLEVA, 1994; AHARONOV and ROTHMAN, 1996). Extreme roughness of the pore walls is commonly observed in sandstones containing diagenetic minerals. Roughness is often associated with a fractal structure of the pore-solid interface at scales lower than a few tens of microns (KROHN, 1988a,b). Such fractal structure is thought to have a notable influence on permeability (e.g., THOMPSON, 1991; AHARONOV *et al.*, 1997; PAPE *et al.*, 1999, 2000). According to AHARONOV *et al.* (1997) the pore space can be decomposed into fractal and Euclidean pore space (see also KROHN, 1988a) of which only the latter conducts fluid flow. This notion is very similar to our concept of effective and non-effective porosity. Roughness produces non-effective porosity because viscous fluid flow cannot penetrate inside the narrowest side cavities of the pore-solid interface. In other words, viscous fluid flow filters out the irregularities of the pore-solid interface with wavelengths shorter than some threshold value (BERNABÉ and OLSON, 2000). This remains true whether the pore-solid interface is fractal (i.e., with a power-law power spectrum) or not. One interesting observation by AHARONOV *et al.* (1997) is that a log-log plot of permeability versus Euclidean porosity in sandstones gives an approximate power law with an exponent of 2, consistent with our assertion that Euclidean and effective pore space are basically identical, and that α_0, the slope of a constant-ξ transformation, is equal to 2 in the case of chemically driven processes. The analysis of PAPE *et al.* (1999, 2000) is quite different. Fractality is expressed in their model by a number of power-law relations with non-integer exponents between the relevant geometrical parameters (e.g., porosity, grain and pore dimensions, and so forth). These power-law relations reflect underlying scaling laws associated with the assumed fractal structure. They naturally lead to power law k–ϕ relationships with non-standard exponents (i.e., large non-integer α). Hence, the model demonstrates that a fractal microstructure implies high values of α. However, as shown in previous sections, the reverse is not necessarily true, i.e., a high α does not necessarily imply the existence of a fractal structure.

7. Conclusions

Our main idea here is that the evolution of permeability and porosity in rocks (EPPR) is dictated by the operating process. Thus, EPPRs are process specific, and although different processes may produce qualitatively similar curves, one expects substantial differences in the corresponding rate kinetics. Because porosity-producing and porosity-destroying processes usually involve very different mechanisms, the EPPRs contained in quadrant I and quadrant III in Figure 2a must be distinguished. Additionally, the EPPR curve generated by a given process depends on the initial state of the material (in particular, its initial pore geometry) and on the loading conditions.

A second important idea is that the various EPPR curves may be interpreted in terms of creation/destruction of effective and non-effective porosity. From our definition of ϕ_e and ϕ_{ne} in Section 2 it is clear that these two quantities are not purely geometrical. They might be called "kinematic" to stress that they depend on the fluid velocity field. One important consequence is that loss of effective porosity is not exclusively related to loss of connectivity. For example, pore wall roughness is a very

Table 1

Summary of the values of α for different processes and materials

Processes	Materials	α
Plastic compaction	Synthetic aggregates	2.5–3 α increasing with decreasing ϕ if disconnection occurs
Sintering	Porous glass	4.5 for $\phi < 0.10$ disconnection at $\phi \approx 0.04$
Semi-brittle compaction	Salt aggregates	5–7
Elastic compaction	Sandstones	1–25 depending on microstructure
Cataclastic compaction (hydrostatic)	Sandstones $\phi > 0.30$	≈ 20
	$0.15 > \phi > 0.30$	10–20
	$\phi < 0.15$	≈ 10
Cataclastic compaction (triaxial)	Sandstones $\phi > 0.30$	5–10
	$0.15 > \phi > 0.30$	10–20
	$\phi < 0.15$	≈ 20
Dilatant microcracking	Dense rocks	7–8 α decreasing with increasing ϕ
Thermal microcracking	Dense rocks	5–7 $\alpha \approx 1$ at very low ϕ
Dissolution	Sedimentary rocks	>20
Precipitation	Sedimentary rocks	≈ 8
Chemical alteration (roughening)	Porous glass Sedimentary rocks	>10 α decreasing with decreasing ϕ
Diagenesis		$\alpha \approx 2$ at $\phi < 0.10$

efficient destroyer of effective pore space. Disconnection and roughness-controlled EPPRs may possibly be distinguished by noting that disconnection normally occurs only at low ϕ, whereas, at high ϕ, roughening may degrade permeability faster. Another source of non-effective pore space is heterogeneity, which can exist at all scales and in the whole porosity range.

When different mechanisms of change of the pore space are considered, a variety of evolution curves are possible (see Table 1 for a summary). Among all the processes reviewed here, plastic deformation is the least capable of reducing ξ, the ratio of effective to non-effective pore space. However it is often combined with mechanisms leading to disconnection; the importance of which grows as porosity decreases. Note that disconnection rates depend on pressure, temperature, and fluid chemistry. Elastic compression produces EPPR curves that strongly depend on the initial microstructure of the rock. If the most compressible pores are parts of the effective pore space (for example, pore/throat microstructure), elastic compression will produce a rapid decrease of the effective porosity fraction and of permeability. Non-elastic, brittle (i.e., cataclastic) deformation of porous rocks also leads to very complex behavior and may generate a wide variety of EPPR curves, depending on the initial microstructure and on stress conditions (i.e., hydrostatic or deviatoric). Dilatant microcracking in dense rocks is easily interpreted in terms of connectivity increase however, if a significant plastic strain is added to the brittle strain, a more complex behavior is produced. Note also that, in contrast to dense rocks, dilatant microcracking in porous rocks may lead to a decrease in effective porosity. Finally chemically driven processes can also generate a rich class of EPPRs, depending upon whether or not roughening occurs. Moreover, chemical reactions have the ability to produce disorder at scales greater than pore scale and may lead to disconnection of the pore space.

Acknowledgements

We thank Sergei Shapiro and an anonymous reviewer for their helpful suggestions. YB is grateful for enlightening discussions with Peter Schutjens, Till Popp, Hansgeorg Pape and many other participants to the 3rd. Euroconference on Rock Mechanics in Bad Hönnef, Germany. YB also expresses his deep appreciation to Joe Walsh, a pioneer of rock physics and a great friend. This work was partially funded by DOE under grant DE-FG02-00ER.

References

AHARONOV, E. and ROTHMAN, D. H. (1996), *Growth of Correlated Pore-Scale Structures in Sedimentary Rocks: A Dynamical Model*, J. Geophys. Res. *101*, 2973–2987.

AHARONOV, E., TENTHOREY, E., and SCHOLZ, C. H. (1998), *Precipitation Sealing and Diagenesis: 2 – Theoretical Analysis*, J. Geophys. Res. *103*, 23,969–23,981.

AHARONOV, E., THOMPSON, A., and ROTHMAN, D. H. (1997), *Transport Properties and Diagenesis in Sedimentary Rocks: The Role of Micro-scale Geometry*, Geology. *25*, 547–550.

ALMOSSAWI, H. I. H. (1988), *Physical Properties of Synthetic Sandstone Rocks*, Geophys. Prospecting *36*, 689–699.

ANGEVINE, C. L. and TURCOTTE, D. L. (1983), *Porosity Reduction by Pressure Solution: A Theoretical Model for Quartz Arenites*, Geol. Soc. Am. Bull. *94*, 1129–1134.

BERNABÉ, Y. (1991), *Pore Geometry and Pressure Dependence of the Transport Properties in Sandstones*, Geophysics. *56*, 436–446.

BERNABÉ, Y., BRACE, W. F. and EVANS, B. (1982), *Permeability, Porosity, and Pore Geometry of Hot-pressed Calcite*, Mech. Mater. *1*, 173–183.

BERNABÉ, Y. and BRUDERER, C. (1998), *Effect of the Variance of Pore Size Distribution on the Transport Properties of Heterogeneous Networks*, J. Geophys. Res. *103*, 513–525.

BERNABÉ, Y., and OLSON, J. F. (2000), *The Hydraulic Conductance of a Capillary with a Sinusoidally Varying Cross-section*, Geophys. Res. Lett. *27*, 245–248.

BLAIR, S. C., BERGE, P. A., and BERRYMAN, J. G. (1996), *Using Two-point Correlation Functions to Characterizing Microgeometry and Estimate Permeabilities of Sandstones and Porous Glass*, J. Geophys. Res. *101*, 20,359–20,375.

BLOCH, S. (1991), *Empirical Prediction of Porosity and Permeability in Sandstones*, Am. Assoc. Petrol. Geol. *75*, 1145–1160.

BOURBIÉ, T., and ZINSZNER, B. (1985), *Hydraulic and Acoustic Properties as a Function of Porosity in Fontainebleau Sandstone*, J. Geophys. Res. *90*, 11,524–11,532.

BRACE, W. F., PAULDING, B. W., and SCHOLZ, C. H. (1966), *Dilatancy in Fracture of Crystalline Rocks*, J. Geophys. Res. *71*, 3939–3953.

BRACE, W. F., WALSH, J. B., and FRANGOS,, W. T. (1968), *Permeability of Granite Under High Pressure*, J. Geophys. Res. *73*, 2225–2236.

BRUNO, M. S. (1994), *Micromechanics of Stress-induced Permeability Anisotropy and Damage in Sedimentary Rock*, Mech. Mater. *18*, 31–48.

DAROT, M., GUÉGUEN, Y., and BARATIN, M. L. (1992), *Permeability of Thermally Cracked Granite*, Geophys. Res. Lett. *19*, 869–872.

DAVID, C., MENÈNDEZ, B., and DAROT, M. (1999), *Influence of Stress-induced and Thermal Cracking On Physical Properties and Microstructure of La Peyratte Granite*, Int. J. Rock Mech. Min. Sci. Geomech. Abstr. *36*, 433–448.

DAVID, C., WONG, T.-f., ZHU, W., and ZHANG, J. (1994), *Laboratory Measurement of Compaction-induced Permeability Change in Porous Rock: Implications for the Generation and Maintenance of Pore Pressure Excess in the Crust*, Pure Appl. Geophys. *143*, 425–456.

DEWERS, T. and ORTOLEVA, P. (1990), *A Coupled Reaction/Transport/Mechanical Model for Intergranular Pressure Solution Stylolites, and Differential Compaction and Cementation in Clean Sandstones*, Geochim. Cosmochim. Acta, *54*, 1609–1625.

EHRENBERG, S. N. (1997), *Influence of Depositional Sand Quality and Diagenesis on Porosity and Permeability: Examples from Brent Group Reservoirs, Northern North Sea*, J. Sedim. Res. *67*, 197–211.

EHRLICH, R., CRABTREE, S. J. , HORKOWITZ, K. O., and HORKOWITZ, J. P. (1991), *Petrography and Reservoir Physics I: Objective Classification of Reservoir Porosity*, Am. Assoc. Petrol. Geol. Bull. *75*, 1547–1562.

FISCHER, G. and PATERSON, M. S., *Measurement of Permeability and Storage Capacity in Rocks During Deformation at High Temperature and Pressure. In Fault Mechanics and Transport Properties of Rocks* (eds. Evans B. and Wong T.-f.) (Academic Press, San Diego, Calif., 1992), pp. 312–253.

FREDRICH, J. T. (1999), *3D Imaging of Porous Media Using Laser Scanning Confocal Microscopy with Application to Microscale Transport Processes*, Phys. Chem. Earth (A), *24*, 551–561.

FREDRICH, J. T., GREAVES, K. H., and MARTIN, J. W. (1993), *Pore Geometry and Transport Properties of Fontainebleau Sandstone*, Int. J. Rock Mech. Min. Sci. *30*, 691–697.

GAVRILENKO, P. and GUÉGUEN, Y. (1993), *Fluid Overpressures and Pressure Solution in the Crust*, Tectonophys. *217*, 91–10.

GELHAR, L. W., *Stochastic Subsurface Hydrology* (Prentice-Hall, Englewood Cliffs, 1993).

KATSUBE, T. J. and WALSH, J. B. (1987), *Effective Aperture For Fluid Flow in Microcracks*, Int. J. Rock Mech. Min. Sci. Geomech. Abstr. *24*, 175–183.

KROHN, C. (1988a), *Sandstone Fractals and Euclidean Pore Volume Distribution*, J. Geophys. Res. *93*, 3286–3296.

KROHN, C. (1988b), *Fractal Measurements of Sandstones, Shales and Carbonates*, J. Geophys. Res. *93*, 3297–3305.

LE RAVALEC, M., DAROT, M., REUSCHLÉ, T., and GUÉGUEN, Y. (1996), *Transport Properties and Microstructural Characteristcs of a Thermally Cracked Mylonite*, Pure Appl. Geophys. *146*, 207–227.

LINDQUIST, W. B., LEE, S.-M., COKER, D. A., JONES, K. W., and SPANNE, P. (1996), *Medial Axis Analysis of Void Structure in Three-Dimensional Tomographic Images of Porous Media*, J. Geophys. Res. *101*, 8297–8310.

LINDQUIST, W. B. and VENKATURAGAN, A. (1999), *Investigating 3D Geometry of Porous Media from High Resolution Images*, Phys. Chem. Earth (A) *24*, 593–599.

MAALØE, S. and SCHEIE, A. (1982), *The Permeability Controlled Accumulation of Primary Magma*, Contrib. Mineral. Petrol. *81*, 350–357.

MCCREESH, C. A., EHRLICH, R., and CRABTREE, S. J. (1991), *Petrography and Reservoir Physics II: Relating Thin Section Porosity to Capillary Pressure, The Association Between Pore Types and Throat Size*, Am. Assoc. Petrol. Geol. Bull. *75*, 1563–1578.

MCCUNE, C. C., FOGLER, H. S., and KLINE, W. E. (1979), *An Experimental Technique for Obtaining Permeability-Porosity Relationships in Acidized Porous Media*, Ind. Eng. Chem. Fundam. *18*, 188–191.

MENÈNDEZ, B., ZHU, W., and WONG, T.-f. (1996), *Micromechanics of Brittle Faulting and Cataclastic Flow in Berea Sandstone*, J. Struct. Geol. *18*, 1–16.

MOK, U., BERNABÉ, Y., and EVANS, B. (2001), *Permeability, Porosity and Pore Geometry of Chemically Altered Porous Silica Glass*, in J. Geophys. Res., in press.

MOORE, D. E., MORROW, C. A., and BYERLEE, J. D. (1983), *Chemical Reactions Accompanying Fluid Flow Through Granite Held in a Temperature Gradient*, Geochim. Cosmochim. Acta., *47*, 445–453.

MORROW, C. A., LOCKNER, D., MOORE, D. E., and BYERLEE, J. D. (1981), *Permeability of Granite in a Temperature Gradient*, J. Geophys. Res. *86*, 3002–3008.

MOWERS, T. T. and BUDD, D. A. (1996), *Quantification of Porosity and Permeability Reduction Due to Calcite Cementation Using Computer-Assisted Petrographic Image Analysis Techniques*, Am. Assoc. Petrol. Geol. Bull. *80*, 309–322.

MÜLLER-LYDA, I., BIRTHLER, H., and FEIN, E., *Ableitung von Permeabilitäts-Porositätsrelationen für Salzgrus* (GRS-148, Ges. Für Anlagen- und Reaktorsicherheit, Braunschweig, Germany, 1999).

ORTOLEVA, P. J., *Geochemical Self-Organization* (Oxford Monographs on Geology and Geophysics 23, Oxford Univ. Press, New York, 1994).

PAPE, H., CLAUSER, C., and IFFLAND, J. (1999), *Permeability Prediction Based on Fractal Pore-Space Geometry*, Geophysics, *64*, 1447–1460.

PAPE, H., CLAUSER, C., and IFFLAND, J. (2000), *Variation of Permeability with Porosity in Sandstone Diagenesis Interpreted with a Fractal Pore Space Model*, Pure Appl. Geophys. *157*, 603–619.

PEACH, C. J., and SPIERS, C. J. (1996), *Influence of Crystal Plastic Deformation on Dilatancy and Permeability Development in Synthetic Salt Rock*, Tectonophys. *256*, 101–128.

POPP, T., KERN, H., and SCHULZE, O. (2001), *Evolution of Dilatancy and Permeability in Rock Salt During Hydrostatic Compaction and Triaxial Deformation*, in J. Geophys. Res., in press.

REIS, J. C., and ACOCK, A. M. (1994), *Permeability Reduction Models for the Precipitation of Inorganic Solids in Berea Sanstone*, J. Geophys. Res. *18*, 347–368.

REVIL, A. and CATHLES, L. M. (1999), *Permeability of Shaly Sans*, Water Resour. Res. *35*, 651–662.

SCHOLZ, C. H., LEGER, A., and KARNER, S. L. (1995), *Experimental Diagenesis: Exploratory Results*, Geophys. Res. Lett. *22*, 719–722.

SPANGENBERG, E., SPANGENBERG, U., and HEINDORF, C. (1998), *An Experimental Study of Transport Properties of Porous Rock Salt*, Phys. Chem. Earth, *23*, 367–371.

STAUFFER, D. and AHARONY, A., *Introduction to Percolation Theory* (Taylor and Francis, Bristol, Pa., 1992).

Stormont, J. C. and Daemen, J. J. K. (1992), *Laboratory Study of Gas Permeability Changes in Rock Salt During Deformation*, Int. J. Rock Mech. Min. Sci. Geomech. Abstr. *29*, 325–342.

Tenthorey, E., Scholz, C. H., Aharonov, E., and Leger, A. (1998), *Precipitation Sealing and diagenesis: 1 - Experimental Results*, J. Geophys. Res. *103*, 951–967.

Thompson, A. H. (1991), *Fractals in Rock Physics*, Annu. Rev. Earth Planet. Sci. *19*, 237–262.

Todd, A. C. and Yuan, M. D. (1992), *Barium and Strontium Sulfate Solid Solution Scale Formation at Elevated Temperatures*, SPE Prod. Eng. *7*, 85–92.

Vaughan, P. J., Moore, D. E. , Morrow, C. A., and Byerlee, J. D. (1986), *Role of Cracks in Progressive Permeability Reduction During Flow of Heated Aqueous Fluids Through Granite*, J. Geophys. Res. *91*, 7517–7530.

Walsh, J. B. and Brace, W. F. (1994), *The Effect of Pressure on Porosity and the Transport Properties of Rocks*, J. Geophys. Res. *89*, 9425–9431.

Wark, D. A. and Watson, E. B. (1998), *Grain-scale Permeabilities of Texturally Equilibrated, Monomineralic Rocks*, Earth Planet. Sci. Lett. *164*, 591–605.

Worden, R. H., Mayall, M., and Evans, I. J. (2000), *The Effect of Ductile-Lithic Sand Grains and Quartz Cement on Porosity and Permeability in Oligocene and Lower Miocene Clastics, South China Sea: Prediction of Reservoir Quality*, Am. Assoc. Petrol. Geol. Bull. *84*, 345–359.

Zhang, S., Paterson, M. S., and Cox, S. F. (1994a), *Porosity and Permeability Evolution During Hot Isostatic Pressing of Calcite Aggregates*, J. Geophys. Res. *99*, 15,741–15,760.

Zhang, S., Cox, S. F., and Paterson, M. S. (1994b), *The Influence of Room Temperature Deformation on Porosity and Permeability in Calcite Aggregates*, J. Geophys. Res. *99*, 15,761–15,775.

Zhu, W., Evans, B., and Bernabé, Y. (1999), *Changes in Permeability and Porosity During Hot Isostatic Pressing in Calcite Aggregates*, J. Geophys. Res. *104*, 25,501–25,511.

Zhu, W., Montesi, L., and Wong, T.-f. (1997), *Shear-enhanced Compaction and Permeability Reduction: Triaxial Extension Tests on Porous Sandstone*, Mech. Mat. *25*, 199–214.

Zhu, W., and Wong, T.-f. (1996), *Permeability Reduction in a Dilating Rock: Network Modelling of Damage and Tortuosity*, Geophys. Res. Lett. *23*, 3099–3102.

Zhu, W., and Wong, T.-f. (1997), *The Transition from Brittle Faulting to Cataclastic Flow: Permeability Evolution*, J. Geophys. Res. *102*, 3027–3041.

Zoback, M. D. and Byerlee, J. D. (1975), *The Effect of Microcrack Dilatancy on the Permeability of Westerly Granite*, J. Geophys. Res. *80*, 752–755.

(Received February 6, 2001, revised June 8, 2001, accepted August 1, 2001)

Pure appl. geophys. 160 (2003) 961–976
0033–4553/03/060961–16

Pure and Applied Geophysics

Slow Crack Propagation and Slip Correlations

J. SCHMITTBUHL,[1] A. DELAPLACE,[2] K. J. MÅLØY,[3]
H. PERFETTINI,[4] and J. P. VILOTTE[5]

Abstract—The propagation of an interfacial crack through a weak plane of a transparent Plexiglas block is studied experimentally. The toughness is controlled artificially by a sand blasting procedure, and fluctuates locally in space like uncorrelated random noise. The block is fractured in mode I at low speed ($10^{-7} - 10^{-4}$ m/s). The crack front is observed optically with a microscope and a high resolution digital camera. During the propagation, the front is pinned by micro-regions of high toughness and becomes rough. Roughness of the crack front is analyzed in terms of self-affinity. The in-plane roughness exponent is shown to be 0.63 ± 0.05. Experimental results are compared to a numerical model. The model reproduces the self-affine behavior of the crack front, i.e., long-range correlations of the roughness. Analogies between mode I and mode III are presented in order to discuss implications of the experimental results for creeping faults. Accordingly, correlations of the slip pattern are shown to exist over scales substantially larger than the asperity sizes.

Key words: Fracture roughness, crack propagation, creeping fault.

1. Introduction

The propagation of rupture front through heterogeneous solids is a central question for numerous mechanical problems. One main evidence of the fluctuating front propagation is the roughness of fracture surfaces. Numerous works have shown the self-affine properties of crack surfaces (BROWN and SCHOLZ, 1985; BOUCHAUD *et al.*, 1990; MÅLØY *et al.*, 1992; COX and WANG, 1993; SCHMITTBUHL *et al.*, 1995b; BOUCHAUD, 1997). The out-plane roughness exponent ($\zeta \approx 0.80$) is found to be very robust over different materials, different fracture modes and a broad range of length scales (BROWN and SCHOLZ, 1985; POWER *et al.*, 1987; SCHMITTBUHL *et al.*, 1993; SCHOLZ, 1990). A possible "universal" self-affine crack geometry of heterogeneous material has been proposed by BOUCHAUD *et al.*

[1] Laboratoire de Géologie, UMR 8538, Ecole Normale Supérieure, Paris, France.
E-mail: Jean.Schmittbuhl@ens.fr
[2] Laboratoire de Mécanique et Technologie, Ecole Normale Supérieure, Cachan, France.
[3] Fysisk Institutt, Universitetet i Oslo, Oslo, Norway.
[4] Laboratoire de Géophysique Interne et Tectonophysique, Grenoble, France.
[5] Département de Sismologie, UMR 7580, Institut de Physique du Globe, Paris, France.

(1990). Studies of fracture surfaces and models of crack propagation are generally linked: the morphology of the crack surface is often assumed as inherited from the geometry of the crack front (BOUCHAUD *et al.*, 1993). Consequently, the mechanisms of the crack-front geometry during its propagation becomes a key question.

Even when the rupture front is interfacial, i.e., restricted to move within a plane (see Fig. 1), the motion results from interactions between non-local elastic coupling, inertia, wave effects, and quenched heterogeneities at different scales (PERRIN and RICE, 1994; RICE *et al.*, 1994; RAMANATHAN and FISHER, 1997; WILLIS and MOVCHAN, 1995; FISHER, 1998). The front is trapped temporarily by local asperities that may be related to local material heterogeneities or residual (i.e., internal) stresses. Depinning from these asperities involves local instabilities. When the elastic coupling is small, the motion is controlled by individual asperities. On the contrary for strong elastic coupling, asperities interact because of elasticity and the dynamics becomes global.

The aim of this paper is to discuss recent experimental results that describe the propagation of a tensile crack front along a heterogeneous weak plane, and to compare them to a simple numerical model. These results may be relevant for other modes of loading and shear fault rupture, at least for model elastodynamic theory that involves a single displacement variable. In particular, the dynamic spatial correlations may have interesting implications for creeping faults, as will be discussed briefly in the last section.

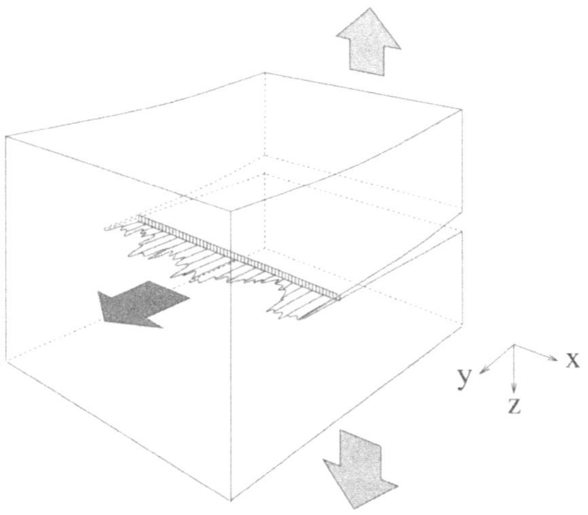

Figure 1
Sketch of the interfacial fracture crack propagation. (from SCHMITTBUHL *et al*, 1995a)

2. *Experimental Setup*

Samples are made of transparent polymethylmethacrylate (PMMA). Each PMMA sample is obtained from the assembling of two plates (SCHMITTBUHL and MÅLØY, 1997; DELAPLACE *et al.*, 1999). The upper plate is 32 cm × 14 cm and 1 cm thick. The lower plate is 34 cm × 12 cm wide and 0.4 cm thick. Each plate is sand-blasted on one side. One important consequence of the sand-blasting is that the transparency of the PMMA is lost, as light scatters from the introduced microstructures. The plates are then annealed together in an oven under a homogeneously normal pressure. The annealing procedure lasts 30 min at a temperature of 205 °C (see Fig. 2).

After annealing both plates together, the newly formed block recovers its transparency. New polymer chains are formed through the rough interface and air bubbles are extracted with the pressure load. During the experiment, the plates are moved apart and a fracture propagates; the polymer chains within the weak plane are broken. The difference of refraction indexes across the interface is recovered and the fracture surface appears opaque due to the light scattering from the microstructures. The transition between the transparent and opaque regions corresponds to the crack front (see Fig. 3).

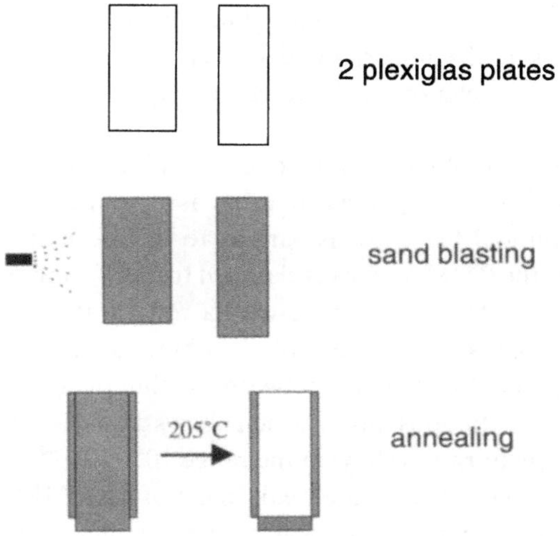

Figure 2

Sample preparation procedure. Two Plexiglas plates of 32 cm × 14 cm and 34 cm × 12 cm are cut and sintered together at 205 °C to form a single new block. Before the annealing process, both plates are sand-blasted with 50 μm steel particles. Sandblasting creates roughness fluctuations that control the local toughness during sintering. Sandblasting also unpolishes the surfaces which induces a loss of transparency. However, transparency is recovered during the sintering process.

Figure 3
Example of a crack front picture using the Kodak DSC420 camera. The resolution is 1536 × 1024 pixels. The image covers an area of 6.5 mm × 4 mm. The crack is propagating from bottom to top. The cracked area is in dark gray. The front obtained from image treatment is superimposed as a black line.

Sandblasting roughens the surfaces of the PMMA plates. The magnitude of the roughness is less than a few micrometers. The random position of the defaults was checked optically using a microscope. Optical signatures of the flaws were observed to be smaller than a cutoff of about 50 μm corresponding to the size of the particle used for the blasting. The height fluctuations introduced by the sandblasting procedure introduce fluctuations in the toughness during the annealing process. We expect the toughness to be uncorrelated on length scales larger than the cutoff length. However, the correlation in the toughness up to this length scale is an open question.

At the final stage, one block with a weak annealing plane is obtained where the crack will propagate. The crack plane is referred as (x, y): the y axis is along the crack propagation direction and the x axis is parallel to the mean crack front.

The thick plate of the PMMA block is clamped to a stiff aluminum frame. A normal displacement is imposed to the thin plate with a rod parallel to the crack front (see Fig. 4). By using an imposed displacement, we obtain a *stable* crack propagation in mode I. Oil is added to reduce friction between the thin plate and the rod. The rod is moved vertically by a continuous motor which allows a quasi-static loading. Typically the imposed displacement rate is slow: in the range $10^{-7} - 10^{-4}$ m/s.

The fracture propagates within the weak plane of the PMMA block. During the propagation, the front is pinned by local regions of high toughness and becomes rough. The geometry of the crack front is observed with a microscope. When static positions of the fronts are analyzed, a digital camera, set on the top of the microscope, provides high resolution images: 1536 × 1024 pixels. Each pixel covers a region of 2.6 μm × 2.6 μm. A sample picture is shown in Figure 3. In the inverted image, the sintered part is seen as light while the dark region represents the fracture

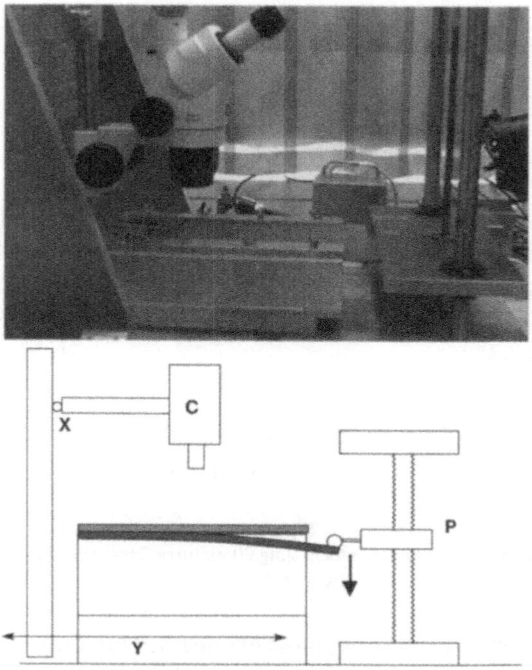

Figure 4
Picture and sketch of the experimental setup. The PMMA block made of two sintered plates is drawn with grays. The lower plate (dark gray) is bent thanks to a mechanical press (P) at constant displacement rate. The crack propagates in the *y* direction and is observed from above using a digital camera (C). The camera is attached to a moving frame that can be translated in both *x* and *y* directions.

aperture. The front is defined as the contrast boundary between the dark and the light areas and is found by using a specially developed software.

3. *Roughness of Pinned Fronts*

Measurements have been performed after a complete stop of the press when the crack front is at rest. The microscope is translated along the front (x direction) and up to 20 pictures of the front are taken. Neighboring pictures overlap one another over one third of the picture width (i.e., over 1.3 mm). Subfronts are extracted from each picture and averaged at overlapping positions. The assembling of 20 pictures provides a high resolution crack front, i.e., full front (up to 2^{14} data points) 45 mm long (see Fig. 5). A complete description of the assembling procedure can be found in DELAPLACE *et al.* (1999).

The in-plane roughness of the crack front is analyzed in terms of self-affinity. A self-affine front line $A(x)$ is statistically invariant under the scale transformation:

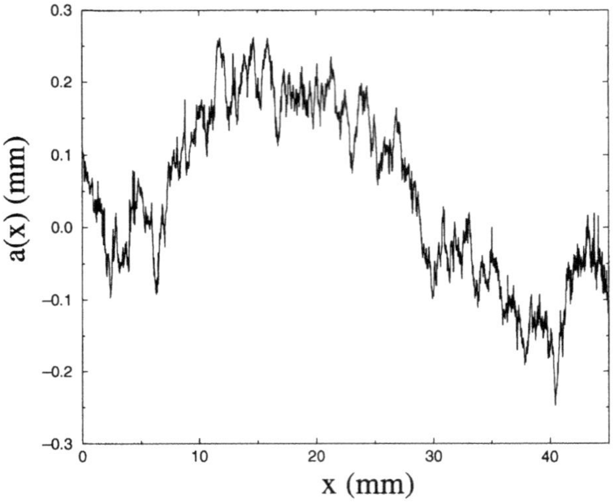

Figure 5
Sample of high resolution front obtained by assembling 20 pictures. Note that the vertical axis is magnified by a factor 75.

$(x, A) \rightarrow (\lambda x, \lambda^{\zeta} A)$ where ζ is the roughness exponent. Numerous techniques exist to probe the self-affine behavior and estimate the roughness exponent. Figure 6 presents an analysis based on a recent technique that uses a wavelet transform: The Average Wavelet Coefficient (AWC) technique (SIMONSEN *et al.*, 1998). A power-law behavior fits nicely the data both for the entire front and the subfronts, which shows the self-affine property of the crack front at rest. The expected slope for an artificial self-affine profile with a roughness exponent ζ is: $\zeta + 1/2(W(a) = \langle W_b(a) \rangle_b \propto a^{\zeta+1/2}$ where a is the scale factor of the wavelet and b the translation factor; an average is performed over the translation factor b). This provides an estimate of the roughness exponent of the crack front: $\zeta \approx 0.60$.

Results are consistent with the power spectrum analysis of the front roughness. Indeed, Figure 7 shows the power spectrum of the full front and the average over each individual subfront. The power-law behavior is also consistent with a self-affine property of the crack front. The slope of the spectrum for an artificial front with an exponent ζ is: $P(k) \propto k^{-2\zeta-1}$. From Figure 7, the roughness exponent of the crack front is also close to $\zeta \approx 0.60$.

In DELAPLACE *et al.* (1999) a full analysis of nine independent full fronts has been performed using four different techniques (SCHMITTBUHL *et al.*, 1995c): the variable bandwidth method, the multiple return probability method, the power spectrum analysis and the AWC method. All methods are consistent, and the final estimate of the roughness exponent is: $\zeta = 0.63 \pm 0.03$.

By performing the full analyses of the self-affinity property of the crack fronts, one observes that the scaling shows an upper cut-off that might be related to the

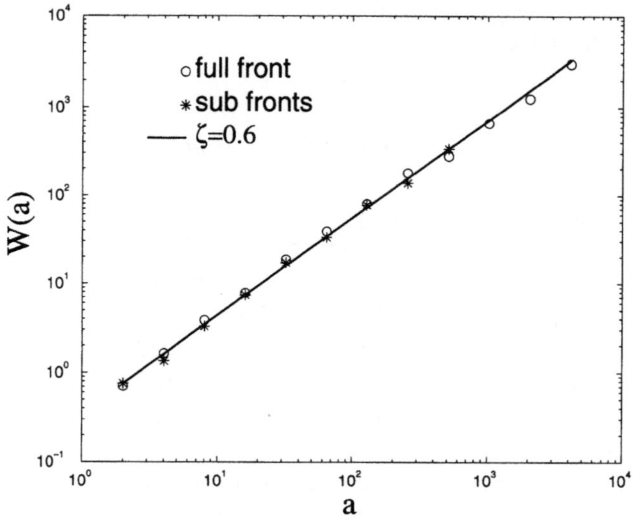

Figure 6
Self-affine analysis of a full-front (i.e., an assembled front from 20 pictures) using the Average Wavelet Coefficient method. The scale a is in pixel unit (i.e., 2.6 μm). The average analysis of each individual subfront is superimposed in the figure. The log-log diagram shows a power-law behavior with a slope of 1.1, consistent with a roughness exponent of 0.6.

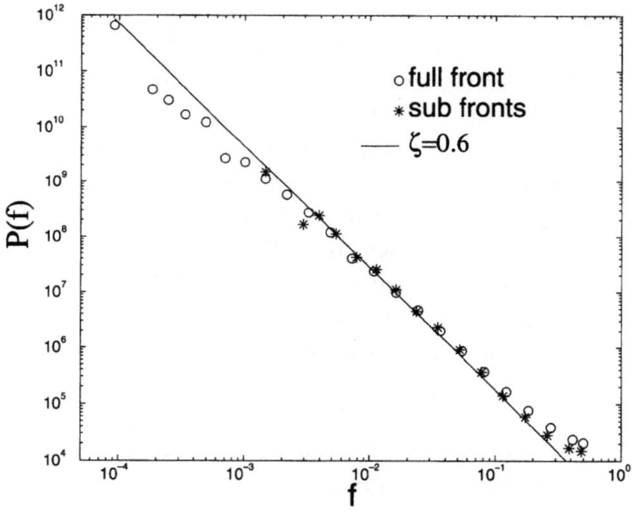

Figure 7
Self-affine analysis of a full-front (i.e., an assembled front from 20 pictures) using the Power Spectrum method. The wavelength unit is in pixel^{-1} where one pixel is 2.6 μm wide. The average analysis of each individual subfront is superimposed in the figure. The log-log diagram displays a power-law behavior with a slope of −2.2, consistent with a roughness exponent of 0.6.

thickness of the Plexiglas plate. Further studies are ongoing for a careful check of the influence of the Plexiglas plate thickness.

4. A Mode I Crack Model

We assume that the roughness magnitude is always significantly smaller than the plate thickness. Accordingly, the experiment is modelled as a half-plane tensile crack propagating through an unbounded elastic solid. The crack is restricted to move in the plane $z = 0$, along the y direction. The crack front position, at time t, is described by a curve $y = A(x, t)$ that is assumed to be single-valued, excluding configurations with overhangs. To first-order, $A(x, t) = v_{cr}t + a(x, t)$ where v_{cr} is a uniform crack extension velocity and $a(x, t)$ the small amplitude deviation around that configuration with zero mean.

The energy per unit area of the interface that must be provided to fracture the solid is denoted $\Gamma(x, A(x, t))$. Interface heterogeneities are considered here as fluctuations in the local energy density: $\Gamma(x, y) = \Gamma_0[1 + \gamma(x, y)]$ where Γ_0 is the mean value of the interface fracture resistance and $\gamma(x, y)$ a dimensionless random quenched fluctuation of zero mean.

The analysis is restricted to a scalar approximation of elasticity (RICE *et al.*, 1994). The displacement field is then a scalar variable $u(x, y, z, t)$, and the associated tensile stress across planes parallel to the crack is $\sigma = (\lambda + 2\mu)(\partial u/\partial z)$, with λ and μ the Lamé coefficients. The problem is to find u satisfying the three-dimensional scalar wave equation: $c^2\nabla^2 u = \partial^2 u/\partial t^2$ with the sound speed $c = \sqrt{(\lambda + 2\mu)/\rho}$ where ρ is the density. The displacement field has a discontinuity across the crack surface $\Delta u(x, y, t) = u(x, y, 0^+, t) - u(x, y, 0^-, t), \quad \forall y < A(x, t)$. The stress must satisfy the boundary condition $\sigma = 0$ on the crack surface, i.e., $(\partial u/\partial z)(x, y, z = 0^\pm, t) = 0$ and meets the remote loading condition, i.e., the static external load.

The fracture energy G is the energy provided to the crack front by the flux of stored elastic energy per unit of new crack area. For small crack-front distortions, the energy release rate G can be expanded about a uniformly extending crack (RICE *et al.*, 1994; WILLIS and MOVCHAN, 1995; RAMANATHAN and FISHER, 1998) and specifically a first-order approximation can be used (SCHMITTBUHL and VILOTTE, 1999):

$$G(x) = G_0(1 + \mathscr{I}(x)) \ , \tag{1}$$

Where G_0 is the energy release rate for a quasi-static straight crack. The non-local kernel is given by (GAO and RICE, 1989):

$$\mathscr{I}(x, a(x, t)) = \frac{1}{\pi} PV \int_{-\infty}^{\infty} \frac{a(x') - a(x)}{(x - x')^2} \, dx' \ , \tag{2}$$

where PV is the principal value.

Accordingly the actual dynamics is replaced by the quasi-static approximation, i.e., elastic waves are neglected, and at each instant t the energy balance between mechanical energy and surface energy leads to:

$$G(x,t) = \Gamma(x, a(x,t)) \ . \tag{3}$$

5. *Numerical Model*

In order to characterize locally the heterogeneity of the interface, we introduce the fracture energy resistance $\Gamma(x,y)$ as a quenched disorder (i.e., a random fluctuation in space that does not evolve with time). To mimic the experiment, the fluctuations of the fracture energy resistance are assumed to be highly correlated over lengths smaller than a characteristic length d, and spatially uncorrelated at higher scales. The cut-off scale d has to be compared to the particle size of the sandblasting procedure ($d = 50\,\mu\text{m}$). Accordingly, the interface is discretized into $N = L/d$ independent segments. The resistance is chosen for each segment from the definition: $\Gamma(x,y) = \Gamma_0[1 + \gamma(x,y)]$ where the average resistance Γ_0 is set to unity, and γ is obtained from a uniform distribution $[\gamma - \delta\gamma; \gamma + \delta\gamma]$.

To avoid edge effects, a L-periodic interface in the x direction is assumed such that the $1/x^2$ kernel in (2) transforms in $(\pi/L)^2/\sin^2(\pi x/L)$ (GAO and RICE, 1989).

Loading results from an imposed displacement. An elementary step (i.e., a time step) corresponds to the motion of only one segment. At each step, the weakest segment is searched for (i.e., event driven dynamics) by assuming that its location x_w corresponds to the least fracture energy $G_{\min}$ required to advance the crack SCHMITTBUHL *et al.*, 1995a; SCHMITTBUHL and VILOTTE, 1999):

$$G_{\min} = \Gamma_0 \min_{\forall x} \left[\frac{1 + \gamma(x, A(x,t))}{1 + \mathcal{I}(x, A(x,t))} \right] \ . \tag{4}$$

The most susceptible segment x_w is advanced by a fixed amount s that gives the unit length in the direction of propagation. At this stage, the crack is unloaded. The procedure is then repeated. The time scale is fixed by the updates of the weakest segments, and not directly related to the real time. After a transient regime, the system organizes into a statistically stationary state.

The behavior of the system is controlled by the competition between local fluctuations of the toughness and the effect of long-range elastic interactions. It can be shown (PERFETTINI *et al.*, 2001) that the stress drop of a point that just slipped by amount s is $2\mu s/(\pi d)$, where s is the slip size of elementary events and d is the discretization length. PERFETTINI *et al.* (2001) showed that the ratio of this stress drop and the rms of the toughness fluctuations $2\delta\gamma/\sqrt{12}$ defines the useful dimensionless parameter $\alpha = \mu s\sqrt{12}/(\pi d \delta\gamma)$. When α is much greater than unity ($\alpha \gg 1$), the elastic interactions screen the effect of the fluctuations of the toughness.

The heterogeneities are not strong enough to pin the front, thus crack advance can never be arrested. In our model under imposed displacement, the runaway of the crack is controlled from the imposed displacement, but this would not be the case with a model at fixed loading stress (see Fig. 8). This propagative regime will be denoted as regime **P**, and results in sidewise propagation of the crack.

When α is considerably smaller than unity ($\alpha \ll 1$), the heterogeneities of the toughness control the front advance (regime **SP** for strong pinning; see Fig. 8). This extreme regime may lead, for instance, to recurrent slip of the same segment, since the stress drop due to elastic interactions is not big enough (compared to $\delta\gamma$) to unload the point. For such a regime, the displacement field is expected to have the same statistical distribution as the fluctuations of the toughness, that is, a white noise in the present study.

For the intermediate regime, the magnitude of the elastic interactions is comparable to the amplitude of the toughness variations. This is a weak-pinning regime **WP** where the interactions between toughness heterogeneities and elastic stress transfers leads to nontrivial spatio-temporal correlations of slip (see Fig. 8) (SCHMITTBUHL *et al.*, 1995a; TANGUY *et al.*, 1998); SCHMITTBUHL and VILOTTE, 1999).

The simulations start with a zero displacement field. Figure 8 shows various slip distributions in steady state related to $\alpha = 10^{-4}$, 10^{-1} and 10^{3} illustrating the three regimes mentioned above. We will focus on the intermediate regime **WP** ($\alpha \simeq 1$) where elastic interactions are comparable to variations of the toughness.

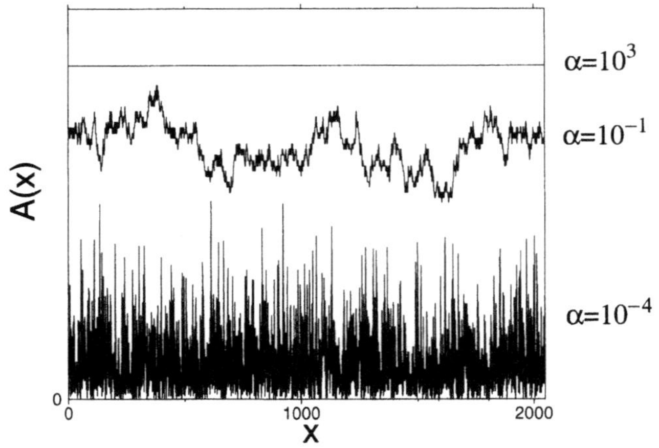

Figure 8

Sample of three fronts obtained from the numerical modeling for various α parameters. The x coordinate is in asperity size unit d. For $\alpha = 10^{3}$, elastic interactions dominate toughness fluctuations and the medium is seen as homogeneous. When $\alpha = 10^{-4}$, elastic coupling is negligible with respect to the influence of toughness fluctuations; roughness of the front is then controlled by the randomness of the toughness. For intermediate α ($\alpha = 10^{-1}$), the crack front develops spatial correlations over large distances.

6. Experiment and Model Comparison

In the numerical model, we let the front evolve from an initially flat slip with increasing displacement. After a transient regime, corresponding to the spreading of slip over the whole length of the sample, the slip correlations reach a stationary regime. The power spectrum (thin line) of the slip distribution, averaged over many fronts, is shown in Figure 9. On the same figure, the power spectrum related to the experimental results (heavy line) is also plotted. The power spectrum of the modeled front exhibits a power-law behavior, characteristics of self-affinity. The roughness exponent is found to be $\zeta \approx 0.35$, a value lower than the $\zeta \approx 0.6$ measured in our experiment.

Even though this model captures the main feature of the experimental observations i.e., the existence of a self-similar front, the value of the roughness exponent we obtain underestimates the value measured in the experiment. This difference in the value of the roughness exponent might come from:

1) *The influence of dynamical effects.* Using an experimental procedure similar to ours, MÅLØY and SCHMITTBUHL (2001) have studied the advance of a mode I crack (in-plane slip). They have observed that the front was advancing due to micro-

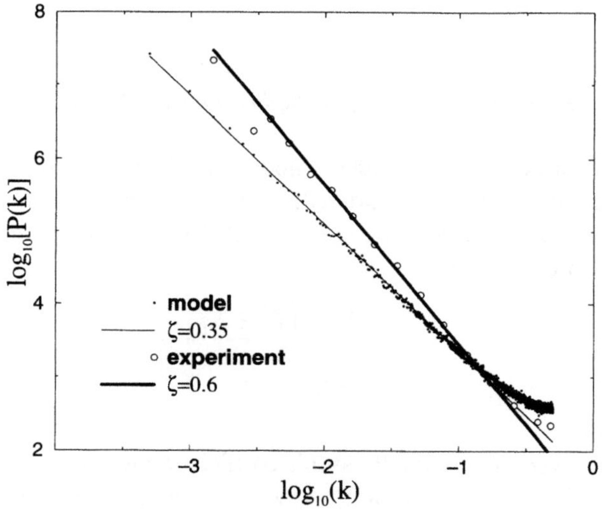

Figure 9
Comparison of the Power Spectra of experimental and numerical fronts. The wavelength unit is in pixel^{-1} (the pixel width is 2.6 μm) for the experimental spectrum. For the numerical model, the unit of wave number is given by the size of the toughness fluctuation (i.e., the discretization size). Both spectra have a power-law behavior on a large range of scales, which shows the self-affine properties of the crack front roughness. However, slopes (that is, roughness exponents) are different: $\zeta \approx 0.35$ for the modeled fronts and $\zeta \approx 0.6$ for experimental fronts.

dynamic events associated with asperity depinning. This result suggests that the quasi-static approximation may be violated. Considering a complete elastodynamic description of this problem, RAMANATHAN and FISHER (1998) found a roughness exponent $\zeta \approx 0.5$, which is significantly higher than the value that is derived using a quasi-static approximation. This difference is very likely due to the presence of stress overshoots which are absent in our model.

2) *The existence of spatial correlations between toughness heterogeneities.* SCHMITTBUHL and VILOTTE (1999) have studied, using a model similar to ours, the influence of various types of these correlations (e.g., Gaussian-like). They showed that, although the self-affine behavior of the front was preserved, the value of the roughness exponent was affected by the presence of spatially correlated asperities. For a Gaussian spatial correlation of the toughness, the crack front shows a roughness exponent: $\zeta = 0.5$. The experimental roughness exponent $\zeta = 0.6$ can be obtained if the toughness fluctuations are spacially correlated using a power-law correlation function with an exponent of -0.4.

3) *The influence of the process zone.* Due to the high stress ahead of the crack tip (crack theory predicts a $1/\sqrt{r}$ singularity, r being the distance from the crack tip), inelastic deformations and micro-fracture interactions may occur and modify locally the response of the medium. A more extended discussion of these type of effects can be found in BOUCHAUD (1997), LOPEZ and SCHMITTBUHL (1998), and ZAPPERI *et al.* (2000).

7. Crack Mode Analogies and Implications for Creeping Faults

The model described above assume a mode I crack. If a shear loaded half-plane crack, on the plane $z = 0$, is considered, it is commonly assumed that fracture processes under mixed mode loading are controlled by the energy release rate or the maximum shear intensity factor:

$$G = \frac{(1-\nu)K_{\mathrm{II}}^2 + K_{\mathrm{III}}^2}{2\mu} \tag{5}$$

$$K = \sqrt{K_{\mathrm{II}}^2 + K_{\mathrm{III}}^2} \tag{6}$$

where ν and μ are respectively the Poisson's coefficient and the shear modulus of the elastic medium, while K_{II} and K_{III} are respectively the mode II and mode III stress intensity factors. For a strike-slip fault, concentration at the straight crack front is only controlled by the mode III loading, while for thrust or normal faults the mode II loading is dominant. GAO and RICE (1986) have derived the first-order perturbation of K_{II} and K_{III} for small deviations in the crack front geometry. This leads to a first-order coupling between mode II and mode III and therefore a second-order coupling effect in G or K. Following GAO *et al.* (1991), one may assume that in the first-order analysis this coupling effect between the shear modes may be neglected, so that the perturbation in the local energy release rate may be written as:

$$G(x) = G^{\infty}\left(1 + \frac{M_{\beta}}{\pi} PV \int_{-\infty}^{\infty} \frac{a(x') - a(x)}{(x - x')^2}\, dx'\right) , \tag{7}$$

where M_{β} is a constant depending on the shearing mode of loading, i.e., β = III for mode III loading and β = II for a mode II loading. This constant can be shown to depend only on the Poisson's coefficient ν with $M_{\mathrm{II}} = (2 - 3\nu)/(2 - \nu)$ and $M_{\mathrm{III}} = (2 + \nu)/(2 - \nu)$. Under this hypothesis, the non-local elastic interactions are of the same type as the model studied before, and our results can to some extent be relevant for mode II and mode III crack experiments.

Another interesting mathematical analogy can be noted here between the mode I perturbed crack front problem and the 2-D problem of an antiplane rupture along a frictional fault (GAO and RICE, 1989; PERRIN and RICE, 1994). In the antiplane rupture problem, the unbounded elastic medium contains a weak plane ($z = 0$) on which frictional sliding can occur. The elastic shear modulus is μ. Only displacement along the y direction is allowed and is presumed to be invariant in the y direction. This defines a model involving a single displacement variable u which satisfies a scalar wave equation and represents the shear slippage with the associated shear stress σ. The slip on the frictional plane is defined by the displacement discontinuity $\delta(x,t) = u(x,0^+,t) - u(x,0^-,t)$ and the shear stress is $\tau(x,t) = \sigma(x_1,0,t)$. Frictional disorder on the fault, such as due to asperities, is introduced through a random pinning force $\eta_p(x,t)$. The quasistatic evolution of the antiplane shear rupture is actually governed by the same equations as the mode I crack problem, equations (1)–(3), provided the identification of the following variables:

$$\begin{cases} a(x,t) \rightarrow \delta(x,t) \\ \dfrac{G - G^{\infty}}{G^{\infty}} \rightarrow \dfrac{\tau}{2\mu} \\ \dfrac{\Gamma - G^{\infty}}{G^{\infty}} \rightarrow \dfrac{\eta_p - \tau^{\infty}}{2\mu} \end{cases} . \tag{8}$$

Since in the experiment the simulated fault is moving at a very slowly imposed rate, it is analogous to a creeping in-plane fault. If one accepts this analogy, our results have interesting implications for low-frequency slip inversions like GPS or InSar measurements, as pointed out by PERFETTINI *et al.* (2001). These instruments are low-pass filters, and will measure an important signal coming from slip on the fault, since the power spectrum of slip decreases with increasing wave number (see Fig. 9). However, the measured slip pattern will be smooth compared to the real one. This feature can be observed in Figure 10 where a synthetic front generated by the model (dark line) and the equivalent filtered one (grey line) are both plotted. The correlations of slip we observe in the experiments are clearly due to small-scale asperities, however these scales are inaccessible to the latter instruments. Therefore, special care should be taken in the interpretation of areas of high slip measured by those instruments since they may not be due, in the view of the results of the experiment, to the presence of heterogeneities of the same size.

8. Conclusion

We have studied experimentally the slow propagation of an interfacial mode I crack in a transparent Plexiglas block. The crack advance has been shown to result from the depinning of asperities or micro-regions of high toughness. Self-affinity was observed, i.e., correlations of slip at all scales. The roughness exponent was found to be $\zeta \approx 0.63 \pm 0.05$. A quasi-static model was developed and was able to reproduce the self-similarity of the front with a roughness exponent $\zeta \approx 0.35$. The discrepancy between the observations and the model may be explained by dynamic effects (due to stress overshots), spatial correlations between asperities or non-elastic processes ahead of the crack tip. We have shown that the results, obtained in mode I, can easily be extended to mode II or III ruptures. These last two modes of rupture are relevant for the description of earthquakes where opening is unlikely to happen. Our results may have interesting applications to the interpretation of low-frequency data such as GPS or InSar.

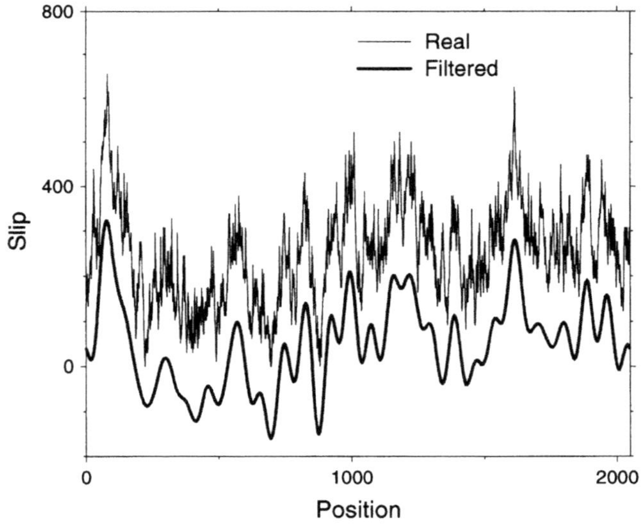

Figure 10
Comparison between the real and measured synthetic slip front of a creeping fault. Because of the low pass filter of the instrument (e.g., GPS or InSar), the small wavelengths are erased from the original signal.

From these results, the hypothesis that the average motion of a creeping fault may arise from a succession of micro-events related to breakage of asperities, is validated. In a mode I experiment similar to ours, MÅLØY and SCHMITTBUHL (2001) demonstrated that these micro-events were dynamical. This could help us view creeping faults differently. Even though these faults appear as moving continuously at large scale, the average slip may result from very localized dynamic instabilities or micro-earthquakes. Some of these events are often observed on creeping faults and one refers to them as micro-seismicity. However, it is possible that only the largest of these dynamic instabilities related to the depinning of large scale (of the order of many meters) asperities, are observed. Continuous slip may result from a series of dynamic events, where some, because of their significant size, are detected, while others, related to depinning of small asperities (such as the ones in the experiment), are beyond the detection threshold of seismic instruments. Such a speculative result, which needs further examination, has the advantage of collectively gathering seismically active and creeping faults, since slip on both of these faults would be due to dynamic instabilities.

Acknowledgements

This work was partially supported by ACI programs "Risques Naturels" and "Jeunes Chercheurs" of the french educational minister the CNRS program ECODEV, the France-Norway program Pics and the CNRS program "Intérieur de la Terre."

REFERENCES

BOUCHAUD, E. (1997), *Scaling Properties of Cracks*, J. Phys.: Cond. Matter *9*, 4319–4344.

BOUCHAUD, E., LAPASSET, G., and PLANÈS, J. (1990), *Fractal Dimension of Fractured Surfaces: A Universal Value?*, Europhys. Lett. *13*, 73–79.

BOUCHAUD, J. P., BOUCHAUD, E., LAPASSET, G., and PLANÈS, J. (1993), *Models of Fractal Cracks*, Phys. Rev. Lett. *71*, 2240.

BROWN, S. R. and SCHOLZ, C. H. (1985), *Broad Bandwidth Study of the Topography of Natural Rock Surfaces*, J. Geophys. Res. *90*, 12,575–12,582.

COX, B. L. and WANG, J. S. Y. (1993), *Fractal Surfaces: Measurement and Application in Earth Sciences*, Fractals *1*, 87–115.

DELAPLACE, A., SCHMITTBUHL, J., and MÅLØY, K. (1999), *High Resolution Description of a Crack Front in an Heterogeneous Plexiglas Block*, Phys. Rev. E *60*, 1337–1343.

FISHER, D. (1998), *Collective Transport in Random Media: From Superconductors to Earthquakes*, Physics Reports *301*, 113–150.

GAO, H. and RICE, J. (1986), *Shear-stress Intensity Factors for a Planar Crack with Slightly Curved Front*, J. Appl. Mech. *53*(4), 774–778.

GAO, H. and RICE, J. R. (1989), *A First-order Perturbation Analysis of Crack Trapping by Arrays of Obstacles*, ASME J. Appl. Mech. *56*, 828–836.

GAO, H., RICE, J., and LEE, J. (1991), *Penetration of a Quasi-statically Slipping Crack into a Seismogenic Zone of Heterogeneous Fracture-resistance*, J. Geophys. Res. *96*(B13), 21,535–21,548.

LOPEZ, J. and SCHMITTBUHL, J. (1998), *Anomalous Scaling of Fracture Surfaces*, Phys. Rev. E *57*, 6999–7005.

MÅLØY, K. and SCHMITTBUHL, J. (2001), *Dynamical Instabilities during Slow Crack Propagation*, Phys. Rev. Lett. submitted.

MÅLØY, K. J., HANSEN, A., HINRICHSEN, E. L., and ROUX, S. (1992), *Experimental Measurements of the Roughness of Brittle Cracks*, Phys. Rev. Lett. *68*, 213–215.

PERFETTINI, H., SCHMITTBUHL, J., and VILOTTE, J. (2001), *Slip Correlations on a Creeping Fault*, Geophys. Res. Lett. *28*(10), 2133–2136.

PERRIN, G. and RICE, J. N. (1994), *Disordering of Dynamic Planar Crack Front in a Model Elastic Medium of Randomly Variable Toughness*, J. Mech. Phys. Solids *42*(6), 1047–1064.

POWER, W. L., TULLIS, T. E., BROWN, S. R., BOITNOTT, G. N., and SCHOLZ, C. H. (1987), *Roughness of Natural Fault Surfaces*, Geophys. Res. Lett. *14*, 29–32.

RAMANATHAN, S. and FISHER, D. (1997), *Dynamics and Instabilities of Planar Tensile Cracks in Heterogeneous Media*, Phys. Rev. Lett. *79*, 877–880.

RAMANATHAN, S. and FISHER, D. (1998), *Onset of Propagation of Planar Cracks in Heterogeneous Media*, Phys. Rev. B *58*, 6026–6046.

RICE, J. R., BEN-ZION, Y., and KIM, K. (1994), *Three-dimensional Perturbation Solution for a Dynamic Planar Crack Moving Unsteadily in a Model Elastic Solid*, J. Mech. Phys. Solids *42*, 813–843.

SCHMITTBUHL, J. and MÅLØY, K. (1997), *Direct Observation of a Self-affine Crack Propagation*, Phys. Rev. Lett. *78*, 3888–3891.

SCHMITTBUHL, J. and VILOTTE, J. (1999), *Interfacial Crack Front Wandering: Influence of Correlated Quenched Noise*, Physica A *270*, 42–56.

SCHMITTBUHL, J., GENTIER, S., and ROUX, S. (1993), *Field Measurements of the Roughness of Fault Surfaces*, Geophys. Res. Lett. *20*, 639–641.

SCHMITTBUHL, J., ROUX, S., VILOTTE, J., and MÅLØY, K. (1995a), *Interfacial Crack Pinning: Effect of Nonlocal Interactions*, Phys. Rev. Lett. *74*, 1787–1790.

SCHMITTBUHL, J., SCHMITT, F., and SCHOLZ, C. (1995b), *Scaling Invariance of Crack Surfaces*, J. Geophys. Res. *100*, 5953–5973.

SCHMITTBUHL, J., VILOTTE, J., and ROUX, S. (1995c), *Reliability of Self-affine Measurements*, Phys. Rev. E *51*, 131–147.

SCHOLZ, C. H., *The Mechanics of Earthquakes and Faulting* (Cambridge Univ. Press, New York, 1990).

SIMONSEN, I., HANSEN, A., and NES, O. M. (1998), *Using Wavelet Transforms for Hurst Exponent Determination*, Phys. Rev. E *58*, 2779.

TANGUY, A., GOUNELLE, M., and ROUX, S. (1998), *From Individual to Collective Pinning: Effect of Long-range Elastic Interactions*, Phys. Rev. E *58*(2), 1577–1590.

WILLIS, J. and MOVCHAN, A. (1995), *Dynamic Weight-functions for a Moving Crack: 1. Mode-I Loading*, J. Mech. Phys. Solids *43*, 319–341.

ZAPPERI, S., HERRMANN, H., and ROUX, S. (2000), *Planar Cracks in the Fuse Model*, Eur. Phys. J. B *17*(1), 131–136.

(Received February 6, 2001, revised June 22, 2001, accepted August 1, 2001)

Pure appl. geophys. 160 (2003) 977–988
0033 – 4553/03/060977–12

Pure and Applied Geophysics

Pressure Oscillation Effects on the Saffman-Taylor Instability

NICOLAS GLAND[1] and DMITRI PISARENKO[1]

Abstract—We studied experimentally the stability of the front between two immiscible fluids in a Hele-Shaw cell, which consists of two parallel glass plates separated by a narrow spacing. The fluid flow in a Hele-Shaw cell is governed by the Laplace's equation (pressure field) and Darcy's law (fluid transport), which makes of this experimental setup a two-dimensional analog of a porous medium (FABER, 1997). SAFFMAN and TAYLOR (1958) demonstrated that when the injected fluid is less viscous than the displaced one, the front between the fluids can be unstable and fingers may develop along the interface. We investigated the possibility of influencing the development of the Saffman-Taylor instability by modulating the injection pressure with weak periodic oscillations of low frequency (5 to 15 Hz). A series of experiments were performed, in which the evolution of the interface between the injected air and the displaced silicon oil was recorded with the help of a numerical photo camera. We observed that the oscillations of the injection pressure inhibit the development of fingers. We quantified the observed stabilization effect by performing a harmonic analysis of the shape of the front, and suggested an interpretation in terms of "stiffening" of the interface.

Key words: Porous medium, two-phase flow, Saffman-Taylor instability, surface tension.

Introduction

SAFFMAN and TAYLOR (1958) observed that, when a fluid is displaced by another less viscous fluid in a Hele-Shaw cell consisting of two closely spaced parallel plates, the interface between the two fluids can be unstable. The injected fluid forms fingers which penetrate into the more viscous fluid. The shape of the interface between the two fluids is controlled by a number of physical parameters (SAFFMAN-TAYLOR, 1958), among which the capillary number $Ca = \eta_1 V/\sigma$ (where η_1 is the viscosity of the fluid that initially fills the cell, σ is the surface tension and V is the fluid velocity) and the ratio of the viscosities of the two fluids η_1/η_2 play the most important role. Higher values of both of these parameters lead to a more pronounced fingering of the interface.

In the limit of low Reynolds numbers, the flow in a horizontal Hele-Shaw cell is governed by Darcy's law according to which the velocity of the fluid is given by

[1] Ecole Normale Supérieure, Laboratoire de géologie, 24, rue Lhomond, 75005 Paris, France.
E-mail: gland@geologie.ens.fr

$$\mathbf{V} = -\frac{b^2}{12\eta}\nabla P, \tag{1}$$

where b is the gap between the plates of the cell, and $K = b^2/12$ represents the effective permeability of the cell. It follows then that for incompressible fluids ($\nabla \cdot V = 0$), the pressure field in the cell satisfies the harmonic equation

$$\nabla^2 P = 0. \tag{2}$$

The Saffman-Taylor instability is the product of a competition of two physical mechanisms: the growth of perturbations on the interface under the effect of increasing gradient of pressure in the vicinity of protrusions of the injected fluid (destabilization), and the smoothing of the interface due to the surface tension (stabilization). Since both effects depend on the local curvature of the interface (Fig. 1), it can be shown that the interface is unstable only within a certain range of wavelengths of perturbations which depends on the ratio of viscosities η_1/η_2 of the two fluids and on the injection rate. CHUOKE *et al.* (1959) calculated the critical wavelength λ_c, below which perturbations vanish for the case of a plane interface. PATERSON (1981) performed a similar stability analysis for the circular geometry.

According to equations (1) and (2), the Saffman-Taylor experiment in a Hele-Shaw cell can be seen as a 2-D analog of two-phase flow through a porous medium. This physical problem is of major importance in a number of geological situations and industrial applications, in particular, hydrocarbon reservoir engineering. A common technology of oil reservoir exploitation consists in injecting water (or gas) in a number of peripheral injection wells in order to displace oil towards the production well. Fingering of the injected fluid into oil can then reduce the injection efficiency and may eventually lead to flooding of the production well.

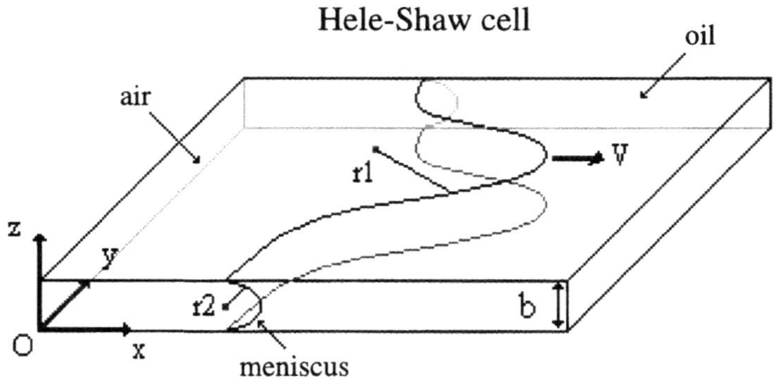

Figure 1
Two-phase immiscible flow in a Hele-Shaw cell; r_1 and r_2 are the curvature radii of the interface in the planes (OXY) and (OXZ).

Heterogeneities of the permeability of porous rocks at different scales, including fractures and faults, make the fluid flow in real conditions far more complex than the experiment in a Hele-Shaw cell. Although in certain cases those heterogeneities may dominate the flow pattern within the geological formation, a realistic description should also take into account the viscous fingering effects which are present even in a homogeneous medium.

In a number of works related to two-phase flows, special attention was paid to the effects of the velocity of fluid motion. It was found in particular that the thickness of the wetting film (BRETHERTON, 1961; DUSSAN, 1987; SCHWARTZ *et al.*, 1986) and the shape of the meniscus between immiscible fluids (PARK and HOMSY, 1984; REINELT, 1987) significantly depend on the rate of displacement. In this study we investigated the possibility of modifying the characteristics of the Saffman-Taylor instability by adding small periodic perturbations to the injection pressure to influence the instantaneous velocity of the flow without changing the average injection rate.

It is known from the work of many authors (BRETHERTON, 1961; DUSSAN, 1987; SCHWARTZ *et al.*, 1986), that at low Reynolds numbers, the average thickness of the wetting film left behind the slow passage of a sufficiently long bubble of fluid (many times the tube diameter) with negligible viscosity in a capillary tube filled with a viscous fluid is found to be a function of bubble speed only. Later works from other authors (PARK and HOMSY, 1984; REINELT, 1987) investigate the interface conditions for two-phase displacement in Hele-Shaw cells; while displacing a viscous fluid from the gap between two closely spaced parallel plates, a thin film of the original fluids remains on the surface of each plate. The shape of the meniscus and the corresponding interface conditions are expressed as an expansion in powers of Ca. Considering those works, it would be relevant to look at possibly modifying the characteristics of the instability of Saffman-Taylor by adding small low frequency oscillations to the injection pressure.

Experimental Setup

We used a horizontal square-shaped (55 × 55 cm) Hele-Shaw cell made of 1 cm, thick polished glass plates (Fig. 2). The constant spacing of 0.5 mm was provided by calibrated mylar separators placed at the corners, where the glass plates were clamped together in order to avoid changes of the spacing under the effect of pressure in the cell. A 5 mm diameter hole was drilled in the center of the top plate in order to inject fluids into the cell, while the gap along the external borders of the plates remained open (i.e., at atmospheric pressure).

In a series of preliminary tests, we verified that the square shape of the cell, which was chosen for technical reasons, induced no measurable anisotropy of the interface shape while the diameter of the latter remained below 10 cm. In the stable regime at slow rates of air injection, the interface thus remained perfectly circular. In all the

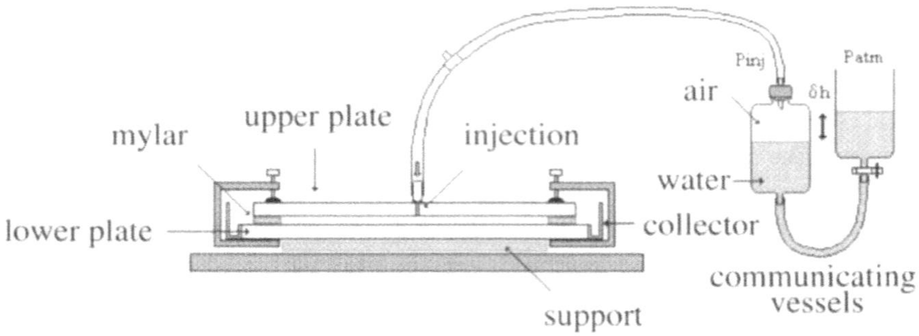

Figure 2
Setup of the Hele-Shaw cell and of the injection system.

experiments described below, we first filled the cell with a dyed silicon oil (Rhodorsyl R47V500) with dynamic viscosity $\eta = 0.5\,\mathrm{Pa} \cdot \mathrm{s}$ at $20\,\mathrm{C}^\circ$ and surface tension $\sigma = 2.11 \times 10^{-2}\,\mathrm{N} \cdot \mathrm{m}^{-1}$ at 20 C°. Before each experiment, the silicon oil was filtered and the cell was cleaned to ensure the absence of dust.

Once the cell was completely filled with oil, we injected air through the central hole, using a system of communicating vessels filled with water in order to create a constant air pressure. We experimentally found that the injection pressure of $P_{inj} = 300$ Pa produced a marginally unstable flow, i.e., such that the Saffman-Taylor instability was developing at a very slow rate for the interface diameter ranging from 1 to 10 cm. All the subsequent experiments were performed with this injection pressure.

The shape of the air-oil front was photographed using a digital camera (Kodak DCS420). The optical contrast between air and oil was enhanced by dying the oil so as to improve the subsequent numerical extraction of the interface curve from the snapshots.

Periodic oscillations of the injection pressure were produced by a piston pump with a differential volume of 7 cm^3 connected through a T-junction to the injection tube. The frequency of periodic motions of the piston could be varied continuously within the range from 1 to 25 Hz.

Pressure Oscillation Effects

Interface Stabilization

We first performed a classical Saffman-Taylor experiment (with a constant injection pressure) in order to use it as a reference (set 1 of snapshots in Fig. 3). After 405 s of injection at a constant rate, a moderate fingering of the interface appeared. Keeping all the parameters unchanged, we then performed several experiments in which we

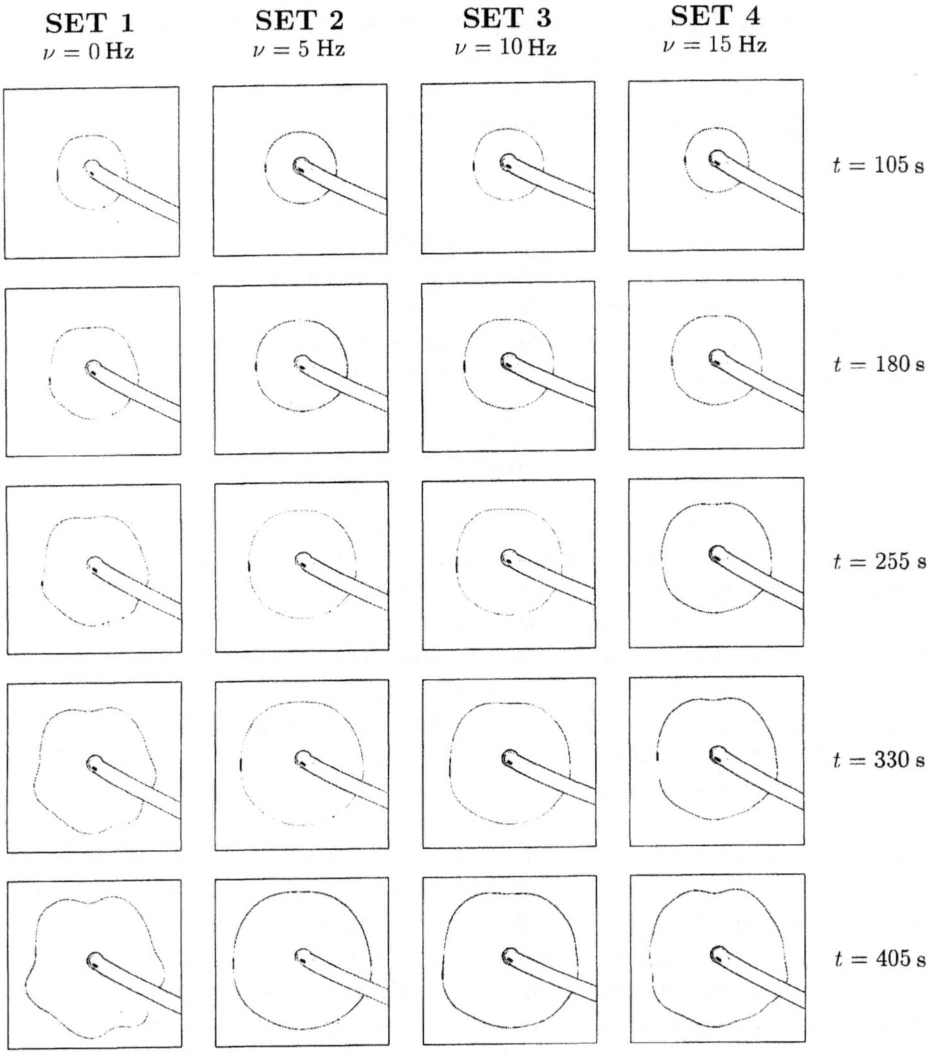

Figure 3
Snapshots of the interface between air (inside) and oil (outside) taken in four different experiments ($\nu = 0$ Hz, 5 Hz, 10 Hz, 15 Hz) at the identical intervals from the beginning of the injection (time given in insets).

added to the injection pressure small periodic oscillations with frequencies ranging from 5 to 15 Hz. The resulting sequences of snapshots of the interface taken at equivalent intervals as in the reference experiment (sets 2 through 4 in Fig. 3), clearly demonstrate a slowdown of fingering with respect to the reference case.

It was observed that the response of the system to the periodic oscillations of the injection pressure depends on the frequency of the oscillations, and that the most significant "stabilization" effect was obtained in the experiment with the oscillation frequency of 5 Hz. This observation can be interpreted in terms of a delay of destabilization or, similarly, as a decrease of the rate of growth of the instabilities.

Harmonic Analysis of the Interface

In order to quantify the observed effect of stabilization, we performed the Fourier analysis of the shape of the air-oil interface in the last snapshot ($t = 405$ s) for each experiment (bottom line of snapshots in Fig. 3). The contour of the interface extracted from snapshots by standard image processing procedures was then digitized in polar coordinates (r, φ) and represented as a truncated Fourier series:

$$R_f(\varphi) = A_0 + \sum_{i=1}^{N} a_i \cos(i\varphi) + b_i \sin(i\varphi), \tag{3}$$

with $N = 50$ harmonics, which was sufficient to reproduce all the shapes of the interface that we observed in the experiments (higher order harmonics had normalized amplitudes smaller than 10^{-3}, which is below the resolution of our analysis). We then calculated the spectra of perturbations of the interface $A_i = \sqrt{a_i^2 + b_i^2}$ with respect to the circular shape and normalized them by the zero-order harmonic A_0 (mean radius of the interface) to compare the spectra corresponding to the experiments with slightly different dimensions of the air-oil interface. For each experiment we present in Figure 4 the final shape of the interface with the corresponding normalized spectrum A_i. The values of the zero-order harmonic A_0, given in insets for each spectrum were very close for all the experiments, which means that the average injection rate was maintained practically constant in all experiments with and without oscillations of the injection pressure. One can also observe that in all the experiments, the amplitude of the leading harmonic A_5 was less than 5% of the zero-order harmonic A_0, which means that the system remained close to the onset of the Saffman-Taylor instability. We do not consider the first harmonic A_1, since its value is strongly influenced by the choice of the origin of the polar coordinates in the digitizing procedure. The harmonic analysis confirms and quantifies our observation that the periodic oscillations, added to the constant injection pressure, decrease the rate of growth of the instabilities of the interface on one hand, and reduce the number of unstable harmonics on the other hand.

In all four experiments the leading harmonic was A_5, which took the following values $A_5(0\,\mathrm{Hz}) = 0.049$, $A_5(5\,\mathrm{Hz}) = 0.014$, $A_5(10\,\mathrm{Hz}) = 0.028$, and $A_5(15\,\mathrm{Hz}) = 0.042$. The interface is thus best "stabilized" at the oscillation frequency of 5 Hz.

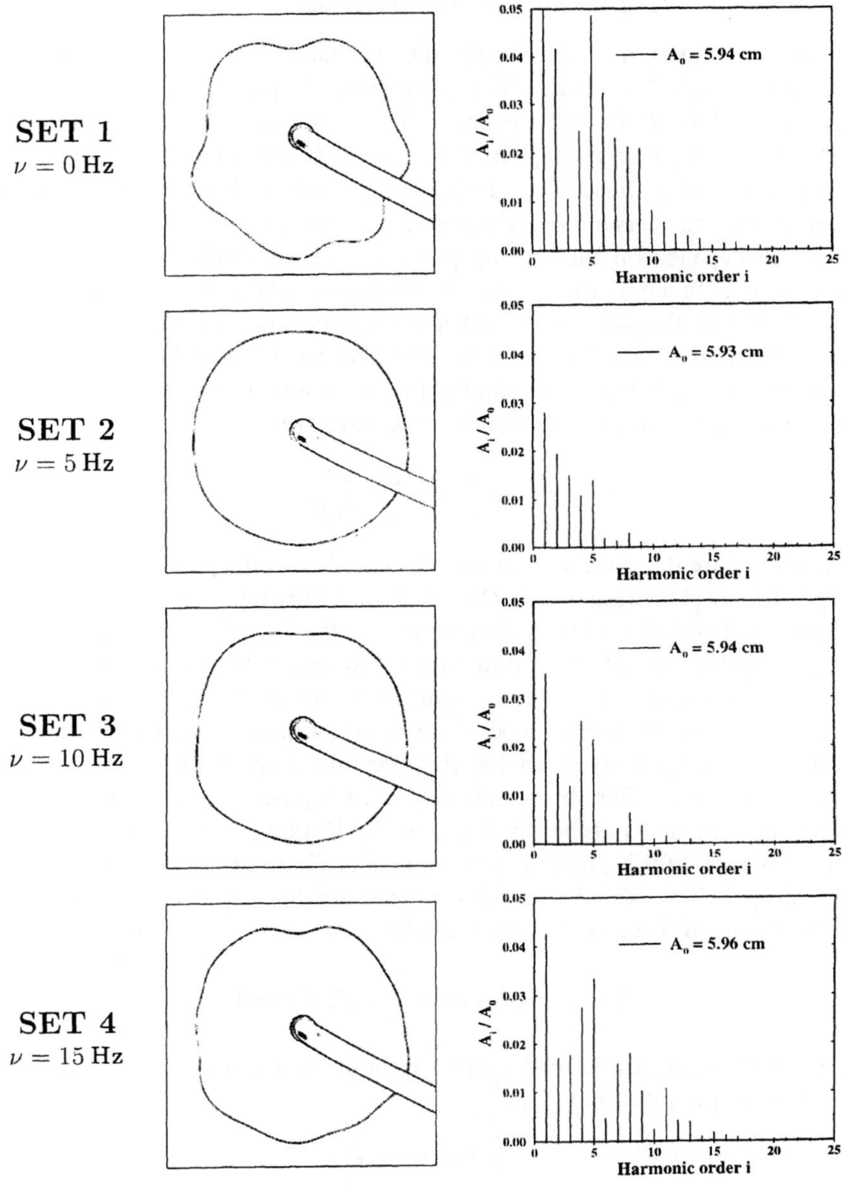

Figure 4

Harmonic analysis of the shape of the air-oil interface in the last snapshot of each experiment (t = 405 s); the spectra $A_i = \sqrt{a_i^2 + b_i^2}$ are normalized by A_0 (mean radius of the interface given in insets). The number of unstable harmonics, as well as their amplitudes, decrease under the effect of oscillation of the injection pressure, compared to the reference experiment. The most significant stabilisation was observed at 5 Hz.

Interface Conditions Modified by Pressure Oscillations

In the previous analysis we reduced the Saffman-Taylor problem to a two-dimensional Darcy's flow driven by a heterogeneous pressure field, by averaging the velocity field over the cell thickness. This frequently used approximation is justified when the transverse shape of the meniscus between the fluids is constant at all the locations and its curvature r_2 is determined exclusively by the cell thickness b. However, if one considers a moving meniscus, a more refined analysis shows that its shape, which depends on the wetting properties of the fluids with respect to the plates, is also a function of its velocity (REINELT, 1987). In the case when the displaced fluid wets the plates of the cell, a thin film of this fluid remains on the plates behind the meniscus. The thickness of this film may not be negligible compared to the cell thickness, in which case it will modify the curvature of the meniscus r_2 and thus directly influence the pressure drop across the interface:

$$P_2 - P_1 = \sigma\left(\frac{1}{r_1} + \frac{1}{r_2}\right), \tag{4}$$

where r_1 and r_2 are the curvature radii of the interface in the plane of the cell and in the perpendicular plane respectively (Fig. 1). It was shown that, in the limit when the viscosity of the injected fluid is small compared to the viscosity of the displaced fluid ($\eta_2 \ll \eta_1$), the dynamics of the wetting film is determined by the relative strength of viscous forces compared to that of capillary forces (PATERSON, 1981), which is characterized by the capillary number Ca. Several authors (PARK and HOMSY, 1984; REINELT, 1987) analytically estimated this pressure drop, taking into account the presence of the wetting film, by performing an asymptotic expansion in the limit of small capillary numbers (typically $Ca < 10^{-3}$). REINELT (1987) also proposed a numerical solution of this problem for the range of Ca up to 1. The expression of the pressure drop, which includes the dependence on the capillary number, can be written in a general form as (REINELT, 1987):

$$P_2 - P_1 = \sigma\left(K_0(Ca)\frac{1}{r_2} + K_1(Ca)\frac{1}{r_1}\right), \tag{5}$$

where K_0 and K_1 are functions of Ca. In the asymptotic limit of $Ca \ll 1$, the functions K_0 and K_1 have the following form:

$$K_0 = 1 + 3.80Ca^{2/3}, \tag{6}$$

$$K_1 = \frac{\pi}{4} - 4.07Ca^{2/3}. \tag{7}$$

In the presence of oscillations of the injection pressure, we propose the following scenario. The instantaneous velocity of the meniscus varies periodically following the variations of the injection pressure, thus inducing periodic changes of the shape of the meniscus and of the associated pressure drop. Since the dependence of the pressure drop on the velocity is nonlinear (REINELT, 1987), the value of the pressure drop *averaged* over the period of oscillations, can therefore be different from its static value. Indeed, the average pressure drop increases in the presence of oscillations, which leads to a stabilizing effect on the interface and consequently reduces the rate of growth of the Saffman-Taylor instability.

In the following we attempt to quantify the observed slowdown of the instability in terms of physical parameters of the problem. The linear stability analysis in the circular geometry (PATERSON, 1981) gives the order of the critical harmonic $n_c = 2\pi R/\lambda_c$ (where λ_c is the corresponding critical wave length and R is the radius of the interface) below which the interface is unstable:

$$n_c = \left(\frac{QR}{2\pi M\sigma} + \frac{1}{4}\right)^{\frac{1}{2}} - \frac{1}{2}, \tag{8}$$

where Q is the surface injection rate and $M = b^2/12\eta$ is the mobility of the oil. This expression can be inverted in order to express the surface tension between the fluids σ as a function of the critical harmonic n_c. Since it was observed that n_c decreases under the effect of pressure oscillations compared to the reference experiment (see spectra in Fig. 4), one can introduce the *effective surface tension* σ^* defined through the value of the highest *observed* unstable harmonic n_c^* in the presence of oscillations:

$$\sigma^*(n_c^*) = \frac{QR}{2\pi M} \frac{1}{\left(n_c^*(\nu) + \frac{1}{2}\right)^2 - \frac{1}{4}}. \tag{9}$$

It is thus possible to determine the value of the effective surface tension σ^*, which allows for the effect of pressure oscillations, by estimating n_c^* through the harmonic analysis of the shape of the interface described above. Physically, the effective surface tension σ^*, which becomes higher than σ for given fluids, reflects the effect of "stiffening" of the interface under the effect of periodic motions of the meniscus superimposed on its slow advance.

By performing the harmonic analysis of the shape of the interface in our four experiments, we determined the values of $n_c^*(\nu)$ for each experiment, and calculated the corresponding values of the effective surface tension $\sigma^*(\nu)$. The surface injection rate in all the experiments was $Q = 5 \cdot 10^{-5}\,\mathrm{m}^2 \cdot \mathrm{s}^{-1}$ and the radius of the interface R was estimated as the value of the zero order harmonic A_0 obtained through Fourier analysis. Figure 5 shows the effective surface tension $\sigma^*(\nu)$ calculated for each experiment and represented as a function of the pressure oscillation frequency ν. The validity of this approach is confirmed by the fact that the value of the effective surface tension σ^* obtained through the harmonic analysis of the interface in the experiment with $\nu = 0$ (the reference experiment without pressure oscillations) $\sigma^*(0) = 1.91 \cdot 10^{-2}\,\mathrm{N} \cdot \mathrm{m}^{-1}$ is quite close to the catalog value of the surface tension for the silicon oil that we used, $\sigma = 2.11 \cdot 10^{-2}\,\mathrm{N} \cdot \mathrm{m}^{-1}$.

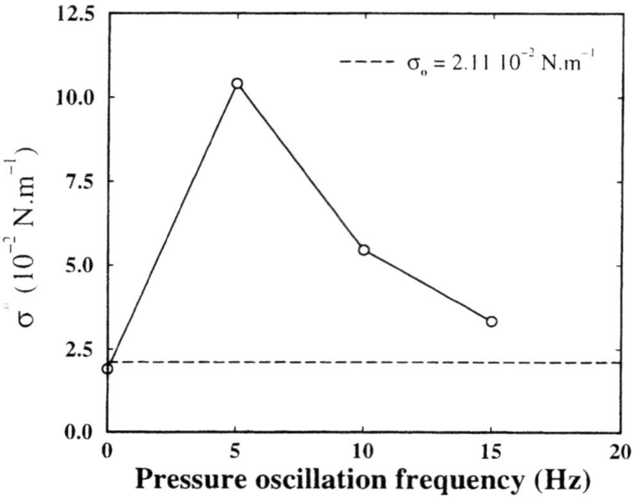

Figure 5
The effective surface tension σ^* obtained from the harmonic analysis of the interface as a function of frequency of pressure oscillations (open circles). Its value σ_0^* Hz obtained through the analysis of snapshots in the reference experiment, is close to the catalog value of the surface tension of the silicon oil σ_o (dashed line). The increase of the effective surface tension with respect to its nominal value can be interpreted as a "stiffening" of the interface under the effect of periodic motions of the meniscus.

The proposed description of the stabilization effect in terms of changes of the effective surface tension σ^* has certain limitations related to uncertainties in determination of the highest unstable harmonic n_c^* from the spectrum of the interface (a small noise is always present in the spectra even for harmonics with $i > n_c$). Nevertheless, it allows us to quantify the effect of stabilization due to the pressure oscillations simply by performing the image analysis of the interface shape.

Discussion and Conclusions

We presented an experimental study of stability of the interface between two immiscible fluids in a Hele-Shaw cell, which showed that the rate of growth of the Saffman-Taylor instability can be reduced by adding small amplitude oscillations to the injection pressure. This effect was clearly visible for marginally unstable flows (with relatively weak rates of growth of instabilities and close to their onset), while it was not possible to stabilize interfaces with a strongly developed fingering.

Since the periodic oscillations superimposed on the constant injection pressure have a zero mean, an additional force responsible for the stabilization effect could result only from a nonlinear mechanism. We suggest that such a mechanism can be related to the variations of the transverse shape (i.e., in the plane perpendicular to that of the cell) of the meniscus between the fluids, and consequently to variations of the associated pressure drop across the interface. This pressure drop of capillary nature is a stabilizing mechanism with respect to the Saffman-Taylor instability, and it is particularly responsible for damping of shortwave length instabilities. A nonlinear response of the meniscus to periodic variations of its velocity could thus lead to an increased *average* value of the pressure drop related to the curvature of the meniscus, and consequently to a stronger stabilization.

We developed a procedure of harmonic analysis of the shape of the interface, based on a truncated Fourier expansion in polar coordinates of the digitized interface curve. This procedure allowed us to analyze the form of the air-oil interface extracted from numerical snapshots, and to quantify the difference between the reference experiment and the experiments with pressure oscillations. The harmonic spectra show that the amplitude of the most unstable harmonic (the 5th order harmonic A_5 in all our experiments) decreased in experiments with pressure oscillations as compared to the reference experiment, the most significant decrease by a factor of 3.5 being observed in the experiment with oscillations at 5 Hz. The stabilization effect was less pronounced for the two other frequencies of pressure oscillations used in our experiments (10 and 15 Hz). The amplitude of pressure oscillations could also importantly contribute to the discussed stabilization effect. However in our experiments, we had no direct control of the pressure variations at the injection point of the cell, since pressure oscillations produced by a piston pump connected to the injection circuit (see Fig. 2) could subsequently be attenuated due to air compressibility and elasticity of connecting tubes. Moreover, this attenuation is likely to be stronger for higher frequencies, which could explain the fact that the observed stabilizing effect of pressure oscillations was weaker at 10 and 15 Hz than at 5 Hz. This dependency on the frequency of oscillations was not extensively studied in the present work, and merits further investigation.

The order of the critical harmonic n_c, and hence the number of unstable harmonics, was also reduced in the presence of pressure oscillations. We used this fact in order to introduce the notion of the effective surface tension σ^*, which allowed us to describe the local changes of the pressure drop across the meniscus in terms of an effective value of the physical parameter of the problem. It is worth noting that the calculation of the effective surface tension by this method using the snapshots of the reference experiment (without pressure oscillations) produced a good estimation of the catalog value of the surface tension of the used silicon oil. The inhibition of the Saffman-Taylor instability can thus be viewed as a result of "stiffening" of the air-oil interface under the effect of periodic motions of the meniscus.

Acknowledgements

We extend thanks to Yves Couder and Hernán Makse for stimulating discussions and helpful suggestions during the course of this work, and Guy Marolleau for valuable technical assistance.

References

BRETHERTON, F. P. (1961), *The Motion of Long Bubbles in Tubes*, J. Fluid. Mech. *10*, 166–188.

CHUOKE, R. L., VAN MEURS, P., and VAN DER POEL, C. (1959), *The Instability of Slow, Immiscible, Viscous Liquid-liquid Displacements in Permeable Media*, Trans. AIME *216*, 188–194.

DUSSAN, V. E. B., *Incorporating the influence of wettability into models of immiscible fluid displacement through a porous media.* In *Physics and Chemistry of Porous Media II* (ed. Schlumberger-Doll Research) (AIP Conference Proceeding 154, 1987), pp. 83–97.

FABER, T. E., *Fluid Dynamics for Physicists*, (Cambridge University Press, Cambridge 1997).

PARK, C. W. and HOMSY, G. M. (1984), *Two-phase Displacement in Hele-Shaw Cells: Theory*, J. Fluid. Mech. *113*, 513–529.

PATERSON, L. (1981), *Radial Fingering in a Hele-Shaw Cell*, J. Fluid. Mech. *113*, 513–529.

REINELT, D. A. (1987), *Interface Conditions for Two-phase Displacement in Hele-Shaw Cells*, J. Fluid. Mech. *183*, 219–234.

REINELT, D. A. (1987), *The Effect of Thin Film Variations and Transverse Curvature on the Shape of Fingers in Hele-Shaw Cell*, Phys. Fluids. *30*, 2617–2623.

SAFFMAN, P. G. and TAYLOR, S. G. (1958), *The Penetration of a Fluid into a Porous Medium or Hele-Shaw Cell Containing a More Viscous Liquid*, Proc. R. Soc. Lond. *245*, 312–329.

SCHWARTZ, L. W., PRINCEN, H. M., and KISS, A. D. (1989), *On the Motion of Bubbles in Capillary Tubes*, J. Fluid. Mech. *172*, 259–275.

(Received February 5, 2001, revised June 1, 2001, accepted August 1, 2001)

Pure appl. geophys. 160 (2003) 989–998
0033 – 4553/03/060989–10

Pure and Applied Geophysics

A Unified Model for Characterisation and Mechanical Behaviour of Rock Fractures

FLAVIO LANARO[1] and OVE STEPHANSSON[1]

Abstract— Roughness and aperture are two important characteristic parameters of rock fractures. It has been demonstrated by many authors that the knowledge of roughness does not directly lead to that of aperture, and aperture is to be handled as a separate geometrical descriptor. For determining normal deformability and flow response of a fracture, the aperture distribution and the mechanical properties of the rock matrix are required. When shearing of joints and fractures is considered, roughness comes into play and affects the evolution of the aperture distribution. The aperture distribution can be evaluated by knowing the correlation between the asperity profiles of the rock walls of a rock fracture. Thus, the distributions of the contact area and void space determine the fracture dilation and hydraulics during shearing. In this study, a mathematical formulation of roughness and aperture of rock fractures is proposed. Both aperture and roughness are interpreted as scale-dependent quantities, and their statistical models are described by fractal theory. The theoretical results are in very good agreement with experimental results and published data. Furthermore, a constitutive model for the hydro-mechanical behaviour of rock fractures is developed based on the new theoretical description of the geometry. The model is based on a fractal-island description of the contact areas and on an approach of micromechanics.

Key words: Aperture, constitutive model, fractals, normal loading, rock fractures, shearing.

1. Introduction

Rock masses are always characterised by some degree of fracturing that affects stability of underground excavations and groundwater flow. Thus, the attention focuses on rock fractures and their geometrical features as the main parameters controlling the mechanical, hydraulic, thermal and chemical properties of rock masses (e.g., GENTIER *et al.*, 2000; PYRAK-NOLTE and MORRIS, 2000). In this research work, samples of a natural fracture were measured by 3-D laser scanning. Aperture and roughness were analysed as fractal objects. All the determined fractal parameters were then introduced into a constitutive model for fractures based on micromechanics. The response of normal loading and shearing of fractures could be simulated.

---

[1] Division of Engineering Geology, Department of Civil and Environmental Engineering, Royal Institute of Technology, Teknikringer 72, SE-10044 Stockholm, Sweden.
E-mails: Flavio@aom.kth.se, Ove@aom.kth.se

2. *Analysis of Aperture and Roughness*

Two samples of size 60 × 60 mm (Samples A and B) from a natural fracture of Stockholm granite have been analysed by means of a 3-D laser scanning technique (accuracy of 50 μm). The surface topography of the walls of the fracture was measured. By knowing the position of the surfaces with respect to each other, also the geometry of the fracture aperture could be determined.

For the analysis of fracture roughness and aperture as a function of the sample size, the fracture samples are subdivided into smaller square samples of 10, 20, 30 and 40 mm side. Points were measured with a density of about 1 point/0.03 mm^2. For each subsample, a statistical analysis of the data gives the frequency distribution of the aperture (Fig. 1) and asperity height. Aperture is evaluated for a very small normal load (only self-weight). For its determination, the distance between the polygonal interpolation of one wall of the fracture and the points of the opposite wall is measured. This technique allows one for an aperture determination that is independent of the assumed coordinate reference system (HAKAMI *et al.*, 1995).

The values of the mean aperture for the two samples are different: 0.108 mm and 0.152 mm, respectively. For different subsample sizes, the mean aperture ranges between 0.055 mm and 0.204 mm. Aperture data have a fairly Gaussian frequency distribution. For subsamples with mean aperture of the same order of magnitude as the equipment accuracy, the aperture frequency turns into a log-normal distribution.

While the range of variation of the aperture mean values reduces by increasing the sample size, the aperture standard deviation increases (LANARO, 2000). The log-log diagram of the standard deviation of the aperture versus the subsample size

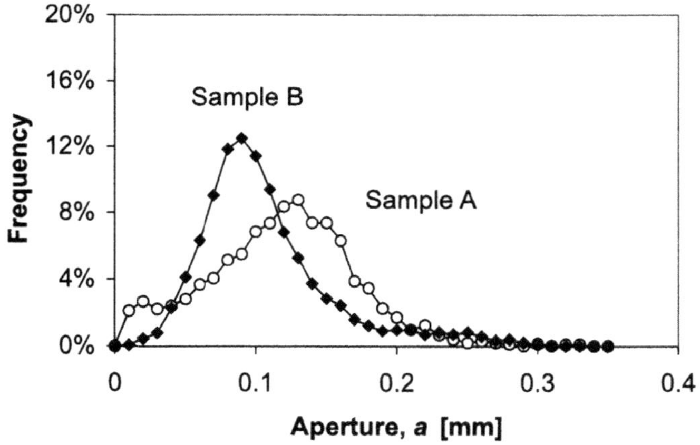

Figure 1
Aperture frequency distributions for two square areas of size 30 mm extracted from Samples A and B (60 × 60 mm).

shows that the standard deviation has an almost linear trend, i.e. a power law is valid for window sizes smaller than 30 mm. The linear trend shown in Figure 2 can be described by the following power law of the sampling size h:

$$\sigma_a^2(h) = G_{\sigma a} h^{2H_a} \ . \tag{1}$$

Objects that exhibit such relation are said to be "self-affine" fractal objects. The power exponent H_a is related to the fractal dimension D_a of the aperture distribution through the relation: $D_a = 3 - H_a$.

The contact area between the surfaces of the fracture can be seen as a cut-off of level zero or relatively small threshold of the aperture (e.g. 50 μm). In other words, the contact areas can be seen as "lakes" on a fractal "landscape" representing the aperture distribution. It is then possible to evaluate the cumulative probability of the contact areas with area S larger than a reference area s. For an isotropic and "self-similar" frequency distribution $f(s)$, the following relation holds (MANDELBROT, 1982):

$$\Pr(S > s) = \int_s^{+\infty} f(s)\, ds = Cs^{(1-D_a)/2} \ , \tag{2}$$

where C is a dimensional proportionality constant. Figure 3 shows the very good agreement between the theoretical formulation (Eq. (2)) and the experimental results for the cumulative probability of the contact areas of sample A.

Many authors have demonstrated that the roughness of rock fractures can be represented as a self-affine fractal object. This implies that the statistics of the asperity topography vary according to power laws of the sample size (BROWN and SCHOLZ, 1985; POWER and TULLIS, 1991; SCHMITTBUHL *et al.*, 1995). A square

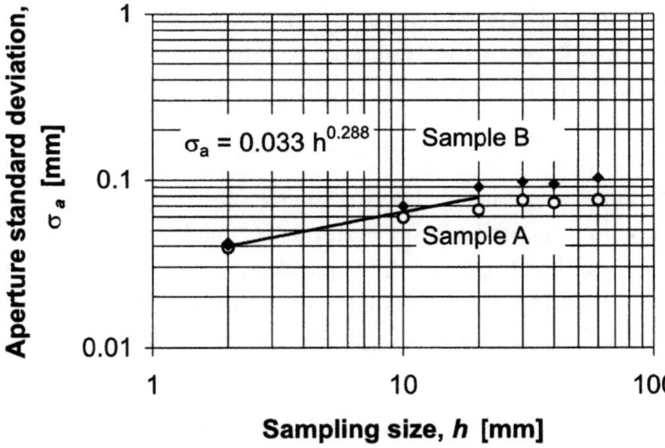

Figure 2
Log-log diagram of the aperture standard deviation σ_a versus sampling size h for Samples A and B.

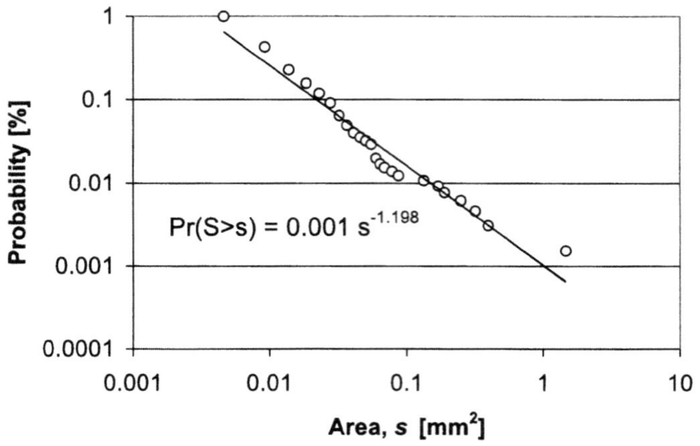

Figure 3
Cumulative distribution of the contact areas measured for Sample A, assuming an aperture threshold of 50 μm for contact areas.

subsample of a fracture surface with edge h can be interpolated with a planar facet. The slope of the facet can be determined with respect to the mean fracture plane. For fractal objects, the standard deviation of the asperity slopes σ_{slope} follows a power law of the sample size:

$$\sigma^2_{\text{slope}} = G_{\gamma z} h^{2H_z - 2} \quad , \tag{3}$$

where $G_{\gamma z}$ is a dimensional proportionality constant and H_z is the Hurst exponent for roughness. Forward and backward slopes are represented by the positive and negative branches of the Gaussian distribution of the asperity slopes with zero mean. The maximum and minimum expected values of the slope would then be related to the standard deviation of the asperity slope (BARTON, 1990). The maximum asperity slope S_{slope} can be related to the standard deviation of the slope distribution by a fractile α (LANARO and STEPHANSSON, 2000):

$$S_{slope}(h) = \pm\alpha\sigma_{\text{slope}} = \pm\alpha\sqrt{G_{\gamma z}} h^{H_z - 1} \tag{4}$$

In this way, the same power law governing the standard deviation of the slopes would apply to the maximum asperity slope ($\pm$ stands for back and forward direction). For small asperity slopes, the tangent of the angle S_{slope} can be replaced by the angle itself in radian. During shearing, the elementary dilation increment due to pure kinematic mechanism will be $S_{slope}(v)dv$, and integrating for a shear displacement of v, it gives:

$$d = \int_0^v S_{slope}(v)\, dv = \alpha\sqrt{G_{\gamma z}} \frac{v^{H_z}}{H_z} \tag{5}$$

that expresses the fracture dilation as a function of the shear dislocation.

3. Joint Normal Deformability

The possibility of detailed investigations of the fracture aperture and roughness with advanced and accurate measurements can be exploited for the prediction of the mechanical behaviour of a fracture during normal loading and shearing.

Assumed that the contact between two fracture asperities behaves as the contact between two ideal spheres, then the Hertzian contact model gives mathematical expressions for the normal load under elastic and elasto-plastic conditions, respectively (BUSHAN, 1999). The relation between the contact area s and the fracture closure δ can be approximated by: $s = \pi R\delta$, where R the equivalent radius of the contact. On the other hand, for a contact undergoing plastic yielding, the mean contact pressure becomes a function of the fracture surface hardness (WILLIAMS, 1994). During normal loading of a rock fracture, small contacts undergo plasticity (MAJUMDAR and BHUSHAN, 1991).

These concepts can be applied to the contact distribution of natural rock fractures. By integrating the normal load P_e acting on the single elastic contact spot of area s and equivalent deformation modulus $E' = \left(\sum_{i=1}^{2}(1-v_i^2)/E_i\right)$, where E_i and v_i are the deformation modulus and Poisson's ratio of the surfaces in contact, respectively, for the normal displacement δ we have:

$$P_e = \frac{4}{3}E'\sqrt{\frac{s}{\pi}}\delta \tag{6}$$

with the cumulative probability in Eq. (2), the following expression can be obtained:

$$P_E = N_{tot}\int_{s_s}^{s_l} P_e f(s)\,ds = \frac{4}{3}\frac{E'}{\sqrt{\pi}}\frac{D_a-1}{2-D_a}s_l^{-\frac{1-D_a}{2}}\left[s_l^{\frac{2-D_a}{2}} - s_s^{\frac{2-D_a}{2}}\right]\delta \tag{7}$$

that applies for a particular fracture closure δ where the smallest and largest contact areas, respectively s_s, and s_l are known. N_{tot} is the total number of measured contacts. By integrating the normal load $P_p = p_p\, s$, where p_p is the plastic contact pressure at the contact that is related to the surface hardness, over all the plastic contacts smaller than s_p, the normal load can be obtained as:

$$P_P = N_{tot}\int_{s_s}^{s_p} P_p f(s)\,ds = -p_p s_l^{-\frac{1-D_a}{2}}\frac{1-D_a}{3-D_a}\left[s_p^{\frac{3-D_a}{2}} - s_s^{\frac{3-D_a}{2}}\right]. \tag{8}$$

If elastic and plastic deformation coexist for the same sample, the total normal load acting onto the contact areas of the fracture can be given by the summation of Eqs. (7) and (8) (where $s_s = s_p$ in Eq. (7)). In these equations, an increase of geometrical complexity of the aperture volume (i.e., increase of D_a) determines more numerous and smaller contact zones. This strongly affects the normal stiffness and frictional properties of the rock fracture, because fractures with small contacts are much stiffer than those with large and clustered contacts for a given percentage of contact area (HOPKINS *et al.*, 1990).

For illustrating the efficiency of the formulation developed, some results by CAPASSO *et al.* (1999) were simulated. In their research work, three replicas of fractures from granite cores of 120 mm diameter were tested in laboratory under normal constraint, and the deformations of three cores of mortar including a fracture were measured. By substituting the mechanical parameters in Eqs. (7) and (8) ($E' = 32$ GPa, $\nu = 0.25$, $p_p = 82$ MPa, $D_a = 2.75$, $s_s = 0.003$ mm^2, $s_l = 2.750$ mm^2, $\delta_0 = 0.015$ mm), the relation between the fracture closure and applied normal load can be determined as shown in Figure 4. A fair to good agreement between applied stress and normal closure is obtained.

4. Joint Shearing

As regards rock joints, any of the classical friction laws seems not to apply. In fact, if the frictional force is proportional to the normal force, their ratio is independent on the apparent contact area (e.g., RABINOWICZ, 1965; MOORE, 1975). Different results were obtained by BARTON and BANDIS (1982) when testing joint

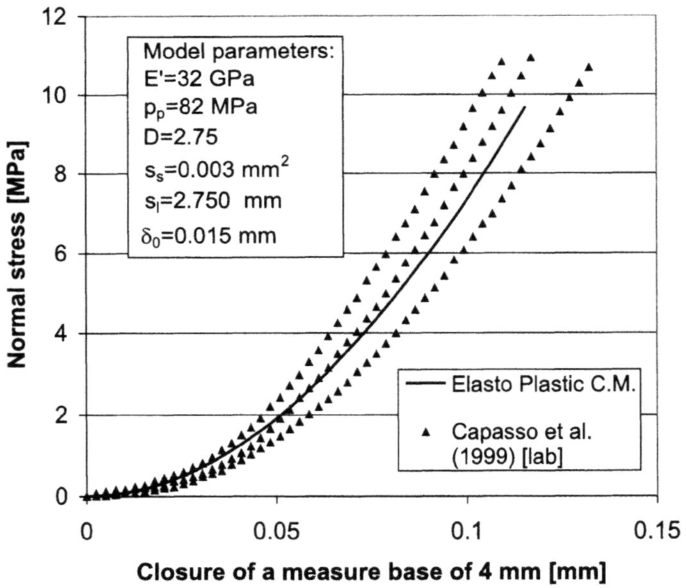

Figure 4
Comparison between three experimental results reported by CAPASSO *et al.* (1999) and the elasto-plastic Hertzian contact model for rock fractures with the listed set of parameters.

samples of different size. In that study, the frictional force seemed to be scale-dependent such that small samples exhibited higher strength per apparent unit area than large samples.

Shearing is the geometrical process that shifts the upper surface with respect to the lower surface of a fracture, under a dilation slope S_{slope} (damage of the asperities is here neglected). After a shearing displacement v, the mean value of the aperture μ_{av}, that was equal to μ_a for the fracture "at rest", becomes: $\mu_{av} = \mu_a + S_{slope}\, v$, while the variance of the aperture σ^2_{av} changes as follows (LANARO, 2000):

$$\sigma^2_{av} = \sigma^2_a + \gamma_z(v) - 2\,\mathrm{cov}(\Delta z^v, a) \tag{9}$$

where Δz^v is the increment of the asperity height between two points distant v on one surface of the fracture. Thus, the aperture variance increases with shearing by a term equal to the variogram of the asperity height γ_z calculated for a distance of the same amount as the shearing displacement v. The deduced term is of importance for small shearing displacements but becomes constant when the delay distance increases to the level of the aperture stationarity (cf. Fig. 2).

POWER and DURHAM (1997) experimentally observed that the size of the contact spots (zones with zero aperture) increased during shearing while their number diminished. The same behaviour is predicted by the analytical model: the growth of the contact spots is justified by the increase of the aperture correlation length. Hence, the reduction of the number of contact spots due to the increase of persistence, and the Hurst exponent of the aperture distribution are also predicted by the model.

Also the mechanics of a fracture during shearing can be modelled by contact mechanics. This solution considers the occurrence of slip in a peripheral annular area around the stick contact between two asperities or equivalent spheres. In this annular area, the tensions due to shearing would be far too high for the materials in contact if the slip would not take place. By assuming this, the relative displacement γ between the walls of the fracture due to shearing can be expressed as (BUSHAN, 1999):

$$\gamma = \frac{3\varphi P}{16\sqrt{\frac{s}{\pi}}}\frac{1}{G'}\left[1 - \left(1 - \frac{Q}{\varphi P}\right)^{2/3}\right] \tag{10}$$

or:

$$Q = \varphi P\left[1 - \left(1 - \frac{16\sqrt{\frac{s}{\pi}}G'\gamma}{3\varphi P}\right)^{3/2}\right], \tag{11}$$

where Q is the shearing force, P the normal load, G' the equivalent shear modulus of the rock and φ the rock friction angle in radian (friction coefficient). By knowing the frequency distribution $f(s)$ of the contact areas, Eq. (10) can be integrated over the whole fracture extension, and the shear response Q_E of the fracture without plastic deformations becomes:

$$Q_E = N_{tot} \int_{s_s}^{s_l} Q_e(s, P_e) f(s)\, ds = -\varphi\left[1 - \left(1 - \frac{4\gamma G'}{\varphi E'\delta}\right)^{3/2}\right] P\ . \qquad (12)$$

The slip-stick condition in Eq. (10) applies only if the shearing force does not exceed the frictional force φP acting on a contact of area s.

In case of contacts under elastic condition, Eq. (10) states that slipping occurs when the ratio between normal and shear displacement exceeds a certain limit that depends only on the mechanical properties of material the fracture surfaces, and not on the contact area s. If plasticity induced by normal loading occurs, then the areas undergoing slipping are smaller. When slipping occurs, the tangential force Q_{slip} is given by surface friction and equals φP. By integrating the tangential force for all the areas under slipping ($s < s_m$), we obtain:

$$Q_{\text{Slip}E} = N_{tot} \int_{s_s}^{s_m} \varphi P f(s)\, ds = -\frac{4}{3}\varphi \frac{1 - D_a}{2 - D_a} \frac{E'\delta}{\sqrt{\pi}} s_l^{\frac{D_a - 1}{2}} \left(s_s^{\frac{2 - D_a}{2}} - s_m^{\frac{2 - D_a}{2}}\right) \qquad (13)$$

for elastic conditions, and:

$$Q_{\text{Slip}P} = N_{tot} \int_{s_s}^{s_m} \varphi P f(s)\, ds = -\varphi p_p \frac{1 - D_a}{3 - D_a} s_l^{\frac{D_a - 1}{2}} \left(s_s^{\frac{3 - D_a}{2}} - s_m^{\frac{3 - D_a}{2}}\right) \qquad (14)$$

for plastic conditions. In these equations, D_a is the fractal dimension of the aperture distribution during shearing, which is a function of the shear dislocation v.

5. *Discussion and Conclusions*

New tools and techniques for roughness and aperture determination and representation have been introduced into constitutive models for rock fractures simulating normal loading and shear behaviour (e.g., YOSHIOKA and SCHOLZ, 1989a, b; BORODICH and MOSOLOV, 1992; WANG, 1994). In this study, micromechanics is applied for determining the behaviour of the contact areas between the rough surfaces of a fracture. The method is proposed as an alternative to empirical methods extrapolated from experimental results, which do not have a strong theoretical base (e.g., BARTON *et al.*, 1985; KULATILAKE *et al.*, 1995). Moreover, this study incorporates the results of recent studies that illustrate the fractal nature of the aperture and contact areas of the natural rock fractures.

The measurements of the topography of the rock surfaces of the samples are obtained by a 3-D laser scanner and relocated in the position they had in the closed fracture samples so that the aperture (or distance between the two opposite rock surfaces) can be calculated. By interpreting the aperture distribution as a fractal object (Cox and Wang, 1993), the frequency distribution of the contact areas, where the aperture is zero, is determined.

The response of the sample to external constrains during normal loading and shearing is given by integrating the forces acting on a single contact over all the contact areas of the sample. This is a major improvement with respect to other constitutive models because, in the proposed model, the contact area is directly measured and used as input parameter while previously the surface roughness parameters were used for characterising the fractures, and thus the spatial correlation of the aperture was not considered (e.g., Yoshioka and Scholz, 1989a; Wang, 1994). In shearing, the conditions of stick and slip at the contacts, occurring at different levels of shear displacements, are also considered and depend on the size of the contact area. The proposed method considers scale-dependency behaviour since the maximum measured area of the contact spots always depends on the sample size, and this aspect is not considered in any of the former models. On the other hand, the deformations inside the rock blocks defining the fracture walls are neglected, as the interaction between neighbour contacts is not considered.

The analytical formulation for aperture during shearing proposed by Lanaro (2000) describes the changes in magnitude and complexity of the aperture observed in the laboratory. In turn, the use of this geometrical model for determining the mechanical behaviour of the fractures makes it possible to consider riding of the surface asperities and increase of the spatial statistical correlation of the aperture during shearing. More experimental efforts should be devoted to the quantification of some of the parameters involved in the simulation of shearing. The knowledge of the frequency distribution of the contact areas during shearing is crucial for obtaining reasonable prediction of shear strength and deformability of rock fractures and joints.

Acknowledgements

The authors particularly thank the Swedish Nuclear Fuel and Waste Management Co. (SKB) for supporting this research work.

References

Barton, N. and Bandis, S. (1982), *Effect of Block Size on the Shear Behaviour of Jointed Rock*, Keynote Lecture, 23rd U.S. Symp. on Rock Mech., Berkeley, California (U.S.A.).

Barton, N. R. *Scale effects or sampling bias?* In *Scale Effects in Rock Masses* (ed. P. da Cunha) (Balkema, Rotterdam 1990) pp. 1–25.

Barton, N., Bandis, S., and Bakhtar, K. (1985), *Strength, Deformation and Conductivity Coupling of Rock Joints*, Int. J. Rock Mech. Min. Sci. and Geomech. Abstr. *22*, 3, pp. 121–140.

Borodich, F. M. and Mosolov, A. B. (1992), *Fractal Roughness in Contact Problems*, J Appl. Maths. Mechs. *56*, 5, 681–690.

BROWN, S. R. and SCHOLZ, C. H. (1985), *Broad Bandwidth Study of the Topography of Natural Rock Surfaces*, J. Geophys. Res. *90*, B14, 12,575–12,582.

BUSHAN, B. *Principles and Applications of Tribology* (John Wiley and Sons, New York 1999).

CAPASSO, G., SCAVIA, C., GENTIER, S., and PELLEGRINO, A. (1999), *The Influence of Normal Load on the Hydraulic Behaviour of Rock Fractures*, Proc. of the 9th Int. Congress on Rock Mech., Paris, France (Balkema, Rotterdam) pp. 863–868.

COX, B. L. and WANG, S. Y. (1993), *Slit-island Fractal Analysis of Single Fracture Aperture Patterns*, Lawrence Berkeley National Laboratory Annual Report, CA, U.S.A., pp. 99–102.

GENTIER, S., RISS, J., ARCHAMBAULT, G., FLAMARD, R., and HOPKINS, D. (2000), *Influence of Fracture Geometry on Shear Behaviour*, Int. J. Rock Mech. Min. Sci. *37*, 1–2, 161–174.

HAKAMI, E., EINSTEIN, H. H., GENTIER, S., and IWANO, M. (1995), *Characterisation of Fracture Aperture – Methods and Parameters*, Proc. of the 8th Int. Congr. on Rock Mechanics, Tokyo, Japan (Balkema, Rotterdam) pp. 751–754.

HOPKINS, D. L., COOK, N. G. W., and MYER, L. R. (1990), *Normal Joint Stiffness as a Function of the Spatial Geometry and Surface Roughness,* Proc. Int. Symp. Rock Joints, Loen, Norway (eds. N. Barton and O. Stephansson) (Balkema, Rotterdam) pp. 203–210.

KULATILAKE, P. H. S. W., SHOU, G., HUANG, T. H., and MORGAN, R. M. (1995), *New Peak Shear Strength Criteria for Anisotropic Rock Joints*, Int. J. Rock Mech. Min. Sci. and Geomech. Abstr. *32*, 7, 673–697.

LANARO, F. and STEPHANSSON, O. (2000), *A Novel Geometrical Model for Rock Joint Description – Roughness, Angularity and Aperture*, Proc. of ISRM Int. Symp. EUROCK 2000, Aachen, Germany (ed. Deutsche Gesellschaft für Geotechnik e.V., DGGT) (Verlag Glückauf, Essen), pp. 287–294.

LANARO, F. (2000), *A Random Field Model for Surface Roughness and Aperture of Rock Fractures*, I. J. Rock Mech. and Min. Sci. *37*, 1195–1210.

MAJUMDAR, A. and BHUSHAN, B. (1991), *Fractal Model of Elastic-plastic Contact Between Rough Surfaces*, Trans. ASME, J. of Tribology, *113*, pp. 1–11.

MANDELBROT, B. B., *The Fractal Geometry of Nature* (W.H. Freeman, New York 1982).

MOORE, A. J., *Principles and Applications of Tribology* (Pergamon Press, New York 1975).

POWER, W. L. and DURHAM, W. B. (1997), *Topography of Natural and Artificial Fractures in Granitic Rocks: Implications for Studies of Rock Friction and Fluid Migration*, Int. J. Rock Mech. Min. Sci. *34*, 6, 979–989.

POWER, W. L. and TULLIS, T. E. (1991), *Euclidean and Fractal Models for the Description of Rock Fracture Roughness*, J. Geophys. Res. *96*, B1, 415–424.

PYRAK-NOLTE, L. J. and MORRIS, J. P. (2000), *Single Fractures under Normal Stress: The Relation Between Fracture Specific Stiffness and Fluid Flow*, Int. J. Rock Mech. Min. Sci. *37*, 1–2, 245–262.

RABINOWICZ, E., *Friction and Wear of Materials* (John Wiley and Sons, New York 1965).

SCHMITTBUHL, J., SCHMITT, F., and SCHOLZ, C. (1995), *Scaling Invariance of Crack Surfaces*, J. Geophys. Res. *100*, B4, 5953–5973.

WANG, J.-A. (1994), *Morphology and Mechanical Behaviour of Rock Joints*, Ph.D. Thesis, Faculty of Mining and Geology, Technical University of Silesia, Gliwice, Poland, 236 pp.

WILLIAMS, J. A., *Engineering Tribology* (Oxford Univ. Press, Oxford 1994).

YOSHIOKA, N. and SHOLZ, C. H. (1989a), *Elastic Properties of Contacting Surfaces under Normal and Shear loads – 1. Theory*, J. Geophys. Res. *94*, B12, 17,681–17,690.

YOSHIOKA, N. and SHOLZ, C. H. (1989b), *Elastic Properties of Contacting Surfaces under Normal and Shear Loads – 2. Comparison of Theory with Experiment*, J. Geophys. Res. *94*, B12, 17,691–17,700.

(Received January 31, 2001, revised September 12, 2001, accepted October 1, 2001)

Pure appl. geophys. 160 (2003) 999–1008
0033–4553/03/060999–10

Pure and Applied Geophysics

Analytical Model for Permeability Evolution in Microcracking Rock

G. D. H. Simpson,[1] Y. Guéguen,[2] and F. Schneider[3]

Abstract—Analytical expressions to predict the enhancement of permeability due to stress-induced microcracking in initially low porosity rock are presented. A fracture mechanical microcrack model is employed to derive integrated effective hydraulic variables as a function of stress, which are then used to calculate the evolution of permeability using the statistically-based Dienes model. The model enables determination of permeability enhancement as a function of two loading parameters and three material parameters. Results are in reasonable agreement with experimental measurements and indicate that appreciable increases in permeability can be anticipated during brittle failure. The analytical nature of the model makes it easily incorporatable into numerical models that require quantification of the permeability evolution as a function of stress, for which there is currently no law.

Key words: Permeability, microcracking, fluid flow.

Introduction

Although classical treatments of fluid flow assume the permeability in a material constant, it is now recognised that deformation can drastically modify the permeability and nature of fluid flow. Stress-induced permeability enhancement (between 0.3 and 5 orders of magnitude) is a feature observed in association with dilatant microcrack deformation in experiments on initially low porosity rocks at relatively low (<200 MPa) confining pressures (Zoback and Byerlee, 1975; Zhang *et al.*, 1994; Peach and Spiers, 1996; Zhu and Wong, 1997), and it has been suggested to play an important role in controlling fluid flow in sedimentary basins, fault zones and metamorphic terrains.

Most models previously employed to investigate the evolution of permeability as a function of stress are based on elastic closure laws, where the hydraulic parameters (aperture and crack length) vary reversibly in response to applied stresses (Walsh and Brace, 1984; Seeburger and Nur, 1984; Gavrilenko and Guéguen, 1989).

[1] Institut Français du Pétrole and Ecole Normale Supérieure, Paris, France. Now at: Institut für Geologie, Sonneggstr. 5, ETH-Zürich, Switzerland. E-mail: simpson@erdw.ethz.ch
[2] Laboratoire de Géologie, Ecole Normale Supérieure, Paris, France.
[3] Institut Français du Pétrole, Paris, France.

These models have been used to investigate pressure effects. A recent study by ZHU and WONG (1999) focussed on quantifying the permeability evolution due to microcrack growth driven by deviatoric stress. These authors simulated dilatancy and the resulting permeability evolution implicitly by incorporating an increasing number of microcracks into a random network model. An alternative approach is to derive crack parameters, and their evolution as a function of stress, explicitly on the basis of fracture mechanics (SIMPSON *et al.*, 2001). Although more complex, such models are attractive in that they are physically derived.

We present a theoretical model developed to predict the evolution of permeability due to stress-induced microcracking. The approach taken is to derive integrated microstructural parameters as a function of stress from a fracture mechanical microcrack model (KACHANOV, 1982), and to calculate the resulting permeability enhancement from a statistical permeability model (DIENES, 1982). The result is a set of analytical expressions from which the permeability can be determined as a function of stress for given material constants. A complementary numerical investigation has been published recently by SIMPSON *et al.* (2001). The aim of this publication is to present for the first time the derivation of a more simplified model which retains the first order effects but which is more easily applied (because it is resolved analytically). As shown by SIMPSON *et al.* (2001), results calculated with the model presented in this paper are comparable to results of the more complex model and reproduce the broad features demonstrated in experiments.

Permeability Model

DIENES (1982) derived an analytical expression for the permeability of cracked media by considering the average fluid flux per crack, the crack density and the connectivity between cracks. Cracks are considered to be thin penny-shaped discs, the flow through which is assumed to be governed by Poiseuille's Law. DIENES (1982) showed that the permeability k, for an isotropic distribution of cracks is

$$k = \frac{8\pi^2}{30}\bar{A}^3 N_o \bar{c}^5 f \ , \tag{1}$$

where $\bar{c}$ is the crack radius, $\bar{A}$ is the aspect ratio ($=\bar{w}/\bar{c}$), $\bar{w}$ is the crack aperture, N_o is the number of cracks per unit volume and f is the fraction of cracks ($0 \leq f \leq 1$) which are hydraulically connected (overbars denote average properties). The above equation is valid if the distribution functions for the various crack parameters are narrow. We follow this assumption in order to simplify the following analysis although we note that some microstructural studies show broad crack parameter distributions (e.g., HADLEY, 1976). The connectivity between cracks can be calculated for simple lattices using percolation theory to yield $f \simeq 9/4(p - 1/3)^2$ for $1/3 < p < 1$, where p ($\simeq \pi^2/4\bar{c}^3 N_o$) is the probability that two cracks intersect

(GUÉGUEN and DIENES, 1989). Note also that f is zero when $p < 1/3$ and f is 1 when $p > 1$. The crack porosity, Φ can be derived as $\pi \bar{c}^2 \bar{w} N_o$.

Equation (1) implies that the modification of permeability can be thought of as a two-step process, the first corresponding to the attainment of connectivity at the percolation threshold, and the second related to increasing c and/or A at constant f. The important parameters controlling the permeability for both regimes are c, A and N_o. We derive dependencies of c and A and f on stress from fracture mechanics, while we assume N_o remains constant (i.e., does not change as a function of deformation).

Crack Growth Model

Elastic fracture mechanics is based on the premise that rocks contain pre-existing microcracks which propagate once the crack tip stresses have exceeded a critical value. Experimental and theoretical studies have demonstrated that cracks loaded by shear tractions grow by the propagation of tensile 'wing cracks' to form a kinked geometry (HORII and NEMAT-NASSER, 1982). Wing cracks nucleate at approximately 70° ($=\omega$) to the shear crack and progressively rotate with subsequent propagation towards the maximum principal stress (Fig. 1). KACHANOV (1982) proposed a model which enables calculation of the equilibrium wing crack length l as a function of applied stresses for a single (non-interacting) crack:

$$\sqrt{\frac{l}{l_o}} = \sqrt{\frac{2}{\pi} \frac{K_{IC}}{\sigma' \sqrt{l_o}} - \frac{2\sqrt{2}}{\sqrt{3}} \frac{F}{\sigma'}} , \tag{2}$$

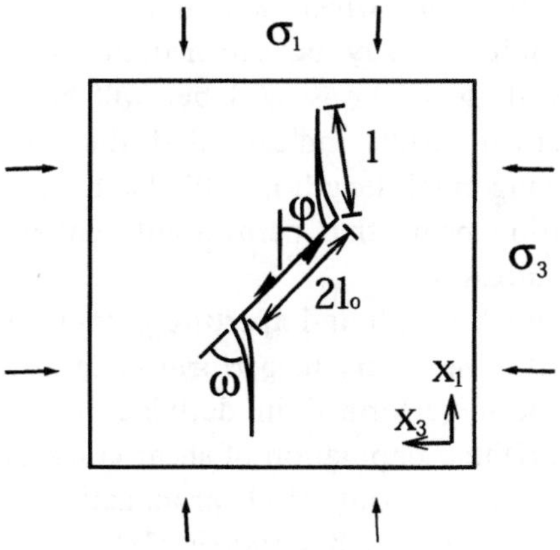

Figure 1
Geometry of tensile wing cracks propagating from the tips of a pre-existing shear crack.

where $2l_o$ is the length of the initial shear crack, K_{IC} is the critical fracture toughness, σ' is the effective stress (negative in compression) resolved normal to the wing crack and F is the driving Coulomb stress:

$$F = \tau + \mu\sigma_n \ , \tag{3}$$

where $\tau = (\sigma_3 - \sigma_1)\sin\phi\cos\phi$, μ is the friction coefficient, $\sigma_n = \sigma_1(\sin^2\phi) + \sigma_3(\cos^2\phi)$, σ_1 and σ_3 are the maximum and minimum Terzaghi effective principal stresses respectively, and ϕ is the angle between the shear crack and the direction parallel to σ_1. Effective stresses are assumed here to be constant with respect to crack growth which, in the presence of fluid, implies perfectly drained conditions. This assumption is only valid for scenarios in which the rate of crack growth is small relative to the rate of fluid flow, which may not be true for low permeability rocks and in cases where the loading rate is such that crack growth is rapid. Equation (2) is accurate for wing crack lengths which are smaller or on the same order as the length of the initial shear crack. KACHANOV (1982) also derived an approximate expression for the average opening of the wing crack v for plane strain conditions:

$$\frac{v}{l_o} = \frac{\sqrt{2}\pi(1-\nu^2)}{E}\left(\frac{4\sqrt{6}}{9}F + \frac{1}{2}\sigma'\sqrt{\frac{l}{l_o}}\right)\sqrt{\frac{l}{l_o}} \ , \tag{4}$$

where E is Young's modulus, ν is Poisson's ratio (both with values corresponding to uncracked rock) and other parameters are defined above. In order to maintain the analytical nature of the following equations, we approximate the normal stress acting on the wing crack (σ') in equations (2) and (4) with the mean stress ($= (\sigma_1 + \sigma_3)/2$) while we retain the ϕ dependency in F. Although both σ' and F are of the form $A + B(\phi)$, the ratio $B(\phi)/A$ is significantly higher for F than for σ' and therefore neglecting the ϕ dependency in σ' appears to be acceptable. This assumption is only strictly valid when $(\sigma_1 - \sigma_3)/2$ is negligibly small or when $\omega - \phi \sim \pi/4$. These conditions may be satisfied at the onset of crack growth (when $\omega = 70$ and $\phi \simeq 30°$ so $\omega - \phi \sim \pi/4$) but will become increasingly less so at high stresses. Numerical results indicate that the error associated with this approximation for the wing crack length is -19.6% at $\sigma_1/\sigma_3 = 10$ and -28.0% at $\sigma_1/\sigma_3 = 20$, the implication being that permeability enhancement is progressively underestimated at high stresses.

To obtain averaged crack length and aperture parameters from the single crack equations ((2) and (4)), it is necessary to perform integration throughout the solid area (in 2-D), as has been performed in deriving the permeability. In a rock containing a randomly oriented population of shear cracks, only favorably oriented cracks are active with respect to wing crack propagation. The angular range over which cracks are active is defined by KACHANOV (1982):

$$\phi_{1,2} = \arctan[(\sigma_3 - \sigma_1)/(2K - 2\mu\sigma_1) \pm K_o] \ , \tag{5}$$

where

$$K_o = \frac{\sqrt{(\sigma_3 - \sigma_1)^2 - 4(K - \mu\sigma_1)(K - \mu\sigma_3)}}{2(K - \mu\sigma_1)}$$
$$K = \left((2 - \nu)\sqrt{6\pi}K_{IC}\right)/\left(8\sqrt{2l_o}\right) .$$

Taking the integral of equations (2) and (4) between ϕ_1 and ϕ_2 and dividing the respective results by $\phi_2 - \phi_1$ yields expressions for the average wing crack length $\bar{l}$:

$$\sqrt{\bar{l}/l_o} = c\left[-2(c_1 + c_2)\sqrt{2}(3S(\lambda + 1))^{-1}\right]_{\phi_1}^{\phi_2} , \qquad (6)$$

where

$$c = (\phi_2 - \phi_1)^{-1}, \quad S = -\sigma_3\sqrt{\pi l_o}/K_{IC}, \quad \lambda = \sigma_1/\sigma_3$$
$$c_1 = -\sqrt{3}\mu S \sin\phi\cos\phi(\lambda - 1)$$
$$c_2 = -\sin\phi^2\sqrt{3}S(\lambda - 1) + \sqrt{3}\phi\left(\sqrt{3} + S\lambda\mu + S\mu\right)$$

and the average wing crack aperture $\bar{v}$

$$\frac{\bar{v}}{l_o} = c\left[\frac{dB}{(1 + \lambda)} + \frac{(c_8 + c_9)B}{S(1 + \lambda)} + c_{10}\right]_{\phi_1}^{\phi_2} , \qquad (7)$$

where

$$c_3 = -2\sin\phi\cos\phi^3(2\mu^2\lambda - \mu^2 + 1 - 2\lambda + \lambda^2)/9$$
$$c_4 = \phi(2\mu^2\lambda + 3\mu^2\lambda^2 - 2\lambda + 3\mu^2 + \lambda^2 + 1)/9$$
$$c_5 = \sin\phi\cos\phi(2\mu^2\lambda - 2\lambda + \lambda^2 + 1)/9$$
$$c_6 = 4\mu\cos\phi^4(\lambda - 1)/9 + \mu^2\sin\phi\cos\phi(1 - \lambda^2)/3$$
$$c_7 = -4\lambda\mu\sin\phi^4(\lambda - 1)/9 - 2\mu^2\lambda^2\sin\phi^3\cos\phi/9$$
$$d = c_3 + c_4 + c_5 + c_6 + c_7$$
$$c_8 = -4\sqrt{3}\mu\phi(\lambda + 1)/9 - 4\cos\phi^2\sqrt{3}(\lambda - 1)/9$$
$$c_9 = 4\sqrt{3}\mu\sin\phi\cos\phi(\lambda - 1)/9$$
$$c_{10} = -2\frac{\phi B}{S^2(\lambda + 1)} \qquad B = -\left(\sqrt{2\pi}(1 - \nu^2)\sigma_3\right)/E .$$

In deriving equation (6) we have assumed that the crack length distribution is narrow so that $\sqrt{l/l_o}$ is close to $\sqrt{\bar{l}/l_o}$. In cases where this condition is not met, equation (2) should be expanded prior to integration. Note also that in taking the integral of equations (2) and (4), the anisotropy in l and v (and permeability) with orientation resulting from deviatoric stresses are averaged-out. SIMPSON *et al.* (2001) calculated the full permeability tensor and showed that the permeability anisotropy (ratio of permeability in the direction of maximum compression to permeability in the direction of minimum compression) is typically less than 50, but that it becomes increasing important as the effective confining pressure approaches zero.

Effective Crack Approximation

Equations (6) and (7) enable the average wing crack length and aperture to be calculated as a function of applied stresses. However, because $\bar{l}$ and $\bar{v}$ are not directly equivalent to c and w, respectively, which are required to determine the permeability, we are forced to make a further simplification. We make the argument that flow through the actual kinked crack geometry, in which the hydraulic aperture varies strongly along the length of the crack, can be described by "effective" hydraulic parameters which are constant along the crack. We proceed by replacing the "actual" kinked crack with an "equivalent" crack (Fig. 2) which has the same surface area but which has a new equivalent length $2c$ defined as

$$c = (\bar{l}\sin\omega)/(\sin\gamma) \ , \tag{8}$$

where ω is the angle (assumed constant at 70°) between the wing crack and the initial shear crack and $\gamma = \arctan\left[(\bar{l}\sin\omega)/(\bar{l}\cos\omega + l_o)\right]$. The approximation also enables the determination of an effective crack aperture w

$$w = (2\bar{v}\bar{l} + S_o)/(2c) \ , \tag{9}$$

where $2\bar{v}\bar{l}$ is the surface area ($=S_a$) created by wing crack propagation and S_o ($=w_0 2l_o$) is the crack surface area prior to crack growth (where the initial effective crack aperture is $w_o = \Phi_o/(\pi N_o l_o^2)$ and Φ_o is the initial porosity). The initial permeability, k_o is defined as $k_o = (8\pi^2)/30 w_o^3 l_o^2 N_o$. Note that $N_o l_o^3$ ($=\rho$) is the

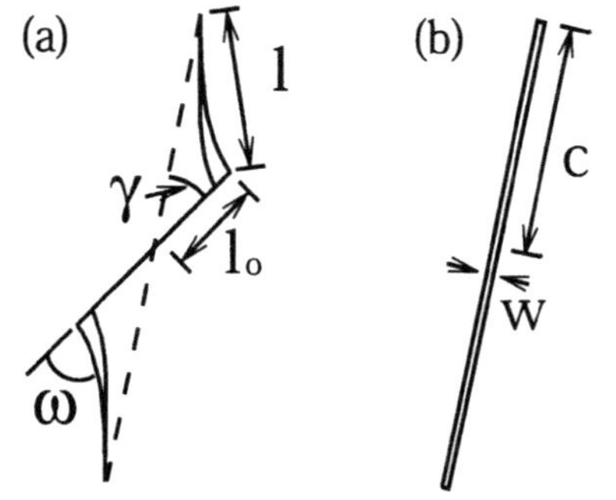

Figure 2
Relationships between actual crack geometry (a) and the corresponding effective crack approximation (b).

dimensionless crack density parameter which controls crack connectivity (GUÉGUEN and DIENES, 1989).

The approximation that has been made to pass between the deformation and permeability models is, by no means, unique. While we argue that the exact details of the replacement process do not govern the model results, it is clear that other equally valid approximations exist. Incidentally, the effective crack approximation is similar to the crack replacement procedure sometimes applied in fracture mechanics problems (see HORII and NEMAT-NASSER, 1982).

It is likely that the effective crack approximation leads to overestimation of the permeability. This is because the permeability is controlled by the narrowest crack apertures which tend to be overestimated. The effective crack approximation is most reasonable when the ratio S_a/S_o is small. When S_a/S_o is zero, as it is prior to wing crack growth (when $c = l_o$), the two crack systems are exactly equivalent. On the other hand, when wing cracks are long but the initial crack sectional area is very small (i.e., low initial porosity), the ratio S_a/S_o can potentially become large. Calculation for reasonable parameter values (i.e., $E = 50000$ MPa, $N_o = 5 \times 10^{10}\,\mathrm{m}^{-3}$, $\Phi_0 = 0.01, \nu = 0.25, K_{IC} = 1\,\mathrm{MPa}\sqrt{\mathrm{m}}$, $l_o = 0.0005$ m) indicate that S_a/S_o is approximately 0.025 for $\lambda = 10$ and 0.15 when $\lambda = 20$. These calculations indicate that, for the parameter ranges of interest, S_a/S_o is small and the effective crack approximation should be acceptable.

Results

Equations (1), (5)–(9) enable the permeability to be calculated analytically as a function of two loading parameters (λ and S) and three material parameters (B/S, ρ and μ). For comparison with model results, we consider permeability evolution data from two different triaxial deformation experiments performed on the Westerly granite containing pore fluid under confining pressure (ZOBACK and BYERLEE, 1975). Values used to estimate model parameters are presented inset in Figure 3. To focus on the effects of permeability enhancement due to crack growth, results are presented in terms of the calculated permeability normalised to the minimum permeability. In the experiments, the initial permeability is slightly higher than the minimum permeability, which is due to elastic closure of preexisting cracks.

Figure 3 presents experimental data for the Westerly granite (ZOBACK and BYERLEE, 1975) and the calculated permeability enhancement as a function of λ for 3 different initial crack densities. The calculations reproduce the form of the experimental data but generally underestimate permeability enhancement for realistic parameter values. This result is reflected by the unrealistically high initial crack densities required to fit the data. On the basis of geometrical considerations, an initial crack density corresponding to ~1.2 (best fit in Fig. 3a) requires the rock to be close to or above the fragmentation threshold prior to loading (see GUÉGUEN *et al.*, 1997).

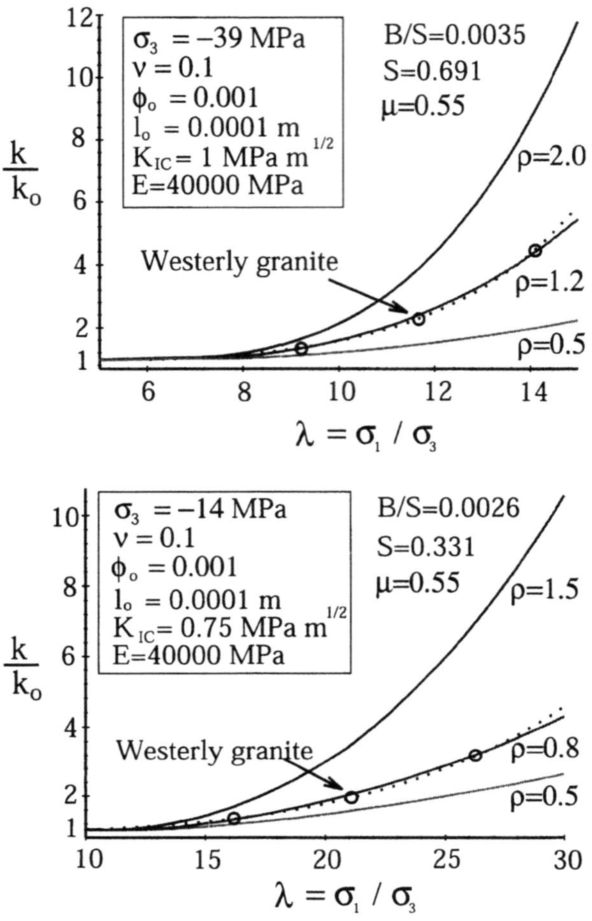

Figure 3

Comparison between results of permeability model (solid lines) and experimental measurements (ZOBACK and BYERLEE, 1975) on the Westerly granite (data in circles, dashed line is the fit, k, permeability; k_o, initial permeability; λ, effective stress ratio (σ_1/σ_3)); σ_1, maximum effective stress; σ_3, minimum effective stress; ν, Poisson ratio; ϕ_o, initial porosity; $2l_o$, initial shear crack length; K_{IC}, stress intensity factor; E, Young's modulus; μ, friction coefficient; $\rho = (N_o l_o^3)$, crack density parameter; N_o, number of cracks per unit volume).

A similar result was found by KACHANOV (1982) who showed that the volumetric strain resulting from a model incorporating frictional sliding and wing crack growth was ~50 times lower than experimental data. In order to explain this discrepancy, KACHANOV (1982) suggested a mechanism whereby there is additional dilatancy associated with frictional sliding over asperities. Allowing for this effect by the introduction of a dilatancy parameter in a numerical model enables data to be fit for realistic initial crack densities (SIMPSON *et al.*, 2001).

Note that the permeability evolution with varying λ in Figure 3 is due solely to variations in crack length and aspect ratio at constant f $(=1)$, due to initial parameter values. In general, the magnitude of permeability enhancement predicted by the model is strongly dependent on λ, S, B/S, μ and ρ.

Concluding Remarks

We have presented the derivation of a simplified version of the more complete numerical permeability model for microcracking rock developed by SIMPSON *et al.* (2001). It has been shown by SIMPSON *et al.* (2001) that the simplified model produces results comparable with the more complete model. The presented analytical model is simple, in that it assumes that the permeability can be represented as a scalar and that deformation is two-dimensional, nevertheless the results are in reasonable agreement with experimental measurements. The advantage of this model is that it is relatively easy to apply because it is resolved analytically. This point makes the model particularly suitable for incorporation into large scale sedimentary basin models which require quantification of how the permeability evolves as a function of stress, for which there is currently no law.

Acknowledgements

We are grateful to M. Kachanov for stimulating discussions during the initial stages of this work. Reviews by J. Heiland and S. Baraka-Lokmane are appreciated.

REFERENCES

DIENES, J. K., *Permeability, percolation and statistical crack mechanics.* In *Issues in Rock Mechanics* (eds. Goodman R. E. and Heuze F. E.) (American Institute of Mining, Metallurgical and Petroleum Engineers, New York 1982) pp. 86–94.

GAVRILENKO, P. and GUÉGUEN, Y. (1989), *Pressure Dependence of Permeability: A Model for Cracked Rocks*, Geophy. J. Int. *98*, 159–172.

GUEGUEN, Y. and DIENES, J. (1989), *Transport Properties of Rocks from Statistics and Percolation*, Math. Geol. *21*, 1–13.

GUEGUEN, Y., CHELIDZE, T., and LE RAVALEC, M. (1997), *Microstructures, Percolation Thresholds, and Rock Physical Properties*, Tectonophys. *279*, 23–35.

HADLEY, K. (1976), *Comparison of Calculated and Observed Crack Densities and Seismic Velocities in Westerly Granite*, J. Geophys. Res. *81*, 3484–3494.

HORII, H. and NEMAT-NASSER, S. (1982), *Compression-induced Nonplanar Crack Extension with Application to Splitting, Exfoliation and Rockburst*, J. Geophys. Res. *87*, 6805–6821.

KACHANOV, M. L. (1982), *A Microcrack Model of Rock Inelasticity. Part II: Propagation of Microcracks*, Mech. Mat. *1*, 29–41.

PEACH, C. J. and SPIERS, C. J. (1996), *Influence of Crystal Plastic Deformation on Dilatancy and Permeability Development in Synthetic Rock salt*, Tectonophys. *256*, 101–128.

SEEBURGER, D. A. and NUR, A. (1984), *A Pore Space Model for Rock Permeability and Bulk Modulus*, J. Geophys. Res. *89*, 527–536.

SIMPSON, G. D. H., GUÉGUEN, Y., and SCHNEIDER, F. (2001), *Permeability Enhancement Due to Microcrack Dilatancy in the Damage Regime*, J. Geophys. Res. *106*, 3999–4016.

WALSH, J. B. and BRACE, W. F. (1984), *The Effect of Pressure on Porosity and the Transport Properties of Rock*, J. Geophys. Res. *89*, 9425–9431.

ZHANG, S., COX, S. F., and PATERSON, M. S. (1994), *The Influence of Room Temperature Deformation on Porosity and Permeability in Calcite Aggregates*, J. Geophys. Res. *99*, 15,761–15,775.

ZHU, W. and WONG, T.-F. (1997), *The Transition from Brittle Faulting to Cataclastic Flow: Permeability Evolution*, J. Geophys. Res. *102*, 3027–3041.

ZHU, W. and WONG, T.-F. (1999), *Network Modeling of the Evolution of Permeability and Dilatancy in Compact Rock*, J. Geophys. Res. *104*, 2963–2971.

ZOBACK, M. D. and BYERLEE, J. D. (1975), *The Effect of Microcrack Dilatancy on the Permeability of Westerly Granite*, J. Geophys. Res. *80*, 752–755.

(Received January 15, 2001, revised May 17, 2001, accepted August 1, 2001)

Pure appl. geophys. 160 (2003) 1009–1022
0033–4553/03/061009–14

Pure and Applied Geophysics

A Simple Model for Deviations from the Cubic Law for a Fracture Undergoing Dilation or Closure

SOURITH SISAVATH,[1] AZZAN AL-YAARUBI,[1]
CHRIS C. PAIN,[1] and ROBERT W. ZIMMERMAN[1]

Abstract—Experimental observations show that flow through a fracture decreases more rapidly than the cube of the mean aperture (COOK, 1992). In order to provide a possible explanation of these experimental findings, we study creeping flow through a fracture of varying aperture that is symmetric about its midplane, using the power series of the stream function obtained by VAN DYKE (1987) for low Reynolds numbers. For the case of sinusoidally-varying walls, a simple expression relating the effective hydraulic aperture of the channel to the mean aperture and to the amplitude and wavelength of the sinusoidal wall profiles is obtained. Comparison is made to previous studies (KITANIDIS and DYKAAR, 1997) and to finite element calculations, and good agreement is obtained. The effect of fracture closure is then modelled as a decrease of the mean aperture without a change in the roughness. A power law relationship can be obtained between the flowrate and the mean aperture, with an exponent as high as 10, thus providing a potential mechanistic explanation of the experimental findings of PYRAK-NOLTE *et al.* (1987).

Key words: Fractures, cubic law, permeability, hydraulic conductivity, Navier-Stokes.

1. Introduction

Fracture-dominated flow is important in many areas, such as naturally fractured petroleum reservoirs and underground disposal of radioactive waste. Flow in these media usually takes place through a network of fractures or, in some cases, through a single fracture. The understanding of flow through a single fracture is therefore essential in order to begin to model transport phenomena in fractured geological media.

The oldest and simplest model of fracture flow was that of viscous flow between two parallel plates, which is governed by the well-known "cubic law" (BEAR, 1972; WITHERSPOON *et al.*, 1980). For large apertures, fracture roughness can be neglected, and a parallel plate model will be fairly accurate. However, when the roughness is of the same order of magnitude as the mean aperture, the complexity of the fracture

[1] Department of Earth Science and Engineering, Imperial College of Science, Technology and Medicine, London SW7 2BP, United Kingdom. E-mail: r.zimmerman@ic.ac.uk

surfaces and the variations in aperture must be taken into account. In order to deal with the variations of the aperture of the fracture, the Reynolds lubrication equation (BROWN, 1987) is often used as an alternative to the considerably more computationally intensive solution of the Navier-Stokes equations. In this approximation, the fluid velocity is assumed to be parabolic along the channel, although the magnitude of the parabola may vary as a function of the $\{x, y\}$ coordinates in the nominal fracture plane. The derivation of the lubrication equations shows (SCHLICHTING, 1968) that the aperture profile must vary "slowly" for this approximation to be valid. However, several recent studies (BROWN *et al.*, 1995; MOURZENKO *et al.*, 1995; NICHOLL *et al.*, 1999) have suggested that the apertures in rock fractures often do not vary slowly enough to ensure the validity of the lubrication approximation. YEO *et al.* (1998) indeed found, using aperture replicas of a natural sandstone fracture, that the measured hydraulic conductances were generally about 30–100% lower than those obtained numerically by solving the Reynolds lubrication equation using the actual aperture distribution. More accurate analytical models need therefore to be investigated.

As a fracture closes up under, say, an increasing normal stress, its mean aperture will decrease, and its hydraulic conductivity will also decrease. If the fracture were bounded by two smooth parallel plates, the conductivity would decrease as the cube of the mean aperture. In fact, this behaviour is observed for fractures having relatively large apertures. However as the aperture continues to close, the conductivity begins to decrease at a faster rate, with a log-log exponent that may be as large as 8–10 (PYRAK-NOLTE *et al.*, 1987). NOLTE *et al.* (1989) argued that this is due, at least in part, to the fact that as the fracture closes, the contact area increases, thus increasing the tortuosity of the flow. In this paper we propose an additional mechanism, related to the increase in *relative* roughness as the fracture closes, which may also give rise to these deviations from the cubic law.

A full understanding of flow through rock fractures is gradually being developed through the work of various research groups. Many different phenomena need to be accounted for, such as anisotropy (MEHEUST and SCHMITTBUHL, 2001), contact area (WALSH *et al.*, 1997), scaling laws (ORON and BERKOWITZ, 1998), stress effects (DURHAM and BONNER, 1994), inertial flow effects (SKJETNE *et al.*, 1999), *etc.* The present paper makes no claims of presenting a unified model for all aspects of fracture flow. Instead, our aim is merely to develop a model that physically captures some of the features of rough-walled rock fractures, and which can qualitatively explain the "deviations from the cubic law" that have been observed by PYRAK-NOLTE *et al.* (1987) and others (WITHERSPOON *et al.*, 1980; DURHAM and BONNER, 1994).

To achieve this, we consider a model of a fracture in which the aperture varies sinusoidally in one direction (say the x-direction), but does not vary in the y-direction. A perturbation approach, following the method of VAN DYKE (1987), is utilised to find an approximate solution of the equations in the Stokes limit (*i.e.*, zero

Reynolds number). A simple expression is obtained for the equivalent hydraulic aperture of the fracture, defined as the aperture of two parallel plates that would have the same hydraulic conductivity as the actual fracture. Finally, the effect of fracture closure is studied by assuming that, to "first-order," when a fracture closes the two opposing walls move towards each other, parallel to the fracture mid-plane, causing a decrease in the mean aperture without change in the "roughness." Although this is clearly an oversimplified mechanical model, this model of fracture closure has been used in many of the key papers of fracture flow (BROWN, 1987), and indeed continues to be invoked (MEHEUST and SCHMITTBUHL, 2000). Moreover, we emphasise that we are not attempting in the present paper to quantify the relationship between normal stress and fracture closure. We instead merely aim to understand the manner in which the conductivity of a fracture decreases as its mean aperture decreases. The fact that the relationship between mean aperture and normal stress is generally nonlinear (COOK, 1992), for example, is not relevant in the present context.

2. Perturbation Method for Flow Through a Symmetric Fracture with Varying Walls

Although real fractures possess a 3-D geometry, we focus, for the sake of mathematical simplicity, on the effect of roughness longitudinal to the flow direction. ZIMMERMAN *et al.* (1991) showed that such a "2-D model," in which aperture varies only in the direction of flow, gives qualitatively the same results as a full 3-D model, in the sense of predicting that the hydraulic aperture is less than the mean aperture by an amount that increases with increasing relative roughness. Recently, such 2-D models have been used (ORON and BERKOWITZ, 1998; SKJETNE *et al.*, 1999) to study the effects of non-zero Reynolds number, for example. We therefore consider a symmetric constricted fracture whose local half-aperture is described by

$$H(x) = H_0 g(x), \tag{1}$$

where H_0 is the mean half-aperture of the fracture, and $g(x)$ is a dimensionless function with a mean value of 1. We consider laminar, steady-state flow of an incompressible, Newtonian fluid, in response to a pressure gradient in the x-direction.

To solve the Navier-Stokes equations for this geometry, we introduce a dimensionless stream function $\tilde{\psi}$, defined by (VAN DYKE, 1987)

$$\tilde{u} = \frac{\partial \tilde{\psi}}{\partial \tilde{z}} \quad \text{and} \quad \tilde{v} = -\frac{\partial \tilde{\psi}}{\partial \tilde{x}}, \tag{2}$$

where $\tilde{u}$ and $\tilde{v}$ are, respectively, the velocities in the x and z directions, non-dimensionalised with respect to the mean velocity $U_0 = q/2H_0W$, where q is the flowrate, W is the width of the channel in the y-direction (*i.e.*, into the page), and $\tilde{x}$ and $\tilde{z}$ are the coordinates normalised with respect to H_0. Using the definition of the stream function in Equation (2), the Navier-Stokes equations reduce to the following differential equation for the stream function (VAN DYKE, 1987):

$$\nabla^4\tilde{\psi} = \mathrm{Re}\left(\frac{\partial\tilde{\psi}}{\partial\tilde{z}}\frac{\partial}{\partial\tilde{x}} - \frac{\partial\tilde{\psi}}{\partial\tilde{x}}\frac{\partial}{\partial\tilde{z}}\right)\nabla^2\tilde{\psi}, \tag{3}$$

where Re is the Reynolds number, defined as $\mathrm{Re} = \rho q/2\mu W$, μ being the viscosity of the fluid. The appropriate boundary conditions are

$$\tilde{\psi} = \pm 1, \quad \frac{\partial\tilde{\psi}}{\partial\tilde{z}} = 0 \quad \text{at the boundary} \quad \tilde{z} = \pm g(x). \tag{4}$$

VAN DYKE (1987) expressed the stream function for this problem as a function of a dimensionless perturbation parameter $\tilde{\varepsilon}$ that characterises the slow variation of the channel wall. For creeping flow (Re = 0), his perturbation solution for the stream function can be written as

$$\tilde{\psi} = \frac{1}{2}(3\eta - \eta^3) + \tilde{\varepsilon}^2\frac{3}{20}(4f'^2 - -ff'')(\eta - 2\eta^3 + \eta^5) + O(\tilde{\varepsilon}^4), \tag{5}$$

where $\eta = \tilde{z}/f$ and $f(\tilde{\varepsilon}\tilde{x}) = H(x)/H_0$. The perturbation series was extended to $O(\tilde{\varepsilon}^4)$ by KITANIDIS and DYKAAR (1997). However, the additional term adds considerable complexity to the results. As we are merely trying to determine whether or not this model is capable of qualitatively explaining the experimental results of PYRAK-NOLTE *et al.* (1987), it was felt that extending the analysis to higher-order terms was not warranted. Moreover, as will be seen below, the second-order perturbation solution has reasonable accuracy for the range of parameters that are of greatest physical interest.

Although the stream function was found by VAN DYKE (1987) and KITANIDIS and DYKAAR (1997), the relation between flow and mean pressure drop was not calculated. For this purpose, we now define a mean pressure difference $\overline{\Delta p}$ between two locations x_1 and x_2 along the fracture:

$$\overline{\Delta p} = \frac{1}{H_0}\left[\int_0^{H(x_2)} p(x_2)\,dz - \int_0^{H(x_1)} p(x_1)\,dz\right]. \tag{6}$$

For a constricted fracture that is periodic with period λ in the x-direction, we will have $H(x_1) = H(x_2)$, if $x_2 - x_1 = \lambda$. Equation (6) can then be simplified to

$$\overline{\Delta p} = \frac{1}{H_0}\int_{x_1}^{x_2}\left(\int_0^{H(x_1)} \frac{\partial p}{\partial x}\,dz\right)dx. \tag{7}$$

For creeping flow, the pressure derivative $\partial p/\partial x$ is given by

$$\frac{\partial p}{\partial x} = \mu\left(\frac{\partial^2 u}{\partial x^2} + \frac{\partial^2 u}{\partial y^2}\right). \tag{8}$$

Using Equation (2) and the perturbation solution (5) for the stream function, the mean pressure drop can be expressed as a function of the variation of the channel wall as

$$\begin{aligned}\overline{\Delta p} = \frac{\mu q}{(2H_0)^3 W}\int_{x_1}^{x_2}\Bigg[&-\frac{12g(x_1)}{g(x)^3} + H_0^2 g'(x)^2\left(-\frac{84}{5}\frac{g(x_1)}{g(x)^3} + \frac{120}{5}\frac{g(x_1)^3}{g(x)^5}\right)\\ &+ H_0^2 g''(x)\left(\frac{12}{10}\frac{g(x_1)}{g(x)^2} - \frac{12}{2}\frac{g(x_1)^3}{g(x)^4}\right) + O(\tilde{\varepsilon}^4)\Bigg]\,dx.\end{aligned} \tag{9}$$

This expression is valid for any twice differentiable function $g(x)$.

A fracture profile is by definition continuous, but not necessarily differentiable. According to the Weierstrass approximation theorem, such a profile can always be approximated arbitrarily closely by a differential function, such as a polynomial. The mathematical treatment of continuum problems in regions bounded by surfaces containing small-scale roughness has been discussed by VANDEMBROUCQ and ROUX (1997).

3. *Case of an Axisymmetric Fracture with Sinusoidally-varying Walls*

One of the simplest models of a fracture wall is that in which the profile varies sinusoidally with x. Although this is obviously an inaccurate representation of any specific actual fracture, it does in some sense capture the oscillatory nature of real fracture profiles. Moreover, it is known that such a model can predict some aspects of real fracture flow behaviour, such as the manner in which conductivity depends on the ratio of roughness to mean aperture (ZIMMERMAN *et al.*, 1991). In this spirit, we now focus our attention on a fracture whose boundaries are described by

$$H(x) = \pm H_0\left[1 + \frac{\delta}{H_0}\sin\left(\frac{2\pi x}{\lambda}\right)\right], \tag{10}$$

where H_0 is the mean half-aperture of the fracture, δ is the amplitude of the constriction, and λ is the wavelength (Fig. 1). An appropriate definition for the perturbation parameter is then $\tilde{\varepsilon} = \delta/\lambda$.

The streamlines obtained from the perturbation solution (5) for the two geometries $\{\tilde{\delta} = 0.8,\ \tilde{\lambda} = 10\}$ and $\{\tilde{\delta} = 0.8,\ \tilde{\lambda} = 5\}$ are displayed in Figure 2, in which the parameters $\tilde{\delta}$ and $\tilde{\lambda}$ are, respectively, the amplitude and the wavelength of the constricted fracture, normalised with respect to the mean half-aperture, H_0. These streamlines are qualitatively in good agreement with those obtained by

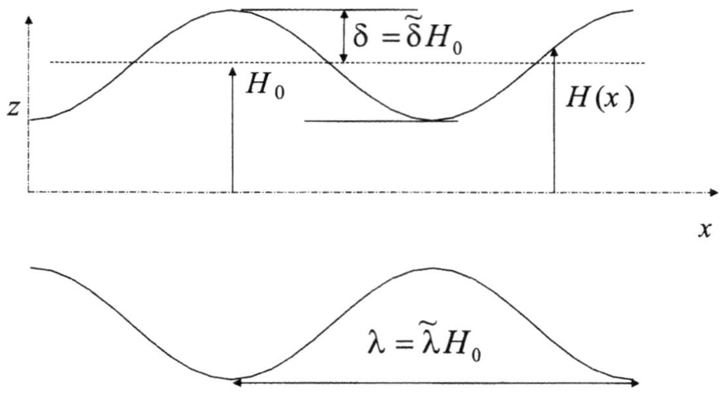

Figure 1
Fracture with sinusoidally-varying walls symmetric about the axis. H_0 is the mean half-aperture, δ is the roughness, and λ the wavelength.

KITANIDIS and DYKAAR (1997), who found the stream function up to the fourth-order term. Most of the flow takes place within a canal from one throat to the next, and for the more abrupt variation, $\{\tilde{\delta} = 0.8,\ \tilde{\lambda} = 5\}$, a stagnant recirculation zone appears near the wall. Note that the "streamlines" that are shown to intersect the wall actually represent "dividing streamlines" (SCHLICHTING, 1968) that separate the main region of fluid flowing through the fracture from small recirculation regions of (macroscopically) stagnant flow. Although such recirculation regions are commonly associated with high-Reynolds number flows, they can also occur in the Stokesian limit of zero Reynolds number (see POZRIKIDIS, 1992, Fig. 6.1.1). These stagnant regions will obviously greatly influence phenomena such as solute transport (RAVEN *et al.*, 1987), discussion of which is beyond the scope of the present paper.

In order to account for the effect of fracture roughness, use is often made of the Reynolds lubrication equation. The Reynolds equation gives an approximate description of flow between slightly non-planar and non-parallel surfaces. In the case of steady-state, incompressible flow though a nominally flat fracture, the Reynolds equation can be written as (e.g., BROWN, 1987)

$$\nabla \cdot [h^3(x,y)\nabla p] = 0, \tag{11}$$

where $h(x, y)$ is the local aperture of the fracture. The Reynolds equation is therefore equivalent to assuming that the "cubic law," which is exact for a laminar flow between two parallel plates, can be applied locally for a constricted channel. For a channel whose aperture varies only in the direction of flow, the Reynolds equation can be integrated directly to yield (SILLIMAN, 1989)

$$\overline{\Delta p} = -\frac{12\mu q}{W} \int_{x_1}^{x_2} \frac{1}{h(x)^3}\, dx. \tag{12}$$

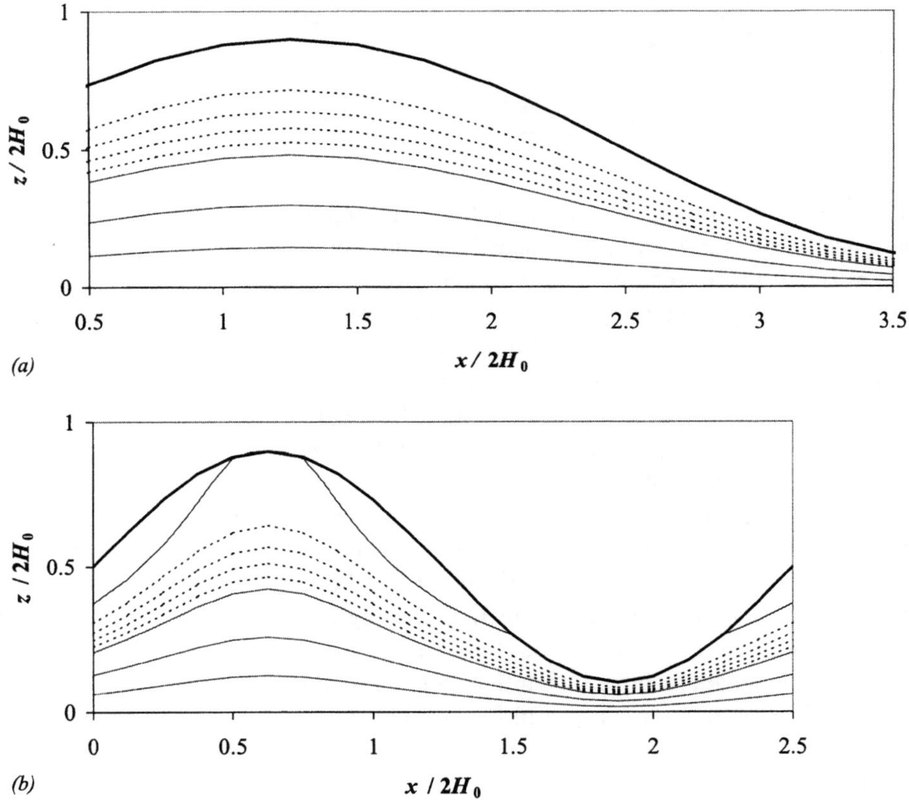

Figure 2
Streamlines for (*a*) $\tilde{\delta} = 0.8$, $\tilde{\lambda} = 10$, and (*b*) $\tilde{\delta} = 0.8$, $\tilde{\lambda} = 5$. The solid and dotted lines delineate streamtubes that convey, respectively, 1/8 and 1/40 of the total flow.

The permeability of a fracture is usually quantified in terms of an equivalent hydraulic aperture (e.g., BROWN, 1987; ZIMMERMAN *et al.*, 1991), which is the aperture of a parallel plate fracture that would have the same hydraulic conductivity as the actual fracture. This equivalent hydraulic aperture h_H can be written (implicitly) in the form of the well-known "cubic law,"

$$q = -\frac{W h_H^3}{12\mu} \frac{\overline{\Delta p}}{\lambda}. \tag{13}$$

For the sinusoidal variation described by Equation (10), the integral appearing in equation (12) can be evaluated to yield (ZIMMERMAN *et al.*, 1991)

$$h_H^3 = \frac{(2H_0)^3 (1 - \tilde{\delta}^2)^{5/2}}{1 + (\tilde{\delta}^2/2)}. \tag{14}$$

The second-order perturbation analysis, Equation (9), gives the hydraulic aperture of the sinusoidal channel as

$$h_H^3 = \frac{(2H_0)^3(1-\tilde{\delta}^2)^{5/2}}{1+(\tilde{\delta}^2/2)}\left[1+\frac{\tilde{\delta}^2}{\tilde{\lambda}^2}\left(\frac{36}{15}\right)\frac{\pi^2(1-\tilde{\delta}^2)}{1+(\tilde{\delta}^2/2)}\right]^{-1}. \tag{15}$$

The second term in the brackets, of $O(\tilde{\varepsilon}^2)$, represents an additional "hydraulic resistance" that is not obtained when using the lubrication theory approximation. Note also that this result (15) is quite similar, except for a difference in the numerical factor multiplying the term $(\tilde{\delta}/\tilde{\lambda})^2$, to the result derived by HASEGAWA and IZUCHI (1983) for flow through a channel bounded by a sinusoidal wall and a plane wall.

In order to assess the validity of Equation (15), numerical calculations were carried out using the finite-element based CFD code Fluidity (PAIN, 2000). The equivalent hydraulic aperture was calculated for the following amplitudes and wavelengths: $\tilde{\delta} = 0.2$, 0.4 and 0.6, and $\tilde{\lambda} = 2$, 4, 6, 8 and 10. A parabolic velocity profile was prescribed at the entrance of the channel. Three wavelengths (unit cells) were used for each simulation, and the overall pressure drop through each wavelength was calculated. It was verified that the flow had become fully-developed by the third wavelength by checking that the pressure drop per wavelength had stabilised between the second and third unit cells. Velocity profiles were also examined, and no appreciable variation could be found between the second and third unit cells. The mesh was refined until further refinement did not affect the results. A sufficiently fine grid was found to contain 40×500 quadratic mesh-elements per wavelength.

Very good agreement was obtained between the numerical results and the perturbation solution (see Fig. 3), provided that the wavelength of the fracture is larger than the mean aperture, i.e., $\tilde{\lambda} > 2$. For shorter wavelengths, $\tilde{\lambda} < 2$, the slowly-varying assumption implicit in the perturbation solution is no longer valid. This regime corresponds to wavelengths less than the mean aperture. We note that the extensive measurements made by YEO *et al.* (1998) on the apertures in a fracture in a Permian sandstone revealed minor roughness of large amplitude at wavelengths less than the mean aperture.

The Reynolds lubrication equation does not account for the wavelength of the corrugations, and consequently its range of acceptable accuracy is restricted to long wavelengths, i.e., $\tilde{\lambda} > 6$. Nevertheless, the Reynolds equation provides a useful upper bound, which is approached as $\tilde{\lambda} \to \infty$. For short wavelengths, $\tilde{\lambda} \to 0$, a plausible (but apparently never rigorously proven) lower bound can be found from the "stagnant limit" (BROWN *et al.*, 1995; BERNABE and OLSON, 2000), in which the flow is stagnant in the troughs of the corrugation and flow takes place only in the center of the fracture. Under such an assumption, the hydraulic aperture of the fracture is given by

$$h_H(\text{stagnant limit}) = 2H_0(1-\tilde{\delta}). \tag{16}$$

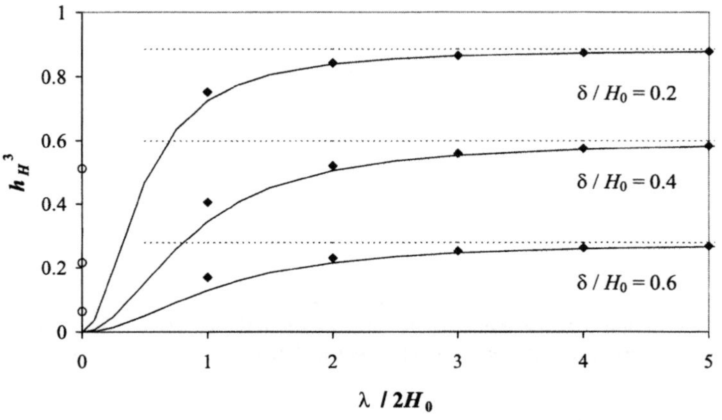

Figure 3
Comparison of the results obtained with Fluidity (◆) with the perturbation solution (—) and the Reynolds lubrication equation (---). The symbols ○ correspond to the stagnant limit (BROWN *et al.*, 1995).

4. Change in Hydraulic Aperture as Mean Aperture Decreases

As mentioned above, it has been found in several experimental studies that, as a fracture closes up under normal stress, the conductivity decreases more rapidly than the cube of the mean aperture (see the review by COOK, 1992). For example, PYRAK-NOLTE *et al.* (1987) made measurements of the flow and the joint aperture as a function of the normal stress, for three different natural joints. For an intermediate range of apertures, they found a power-law relationship between the hydraulic conductivity and the aperture of the fracture, with an exponent n that ranged between 7.6 and 9.8. WITHERSPOON *et al.* (1980) found similar behaviour for flow in a fracture in marble.

Following BROWN (1987) and TSANG and WITHERSPOON (1981), we will assume that, to a first approximation, the displacement of the fracture wall due to application of normal stress is in the direction parallel to the load, so that the sinusoidal shape of the constriction is preserved, along with its amplitude and wavelength. The effect of the normal stress σ is therefore modelled as a decrease of the mean aperture H_0, without a change in the roughness (see Fig. 4). No attempt is made to quantify the *rate* of aperture closure with applied stress, since the focus of this study is on the relationship between geometrical and hydraulic properties of the fracture. As the wall of the fracture is still sinusoidal under the assumptions considered, we again use Equation (15) to calculate the hydraulic aperture. For $H_0 \gg \delta$ and small δ / λ, the relation between the mean aperture and the specific flow is cubic. However, as the mean aperture H_0 decreases with stress, the conductivity does not actually decrease as the third power of H_0, since the dimensionless amplitude $\tilde{\delta}$ and wavelength $\tilde{\lambda}$ vary with H_0. The amplitude of the roughness $\tilde{\delta}$ becomes

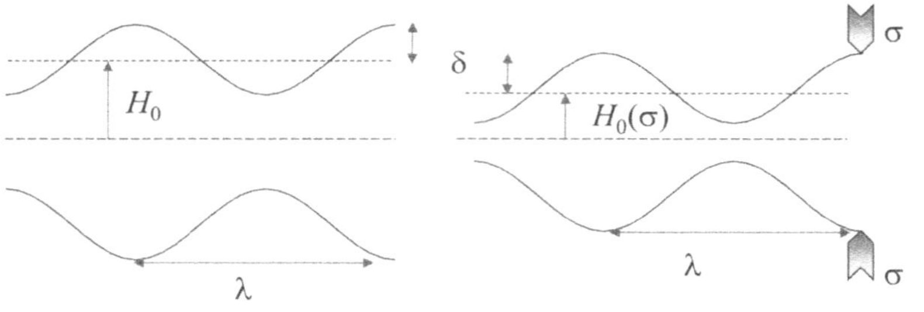

Figure 4
(left): Fracture with sinusoidally-varying walls symmetric about the axis; (right): under normal stress σ, the mean aperture is assumed to decrease, nonetheless roughness δ and wavelength λ remain unchanged.

significant, and a cubic law between the mean aperture and the conductivity no longer applies. As shown in Figure 5, log-log exponents as large as 10 can be found in the power law as the mean aperture decreases. It is indeed possible to obtain the same exponents as found by PYRAK-NOLTE *et al.* (1987), by choosing appropriate values for the amplitude δ and wavelength λ.

The question may be raised as to whether Equation (15) maintains its accuracy as the mean aperture goes to zero. Figure 6 shows the normalised hydraulic conductivity, $(h_H/2H_0)^3$, as a function of the relative roughness, δ/H_0, for a fracture having

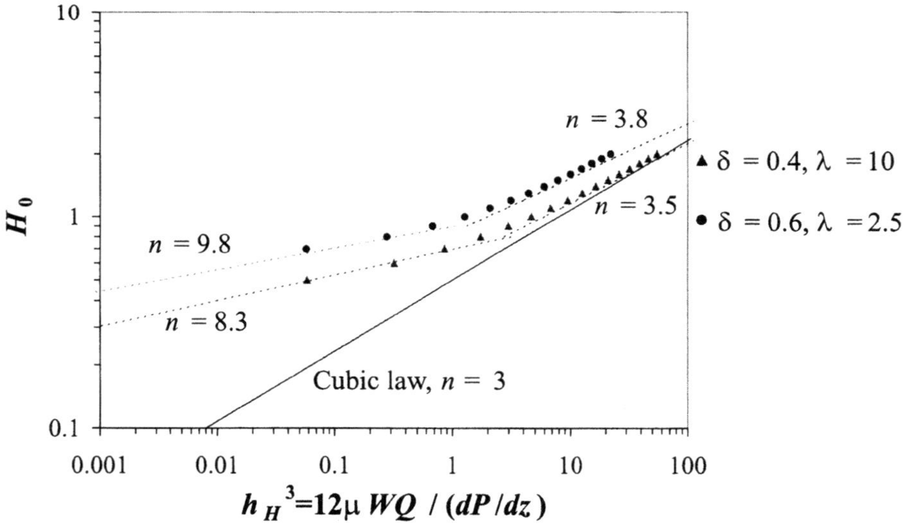

Figure 5
A plot of the hydraulic conductivity as a function of the mean aperture for two different constricted channels. The dotted lines correspond to power laws obtained to fit the results obtained with Equation (15). The power law relating these two parameters can have an exponent well exceeding 3.

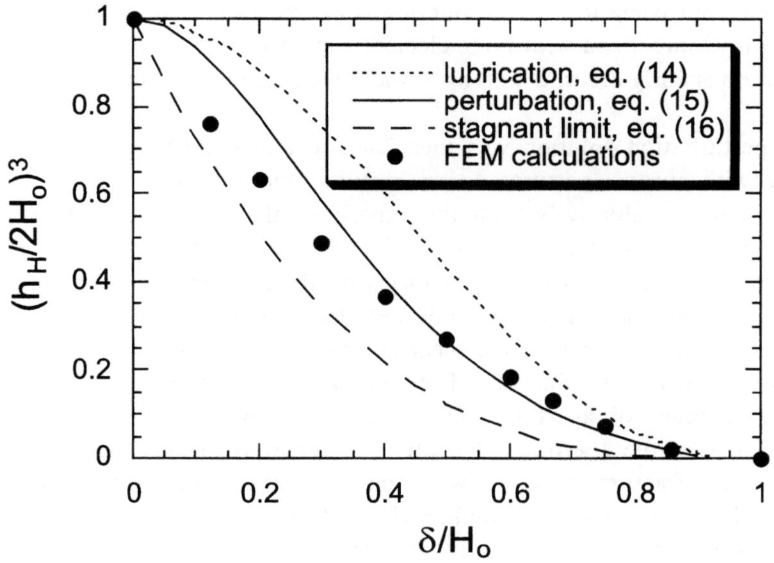

Figure 6
Normalised hydraulic conductivity as a function of relative roughness, for the case $\delta/\lambda = 0.24$. The accuracy of the perturbation solution does not degrade as the mean aperture decreases, which corresponds to moving to the lower left in Figure 5 or the lower right in Figure 6.

$\delta/\lambda = 0.24$, which corresponds to the "worst" of the two cases shown in Figure 5. The four curves show the Reynolds equation prediction, Equation (14), the present second-order perturbation solution, Equation (15), the stagnant limit, Equation (16), and the finite-element results. The perturbation solution is seen to provide a substantial improvement over the lubrication approximation derived by ZIMMERMAN *et al.* (1991). More importantly, the accuracy of the perturbation solution does not degrade as the mean aperture decreases (corresponding to moving to the left along the curves in Fig. 5). These results are consistent with those of Figure 3, which show that a decrease in λ will detrimentally effect the accuracy of the perturbation solution, whereas the accuracy is much less sensitive to changes in the parameter δ/H_0. Finally, we point out that the leftmost points shown in Figure 5, calculated from Equation (15), correspond to the minimum possible value of $H_0 = \delta$, at which point closure occurs. The dotted lines are extrapolated further merely to allow their slopes to be more clearly seen.

5. *Discussion*

A perturbation approach has been used to study flow through a fracture with sinusoidally-varying walls. This solution improves the results obtained using the lubrication approximation, by accounting (to some extent) for the effect of finite wavelength. Comparison with finite element simulations showed that the perturbation solution is fairly accurate as long as the wavelength is not smaller than the mean aperture.

The perturbation solution was then used to study the variation of hydraulic aperture with mean aperture, as a fracture "closes." The results, shown in Figure 5, agree qualitatively with the experimental results obtained by previous investigators such as PYRAK-NOLTE *et al.* (1987) and WITHERSPOON *et al.* (1980). These experimental results can be divided into three regimes (Figure 7). For large apertures, presumably substantially larger than the roughness (regime III), the flow decreases as the cube of the mean aperture. This is also consistent with the recent experimental results of MEHEUST and SCHMITTBUHL (2001). For an intermediate range of apertures (regime II), the flow decreases decidedly faster than the cube of the aperture. The ratio of roughness to mean aperture becomes larger as the fracture closes down, and the development of stagnant zones enhances this decrease in the conductivity. Of course, in the case of a real 3-D fracture, other mechanisms, such as the tortuosity caused by the fluid circumventing regions of contact between the two walls, will also become increasingly relevant as the fracture closes (NOLTE *et al.*, 1989). At very low mean apertures (regime I), the conductivity levels off to some nearly constant value. This regime, which has been discussed by PYRAK-NOLTE *et al.* (1987), cannot be treated with the present model, which does not account for 3-D flow effects such as tortuosity.

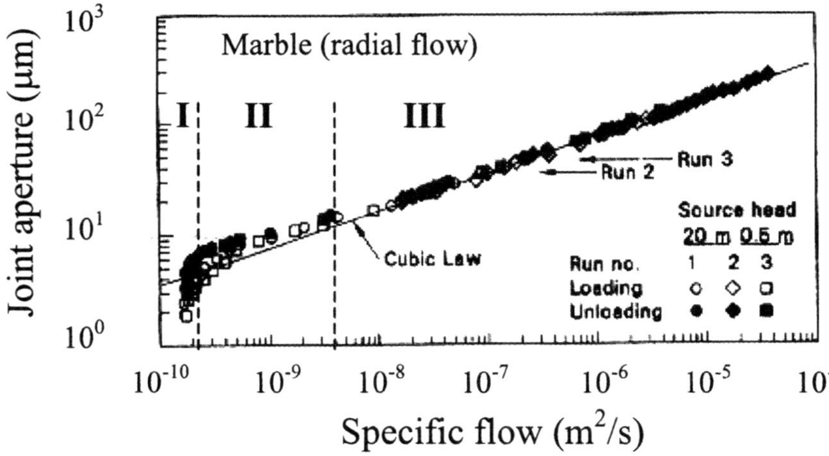

Figure 7

Experimental results showing the relations between specific flow and joint aperture (after WITHERSPOON *et al.*, 1980). Aside from a constant factor, the "specific flow" is equal to the hydraulic conductivity.

Acknowledgements

The authors thank Jean Schmittbuhl and another (anonymous) reviewer for their many incisive comments and criticisms of an earlier draft of this paper.

References

BEAR, J., *Dynamics of Fluids in Porous Media* (Elsevier, New York, 1972).

BERNABE, Y. and OLSON, J. F. (2000), *The Hydraulic Conductance of a Capillary with a Sinusoidally Varying Cross-section*, Geophys. Res. Lett. *27*, 245–248.

BROWN, S. R. (1987), *Fluid Flow through Rock Joints: The Effect of Surface Roughness*, J. Geophys. Res. B *92*, 1337–1347.

BROWN, S. R., STOCKMAN, H. W., and REEVES, S. J. (1995), *Applicability of the Reynolds Equation for Modelling Fluid Flow between Rough Surfaces*, Geophys. Res. Lett. *22*, 2537–2540.

COOK, N. G. W. (1992), *Natural Joints in Rock: Mechanical, Hydraulic and Seismic Behaviour and Properties under Normal Stress*, Int. J. Rock Mech. *29*, 198–223.

DURHAM, W. B. and BONNER, B. P. (1994), *Self-propping and Fluid-flow in Slightly Offset Joints at High Effective Pressures*, J. Geophys. Res. *99*, 9391–9399.

HASEGAWA, E. and IZUCHI, H. (1983), *On Steady Flow through a Channel Consisting of an Uneven Wall and a Plane Wall – Part 1. Case of No Relative Motion in Two Walls*, JSME Bull. *26*, 514–520.

KITANIDIS, P. K. and DYKAAR, B. B. (1997), *Stokes Flow in a Slowly Varying Two-Dimensional Periodic Pore*, Transp. Porous Media *26*, 89–98.

MEHEUST, Y. and SCHMITTBUHL, J. (2000), *Flow Enhancement of a Rough Fracture*, Geophys. Res. Lett. *27*, 2989–2992.

MEHEUST, Y. and SCHMITTBUHL, J. (2001), *Geometrical Heterogeneities and Permeability Anisotropy of Rough Fractures*, J. Geophys. Res. *106*, 2089–2102.

MOURZENKO, V. V., THOVERT, J.-F., and ADLER, P. M. (1995), *Permeability of a Single Fracture—Validity of the Reynolds Equation*, J. Phys II *5*, 465–482.

NICHOLL, M. J., RAJARAM, H., GLASS, R. J., and DETWILER, R. (1999), *Saturated Flow in A Single Fracture: Evaluation of the Reynolds Equation in Measured Aperture Fields*, Wat. Resour. Res. *35*, 3361–3373.

NOLTE, D. D., PYRAK-NOLTE, L. J., and COOK, N. G. W. (1989), *The Fractal Geometry of Flow Paths in Natural Fractures in Rock and the Approach to Percolation*, Pure Appl. Geophys. *131*, 111–138.

ORON, A. P. and BERKOWITZ, B. (1998), *Flow in Rock Fractures: The Local Cubic Law Assumption Re-examined*, Wat. Resour. Res. *34*, 2811–2824.

PAIN, C. C., *Brief Description and Capabilities of the General Purpose CFD Code: Fluidity* (Internal Report, Imperial College, London, 2000).

POZRIKIDIS, C., *Boundary Integral and Singularity Methods for Linearized Viscous Flow* (Cambridge U. P., Cambridge, 1992).

PYRAK-NOLTE, L. J., MYER, L. R., COOK, N. G. W., and WITHERSPOON, P. A., *Hydraulic and mechanical properties of natural fractures in low permeability Rock*. In *Proc. Sixth Int. Congr. Rock Mech.* (eds. Herget G. and Vongpaisal S.) (Balkema, Rotterdam 1987) pp. 225–231.

RAVEN, K. G., NOVAKOWSKI, K. S., and LAPCEVIC, P. A. (1987), *Interpretation of Field Tracer Tests of a Single Fracture using a Transient Solute Storage Model*, Wat. Resour. Res. *24*, 2019–2032.

SCHLICHTING, H., *Boundary-Layer Theory* (McGraw-Hill, New York, 1968).

SILLIMAN, S. E. (1989), *An Interpretation of the Difference between Aperture Estimates Derived from Hydraulic and Tracer Tests in a Single Fracture*, Wat. Resour. Res. *25*, 2275–2283.

SKJETNE, E., HANSEN, A., and GUDMUNDSSON, J. S. (1999), *High-Velocity Flow in a Rough Fracture*, J. Fluid Mech. *383*, 1–28.

TSANG, Y. W. and WITHERSPOON, P. A. (1981), *Hydromechanical Behaviour of a Deformable Rock Fracture Subject to Normal Stress*, Wat. Resour. Res. *86*, 9287–9298.

VAN DYKE, M. (1987), *Slow Variations in Continuum Mechanics*, Adv. Appl. Mech. *25*, 1–43.

VANDEMBROUCQ, D. and ROUX, S. (1997), *Conformal Mapping on Rough Boundaries 1: Applications to Harmonic Problems*, Phys. Rev. E *55*, 6171–6185.

WALSH, J. B., BROWN, S. R., and DURHAM, W. B. (1997), *Effective Medium Theory with Spatial Correlation for Flow in a Fracture*, J. Geophys. Res. *102*, 22,587—22,594.

WITHERSPOON, P. A., WANG, J. S. Y., IWAI, K., and GALE, J. E. (1980), *Validity of Cubic Law for Fluid Flow in a Deformable Rock Fracture*, Wat. Resour. Res. *16*, 1016–1024.

YEO, I. W., DE FREITAS, M. H., and ZIMMERMAN, R. W. (1998), *Effect of Shear Displacement on the Aperture and Permeability of a Rock Fracture*, Int. J. Rock Mech. *35*, 1051–1070.

ZIMMERMAN, R. W. and BODVARSSON, G. S. (1996), *Hydraulic Conductivity of Rock Fractures*, Transp. Porous Media *23*, 1–30.

ZIMMERMAN, R. W., KUMAR, S., and BODVARSSON, G. S. (1991), *Lubrication Theory Analysis of Rough-Walled Fractures*, Int. J. Rock. Mech. *28*, 325–331.

(Received December 15, 2000, revised July 9, 2001, accepted October 1, 2001)

Pure appl. geophys. 160 (2003) 1023–1050
0033 – 4553/03/061023–28

Pure and Applied Geophysics

Scale Effects Related to Flow in Rough Fractures

Y. MÉHEUST[1,2] and J. SCHMITTBUHL[1]

Abstract—A numerical fracture flow simulation based on the lubrication approximation is used to investigate the influence of roughness on the flow inside a rough fracture, at low Reynolds number. Facing surfaces are described as self-affine topographies with identical roughness magnitude. Resolution of the Reynolds equation is achieved using two distinct numerical schemes, with consistency. Fracture closure is studied assuming perfect plastic contact between facing surfaces. Long-range correlations are shown to exist in the local aperture field due to the fracture geometry and subsequently in the local fluxes inside the fracture. Flow channeling is the result of these correlations in terms of spatial distribution of the flow, and is responsible for either flow-enhancing or flow-inhibiting behavior of the fracture. Matching between the two surfaces at scales larger than a mismatch scale is studied. The mismatch scale is the upper limit scale for the local apertures scale invariance. It appears to control flow channeling and the related dispersion of the possible behaviors over a large statistics of fractures with identical statistical features. Hydraulic anisotropy of a given fracture is investigated: the dependence of the fracture transmittivity on the pressure drop orientation is proved to be sinusoidal, with an amplitude that is controlled by the mismatch scale.

Key words: Fracture, flow, roughness, surface matching, hydraulic aperture.

1. Introduction

Flow in crustal rocks with a low permeability is controlled by the hydraulic behavior of the fracture networks they contain (BEAR *et al.*, 1993; SAHIMI, 1993). Two characteristics of the fracture network control the rock permeability: the fracture connectivity (BERKOWITZ and ADLER, 1998), and the permeability of individual fractures. The present work is focused on the latter feature.

The permeability of a fracture depends both on the fracture mean opening and the geometry of the facing surfaces. A number of experimental analyses, both on synthetic (SCHMITTBUHL *et al.*, 1995a; BOUCHAUD, 1997; BOFFA *et al.*, 1998) and natural fractures (BROWN and SCHOLZ, 1985; SCHMITTBUHL *et al.*, 1993), show that fracture surface topographies are rough and exhibit scale invariance over a large range of wavelengths. These fluctuations of the geometric boundaries imposed to flow can strongly influence the permeability of the fracture. For fracture separations large in comparison to the vertical extent of roughness, the simple model of a parallel

[1] École Normale Supérieure, Paris, France. E-mail: meheust@geologie.ens.fr
[2] Present address: Laboratoire de Géologie, 24 rue Lhomond, 75231 Paris CEDEX 05, France

plate with an aperture equal to the fracture mean opening d_m provides a good estimate for the fracture permeability. The volumetric flow $\boldsymbol{Q}$ through the fracture is proportional to the macroscopic pressure gradient, $\overline{\nabla P}$, by a factor that is a cubic law of the fracture opening (IWAI, 1976; COOK, 1992; ZIMMERMAN and BODVARSSON, 1996):

$$\boldsymbol{Q} = -L_y \frac{d_m^3}{12\eta} \overline{\nabla P} \ , \tag{1}$$

where L_y is the dimension of the fracture transverse to flow, and η is the fluid viscosity. For fractures that are close to surface contact, Equation (1) still holds under the condition that the mean aperture d_m be replaced by another characteristic parameter, the hydraulic aperture d_h. The latter is the opening of a parallel plate with a permeability equal to that of the rough fracture. The deviation of d_h from d_m accounts for the difference between the permeability of the fracture and that of a parallel plate with identical mean opening. Thus it is an estimate of the influence of fracture wall geometry on fracture flow.

The effect of roughness on fracture flow has been studied in terms of fracture permeability in a number of experimental works (DURHAM and BONNER, 1994; DURHAM, 1997; YEO *et al.*, 1998). Recent studies also provide an experimental description of the flow paths (BROWN *et al.*, 1998; DIJK *et al.*, 1999). The relation between local features of the flow inside the fracture and the permeability at the fracture scale has been studied using numerical finite difference simulations based upon the Reynolds equation (see Section 2.3) (BROWN, 1987, 1989; THOMPSON, 1991; THOMPSON and BROWN, 1991; AMEDEI and ILLANGASEKARE, 1994). The validity of this equation has been discussed using a comparison with results from a lattice-gas automaton (BROWN *et al.*, 1995), from a complete 3-D simulation of the flow (MOURZENKO *et al.*, 1995), and by comparing experimental measurements and numerical results obtained by application of the Reynolds equation to a recording of the experimental geometry (NICHOLL *et al.*, 1999). Models based on electrical analogs have also been developed (TSANG, 1984; MORENO *et al.*, 1988).

In a previous work (MÉHEUST and SCHMITTBUHL, 2000), we had presented an experimental study of the flow through a plane-rough fracture model, which exhibited a strong hydraulic anisotropy. Anisotropy of the smooth-rough fracture behavior toward flow has been explained in terms of channeling in a complementary work (MÉHEUST and SCHMITTBUHL, 2001), and related to the long-range correlations in the local apertures at the fracture scale, using a numerical model based on the lubrication approximation. We present here the model of a fracture made of two rough surfaces of identical vertical extent, allowing moderate plastic contact between them, and using the same method for simulating fracture flow as in (MÉHEUST and SCHMITTBUHL, 2001). We show that in the limit of slow viscous flows, this configuration is equivalent to that studied experimentally in (MÉHEUST and

SCHMITTBUHL, 2000). We demonstrate how long-range correlations existing in the local apertures are transmitted to the local fluxes inside the fracture, which results in flow channeling. For facing surfaces that are matched at large scales, we show how the latter phenomenon is controlled by local aperture fluctuations at the mismatch length scale.

2. A Numerical Model for the Flow Between Two Rough Walls

2.1. Geometry of the Model

We consider a fracture made of two rough walls with parallel horizontal mean planes. The latter are separated by the mean fracture separation, or mechanical aperture, d_m. Local fluctuations of the rough topographies with respect to their respective mean planes are denoted $h_u(x,y)$ (upper surface) and $h_l(x,y)$ (lower surface) respectively, where (x,y) denotes the position on the horizontal grid. We assume that the height distributions h_u and h_l have the same root-mean-square, σ, which accounts for the vertical extent of the rough topographies. The fracture projection on its horizontal mean plane is assumed to be square, of lateral length L. A steady-state flow is forced between two of its parallel sides. A sketch of the model is shown in Figure 1.

2.1.1. Model for the rough wall topographies

In good agreement with a number of measurements on artificial (SCHMITTBUHL *et al.*, 1995a; BOUCHAUD, 1997) and natural (SCHMITTBUHL *et al.*, 1993; BOUCHAUD, 1997) rock fracture surfaces, we describe each rough surface as an isotropic self-affine fractal. The probability for two points of the topography with horizontal positions distant by l to have altitudes different by a vertical length in the interval $[h; h+dh]$ is $\mathscr{P}(h,l)\,dh$, where the density probability function $\mathscr{P}$ can be written as

$$\mathscr{P}(h,l) = l^{-\zeta}\Phi\left(\frac{h}{l^{\zeta}}\right) , \tag{2}$$

where Φ is an appropriate scaling function (SCHMITTBUHL *et al.*, 1995a). The Hurst exponent ζ accounts for the scale invariance of the topography: a sub-part of the surface of lateral size L/α is statistically identical to the whole surface, provided that the vertical axis be rescaled by $\alpha^{-\zeta}$. Exponent ζ has been chosen to be 0.8, a value that has been proposed as universal (BOUCHAUD *et al.*, 1990).

Such a two-dimensional topography exhibits long-range spatial correlations. Its 2-D Fourier spectrum has an amplitude that scales as a power law of the wave number with an exponent $-2(1+\zeta)$ (SCHMITTBUHL *et al.*, 1995a). We assume its phase to be random. We generate suitable topographies h_i in the Fourier domain: distributions with no spatial correlations are first generated using a Gaussian

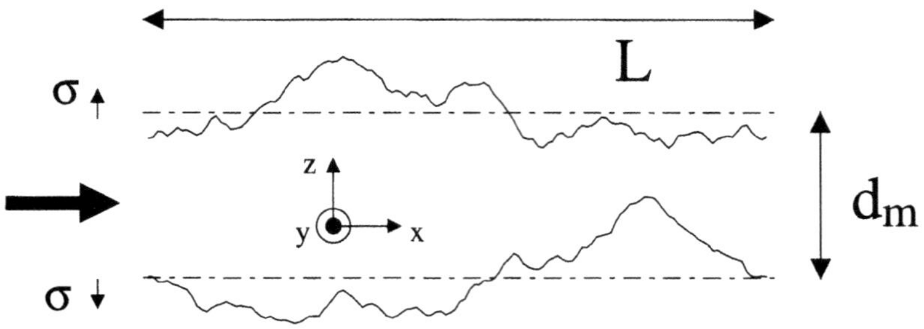

Figure 1
Geometry of the model. The mean fracture separation, or mechanical aperture, is denoted d_m; h_u and h_l denote the local fluctuations of the rough topographies with respect to their respective mean planes. Distributions h_u and h_l have identical root mean square, σ. The horizontal extent of the fracture is $L \times L$. The large arrow denotes the direction of the flow.

probability density function. Correlations are then introduced by multiplying their Fourier modes by the power law $(k_x^2 + k_y^2)^{-(1+\zeta)/2}$, where k_x and k_y stand for the components of the two-dimensional wave vector corresponding to the mode in question.

2.2. *Characteristics of the Local Apertures*

Local apertures inside the fracture are defined as

$$d(x,y) = d_m + h_u(x,y) - h_l(x,y) \quad . \tag{3}$$

Due to the linearity of the Fourier transform, the self-affinity of the rough topographies results in the local apertures field also being an isotropic self-affine fractal (see Appendix A). Its root-mean-square Γ only depends on the root-mean-square of the boundary surfaces, σ, and not on the mean aperture d_m. For example, if the height distributions h_i are perfect Gaussian, Γ is $\Gamma = \sqrt{2}\sigma$. Thus Γ and σ are two parameters that can be used equivalently when comparing the extent of local aperture fluctuations to the mean aperture. In what follows, we use the root-mean-square of the surfaces, σ.

2.3. *Stokes Flow and Lubrication Approximation*

We consider a steady-state flow where viscous forces dominate inertial effects, i.e., with a low Reynolds number: $\mathcal{R}e \ll 1$. In addition, we assume the lubrication approximation, according to which local apertures are slowly varying within the fracture, that is, $\| \nabla d \| \ll 1$. Under these assumptions, the flow through the system can be studied by use of a two-dimensional equivalent system, where the 2-D local velocities are the local fluxes inside the fracture: $\boldsymbol{q} = \int_{d(x,y)} v dz$, and are related to local apertures through a local cubic law identical to (1) (ZIMMERMAN and YEO, 2000; MÉHEUST and SCHMITTBUHL, 2001)

$$q = -\frac{d^3(x,y)}{12\eta}\nabla P \ . \tag{4}$$

The pressure field P inside the fracture, and subsequently, local fluxes, can be obtained through the Reynolds equation, which is simply the conservation equation for the conservative two-dimensional velocity field q:

$$\nabla.(d^3(x,y)\nabla P) = 0 \ . \tag{5}$$

An important consequence of this approximation is that P is derived from the only knowledge of the local apertures, whatever may be the rough surfaces responsible for it. Flow lines can also be obtained using a stream function, as explained in (MÉHEUST and SCHMITTBUHL, 2001).

For local apertures that are self-affine up to the fracture scale, L, the typical local slope along local apertures, s, estimated using the smallest scale described in the model, l_c, scales as (MÉHEUST and SCHMITTBUHL, 2001)

$$s(l_c) \sim l_c^{\zeta-1}\frac{\Gamma}{L^\zeta} \ , \tag{6}$$

where Γ is the root-mean-square of the local aperture field, or, equivalently,

$$s(l_c) \sim l_c^{\zeta-1}\frac{\sigma}{L^\zeta} \ , \tag{7}$$

where σ is the root-mean-square of the rough topographies. Relation (7) diverges with decreasing l_c (since $\zeta = 0.8$). The lubrication approximation requests that $s(l_c)$ be much smaller than 1. For a given fracture, i.e., fixed σ and L, it imposes a lower limit to parameter l_c, i.e., to the self-affine description. From the numerical point of view it imposes an upper limit to the size N of the grid used for sampling local apertures: $N = L/l_c$.

2.4. Numerical Techniques

We use two independent finite difference schemes to solve Equation (5). They are described in detail in (MÉHEUST and SCHMITTBUHL, 2001).

The first method is based on the inversion of the linear system that associates the local aperture values at the nodes of the grid to those of the pressure field. The inversion scheme is based on a biconjugate gradient method. This scheme can be used with any kind of boundary conditions (periodic, semi-periodic, no-flow lateral boundaries), and allows values of the local apertures very close to or equal to 0.

The second one is a perturbation method developed by PLOURABOUÉ *et al.* (1998). It requires periodic boundary conditions, and allows a continuous rotation of the pressure head with respect to the fracture (see MÉHEUST and SCHMITTBUHL, 2001). It uses FFT algorithms, which makes it much faster than the inversion method. From a theoretical point of view, the perturbation method cannot converge for fractures that are very close to surface contact (see MÉHEUST and SCHMITTBUHL, 2001, for the convergence criterion).

For configurations where they are both efficient, the two methods provide results that are different by no more than 10^{-5}. In what follows, we have used periodic boundary conditions enabling the use of one method or the other, alternatively. The spectral method was used preferentially. In cases where it was not appropriate, the biconjugate gradient method was used.

We have used various grid sizes in these simulations, ranging from 256 to 2048. When simulating a fracture with a given size, L, increasing the grid size means diminishing the smallest scale considered in the simulation, l_c. Large grid sizes allow a more precise description of local flow features. However, as shown in Section 2.3, the maximum possible value for the grid size is limited by the lubrication approximation. All grid sizes used throughout the study meet this upper limit requirement. On the other hand, small-scale features do not contribute significantly to the fracture permeability as long as the ratio between L and l_c is as large as 256, a condition that is always met in our simulations. Within this range of allowed N values, relatively small grid sizes have been used for the statistical studies, to gain computation time.

3. Characterizing Correlations Along Local Apertures and Fluxes

As mentioned in Section 2.1.1, our synthetic rough fracture surfaces display isotropic long-range correlations that result in their Fourier power spectrum scaling as a power law of the wave number. As a consequence, the local apertures exhibit similar spectral features (see Appendix A). In order to check this, and also to examine whether local fluxes also display long-range correlations, we use three methods.

Computing the 2-D Fourier spectrum of a distribution provides a way of determining whether the distribution is isotropic or not. A cylindrical shape of the 3-D representation of the spectrum denotes isotropy. If so, the spectrum can be averaged over all directions in the Fourier space to obtain its dependence on the wave number. A self-affine behavior is detected as a power law with an exponent $-2(1+\zeta)$ (see Fig. 6).

Two methods are carried out on linear profiles extracted from the two-dimensional distributions. The results are averaged over many profiles to improve the accuracy. These methods only examine correlations along a given direction. They are used for examining correlations along distributions that are known to be isotropic, for example for determining the Hurst exponents of self-affine fractal distributions. One method consists in computing the 1-D Fourier spectrum of the profiles. An isotropic self-affine distribution exhibits a power law with an exponent $-1-2\zeta$ (SCHMITTBUHL *et al.*, 1995b), as shown in Figure 3. The other method is the average wavelet coefficient method, which consists in computing the wavelet transform of the profiles and averaging it over all possible values for the translation parameter (SIMONSEN *et al.*, 1998). The curve obtained for a self-affine distribution is a power law with an exponent $\zeta + 1/2$. The two methods are used in parallel for consistency.

All methods have been tested successfully on numerically generated distributions. They have also been used successfully to analyze recordings of granite fracture surfaces (see MÉHEUST and SCHMITTBUHL, 2000).

4. Closure of a Rough Fracture

In this chapter we consider a fracture with two independent rough topographies generated on a 512×512 grid. We vary the ratio d_m/σ and observe how the flow evolves as the vertical extent of roughness increases relative to the fracture separation.

In contrast to a number of works that considered elastic surface contact (UNGER and MASE, 1993; MOURZENKO *et al.*, 1997; TSANG and WITHERSPOON, 1981), we allow fracture closure to go beyond surface contact by assuming perfect plastic contacts: parts of the rough topographies that overlap each other simply vanish (BROWN, 1987). Thus, all local apertures that relation (3) would define as negative are set to zero. The overall area of the contact zones, S_c, is kept smaller than 10% of the grid surface, L^2. When contact occurs, some of the fluctuations in the local apertures are lost, which results in a decrease of the root-mean-square of the local apertures, Γ. This change is always smaller than 5%.

4.1. Evolution of the Local Apertures

In Figure 2a, maps of the local apertures $d(x, y)$, normalized by the mean fracture separation, d_m, are drawn for different values of d_m/σ: 14.0, 8.0, 6.0, 4.0, and 3.0, from top to bottom. Black shading indicates closed areas ($d = 0$). Brighter shading denotes larger apertures. The same linear gray scale is used for all maps, which allows a visual comparison between them. A uniform gray scale denotes aperture fluctuations that are negligible in comparison to the mean aperture. As the fracture mechanical aperture is decreased (i.e., from top to bottom), the contrast of the maps grows, indicating that the local aperture fluctuations increase relative to the mean aperture.

We have checked the self-affinity of the two rough topographies. One-dimensional Fourier spectra computed as explained in Section 3 are shown in Figure 3. The self-affine properties of the resulting local apertures for the normalized mechanical apertures 10.0, 3.0, and 2.0 are also checked in this figure. All spectra exhibit a slope of -2.60 ± 0.01, corresponding to the roughness exponent $\zeta = 0.8$ that has been imposed when generating the rough surfaces. The result has been verified by use of the average wavelet coefficient method (not shown here). It is

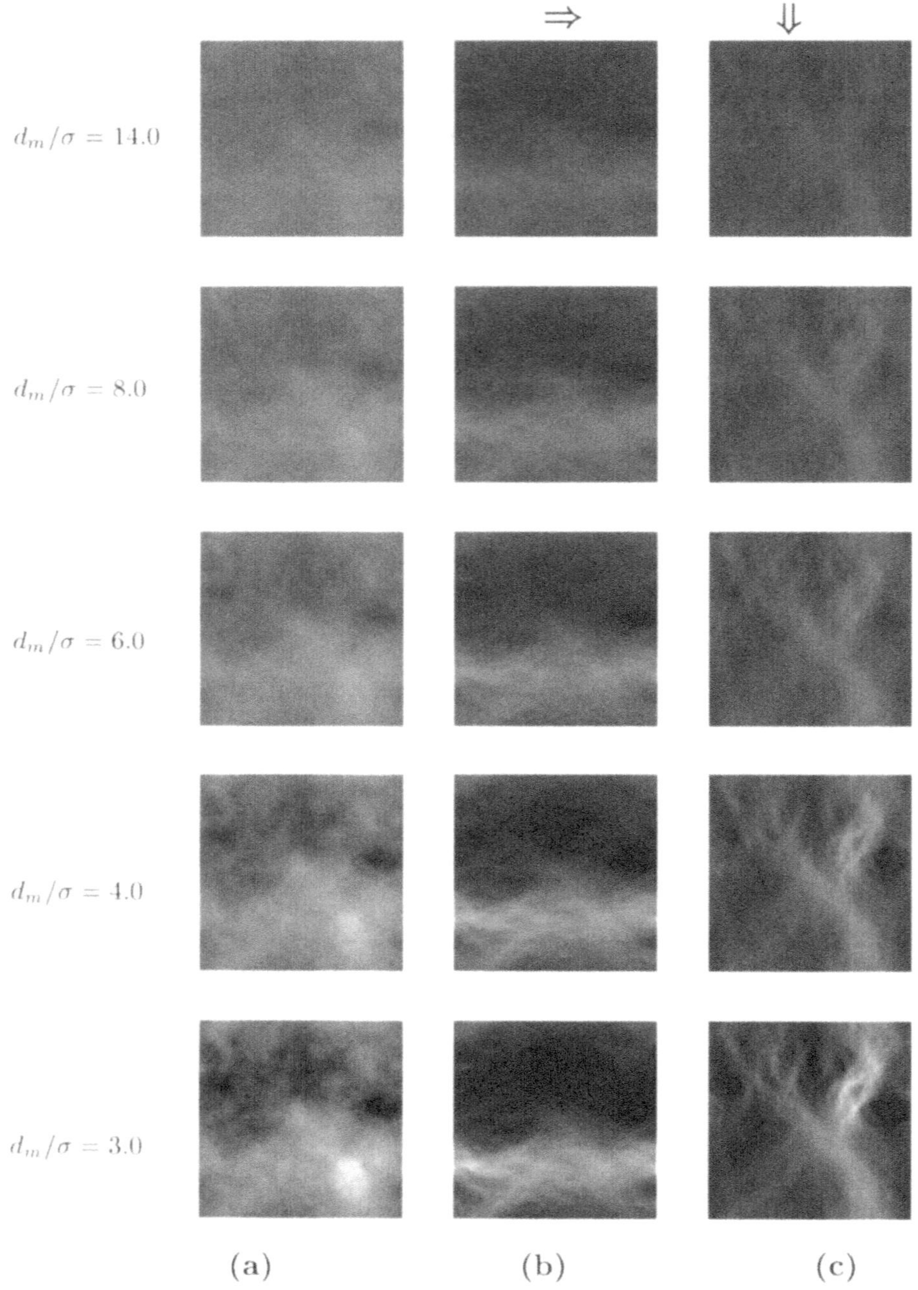
⇒
⇓
$d_m/\sigma = 14.0$
$d_m/\sigma = 8.0$
$d_m/\sigma = 6.0$
$d_m/\sigma = 4.0$
$d_m/\sigma = 3.0$
(a)
(b)
(c)

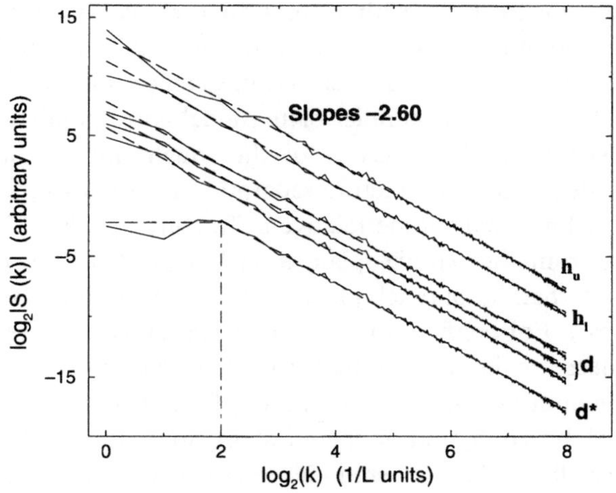

Figure 3
Fourier spectra of the rough topographies: top surface h_u and bottom surface h_l, and of the resulting local apertures, d, for three different fracture separations, $d_m/\sigma = 10.0$, 3.0 and 2.0 (from top to bottom). These values correspond respectively to 0, 0.88 and 6.93% of closed area inside the fracture. The slopes all equal -2.60 ± 0.01, thus the aperture distributions are self-affine with a Hurst exponent identical to that of the topographies: $\zeta = 0.8 \pm 0.01$. — The Fourier power spectrum $S(k)$ of the local apertures (d^*) in the case of facing surfaces that are matched at length scales larger than $L/4$ is shown for comparison ($d_m/\sigma = 10.0$). It displays the same self-affine behavior as d for wave numbers larger than $1/L_c = 4/L$, and is constant otherwise.

straightforward for the rough topographies. For the local apertures in an open fracture, it is also expected, as explained in 2.2 (see also Appendix A). It also holds along local apertures when a small part of the fracture is closed: the self-affinity of the surfaces is transmitted to the resulting local apertures. It is of significant consequence that there is no other characteristic scale available for the geometric description of the fracture than its size L itself.

4.2. Evolution of the Hydraulic Behavior

We have simulated the flow in the space between the two topographies corresponding to the aperture distributions shown in Figure 2a for two orthogonal

◄

Figure 2
(a) Maps of the ratio $d(x,y)/d_m$, for $d_m/\sigma = 14.0$, 8.0, 6.0, 4.0, and 3.0 (from top to bottom). The linear gray scale is the same for all maps. The two maps for $d_m/\sigma = 4.0$ and 3.0 exhibit closed areas, in black. Relative fluctuations of the aperture increase as the fracture separation is decreased. — **(b)** Maps of local fluxes (2D velocities) normalized by the mean local flux; the pressure head is applied on the corresponding aperture distribution in (a) according to the direction of the arrow above the maps; this configuration is labeled ⇒ — **(c)** Same as (b), with a macroscopic flow in the orthogonal direction, and a label ⇓ — Linear gray scales are identical in (b) and (c). This makes a visual comparison of channeling in the two cases possible.

directions of the pressure head with respect to the fracture, labeled ⇒ and ⇓, respectively. The evolution of the ratio of the hydraulic aperture to the mean aperture d_h/d_m when varying the normalized mechanical aperture, d_m/σ, is shown in Figure 4. The behavior of a parallel plate model corresponds to the horizontal line $d_h/d_m = 1$. The hydraulic behavior of the rough fracture deviates monotonously from this ideal model as fracture separation is decreased, as a result of the increase of the roughness magnitude relative to fracture opening. For configuration ⇒, the deviation from the parallel plate model is positive: the fracture is more conducive to flow than a parallel plate with identical mean separation (flow-enhancing behavior). For configuration ⇓, the deviation is negative (flow-inhibiting behavior). The decrease in the permeability of a fracture due to roughness, as displayed by configuration ⇒, is a well-known effect. The possible flow-enhancing behavior exhibited by configuration ⇓ is less obvious: the fracture can be more conducive toward flow than a smooth fracture, or less, depending on the orientation of the macroscopic pressure gradient. Such a hydraulic anisotropy has been observed experimentally for an equivalent smooth-rough fracture by MÉHEUST and SCHMITTBUHL (2000).

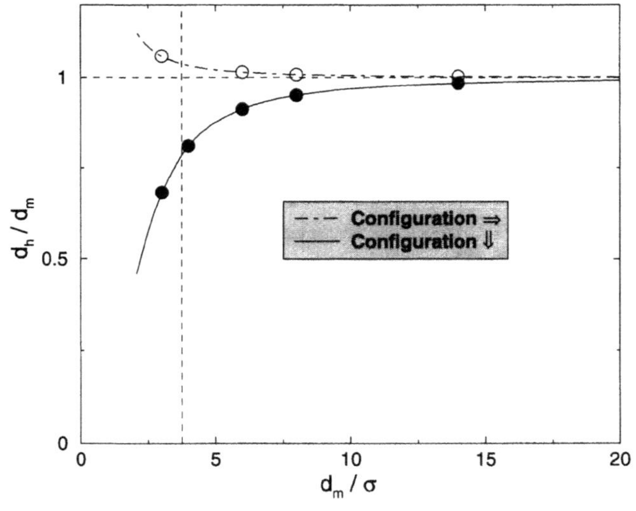

Figure 4
Evolution of the ratio of the hydraulic aperture, d_h, to the mechanical aperture, d_m, as a function of the normalized fracture separation, for the two configurations ⇒ and ⇓, corresponding to orthogonal orientations of the imposed pressure head with respect to the fracture described in Figure 2. d_h/d_m is computed for d_m/σ between 2 and 20, by steps of 0.05. Configuration ⇒ is flow-enhancing, while configuration ⇓ is flow-inhibiting. The circles (plain and white) correspond to aperture distributions shown in Figure 2a. Maps of local fluxes corresponding to the white circles are shown in Figure 2b. Those corresponding to the plain circles in Figure 2c. Mechanical apertures that are on the left of the vertical dashed line correspond to plastic contact between the two surfaces.

4.3. Evolution of the Local Fluxes

In a previous work (MÉHEUST and SCHMITTBUHL, 2001), we have explained how this possible hydraulic anisotropy is related to flow channeling: when the protruding of roughness into the void between the rough surfaces becomes significant, flow tends to avoid regions of small local apertures and find its way through preferential paths along tortuous large aperture channels. The existence of large aperture channels across the whole fracture results from the long-range correlations that exist along local apertures. Flow channeling grows more important as the fracture is closed. This is shown in Figures 2b and 2c, where maps of local fluxes q are shown for configurations $\Rightarrow$ and $\Downarrow$, respectively. In each configuration, the maps are shown for $d_m/\sigma = 14.0$, 8.0, 6.0, 4.0, 3.0 (from top to bottom), that is, for fracture separations considered in Figure 2a. Local fluxes are normalized by the mean local flux, which corresponds to the local flux in the parallel plate model of identical permeability, i.e., of mechanical aperture $d_m = d_h$. All maps in Figures 2b and 2c are drawn using the same linear gray scale. This allows observation of the increase of channeling as the fracture mean aperture is diminished (from top to bottom), and to compare flow channeling between the two configurations. In Figure 2b, bottom map, the flow mainly uses a large aperture channel along direction $\Rightarrow$, that can be seen across the whole fracture in Figure 2a (bottom map). This channel is large and oriented parallel to the applied pressure head which constitutes a configuration decidedly favorable to flow. This is consistent with the flow enhancing behavior observed in Figure 4 (dot-dashed line). In Figure 2c, flow is forced to use a very narrow pass of intermediate aperture, also visible on the right of the bottom map in Figure 2a. This is consistent with the flow-inhibiting behavior displayed in Figure 4 (plain line).

Visual inspection of Figure 2 makes clear that local fluxes and local apertures are correlated to each other. Local fluxes are the result of hydraulic boundary conditions at the fracture scale, coupled with long-range spatial correlations in the geometric boundary conditions (i.e., fracture surfaces). Thus, the distribution of local fluxes should exhibit a trace of the correlations found along local apertures. More precisely, we searched for such a trace in the field $\boldsymbol{\delta q}(x, y)$ that is the deviation between the local fluxes observed inside the fracture, $\boldsymbol{q}(x, y)$, and those observed inside a parallel plate with equal permeability, $\boldsymbol{q}_{pp}$ (uniform vector field). The Fourier power spectrum of the $\|\boldsymbol{\delta q}\|$ field is shown in Figure 5, for a normalized aperture $d_m/\sigma = 4.0$, that is, very close to surface contact, and in configuration $\Rightarrow$. It exhibits an anisotropy that does not exist in the distribution of local apertures. The deviation of the spectrum from a cylindrical shape occurs in a small defect region along the direction of macroscopic flow (direction $\Rightarrow$, or along x), or equivalently of the pressure head: the Fourier spectrum is small for wave vectors with a large magnitude and a direction close to the x direction. In other words, the field $\|\boldsymbol{\delta q}\|$ exhibits few small-scale fluctuations along direction $\Rightarrow$. This is the sign of a "close to-uniform"

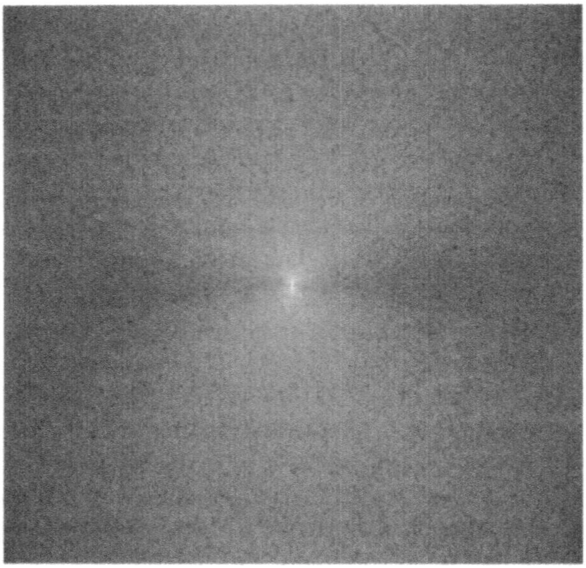

Figure 5
Map of the Fourier spectrum of $\| \boldsymbol{q} - \mathbf{q}_{pp} \|$, where $\boldsymbol{q}_{pp}$ is the uniform field of local fluxes in the equivalent parallel plate. The fracture considered here is the same as in Figure 2. In contrast to the local aperture spectrum (not shown here), this spectrum exhibits an anisotropy of the distribution.

movement along a large aperture pathway in the direction $\Rightarrow$ (see Figure 2b). Distant from the x axis, the spectrum goes back to a cylindrical shape. The defect region represents a small part of the whole spectrum. We have computed the average of the spectrum over all directions, assuming that the defect region only contributes marginally to it. These computations have also been done for the power spectra of the components $\boldsymbol{\delta q}_x$ and $\boldsymbol{\delta q}_y = \boldsymbol{q}_y$. Corresponding curves are shown in Figure 6. They are power laws, which indicate long-range correlations similar to those existing along local apertures. The slopes of the log-log plots are -3.60 ± 0.02 for the component of the local flux transverse to macroscopic flow, $\boldsymbol{\delta q}_y$, -3.41 ± 0.02 for that parallel to it, $\boldsymbol{\delta q}_x$, and -3.42 ± 0.02 for $\| \boldsymbol{\delta q} \|$. The corresponding Hurst exponents are 0.8 ± 0.01, 0.70 ± 0.01, and 0.71 ± 0.01 for $\boldsymbol{\delta q}_y$, $\boldsymbol{\delta q}_x$, and $\boldsymbol{\delta q}$, respectively. This means that the self-affinity of the local aperture field is transmitted to the component of $\boldsymbol{\delta q}$ transverse to the pressure head (same exponent 0.8 ± 0.01). Long-range correlations are also transmitted to the component parallel to macroscopic flow, and subsequently to the amplitude of $\boldsymbol{\delta q}$. However, the resulting correlations are weaker. This is due to the contribution of the defect region mentioned above.

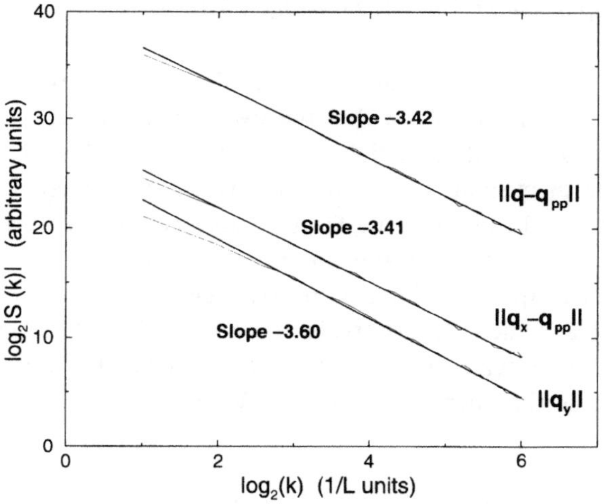

Figure 6

Average over all possible directions of two-dimensional Fourier spectra for the magnitudes of vector $\boldsymbol{\delta q} = \boldsymbol{q} - \boldsymbol{q}_{pp}$ and its components along x and y. Measured slopes correspond to Hurst exponents of 0.80 ± 0.01, 0.70 ± 0.01, and 0.71 ± 0.01 for $\|\boldsymbol{\delta q}_y\|$, $\|\boldsymbol{\delta q}_x\|$ and $\|\boldsymbol{\delta q}\|$, respectively. The component transverse to flow exhibits long-range correlations identical to those found in the local apertures ($\zeta = 0.8$). The component parallel to flow, and subsequently, the vector $\boldsymbol{\delta q}$ itself, are anisotropic. The averaging over all directions provides a power law spectrum that indicates long-range correlations. However these correlations are weaker than those found along local apertures.

4.4. Discussion

We see that the effect of roughness on flow results directly from its self-affinity. The long-range correlations existing in the rough topographies at all scales including fracture scale also exist in the local aperture field. A consequence of this is the existence of a network of large aperture correlated paths across the entire fracture. Depending on the orientation of the macroscopic flow with respect to these channels, channeling occurs inside the fracture in a more or less favorable way. Channeling can be understood as the result of the correlations in the geometric boundary conditions imposed to flow being transmitted to the flow itself. As there is no horizontal characteristic scale smaller than the fracture size for describing the geometry of the fracture, there is no such scale for describing the flow inside the fracture. An important consequence of this is that there is no homogenization scale suitable for the definition of a permeability tensor inside a rough fracture with self-affine local apertures. The hydraulic behavior is controlled by fluctuations of local apertures at all scales up to the fracture scale. Moreover, large scale fluctuations have the largest contribution to it.

What happens then if the upper limit to the local apertures self-affinity is not the fracture size L, but a scale smaller than L? This condition situation occurs for real fractures when the two facing topographies are matched at large scale. This is the topic of the next section.

5. *Influence of Surface Matching on Hydraulic Properties*

Experimental studies carried out on natural fracture surfaces (Brown, 1986; Glover *et al.*, 1998b) have shown that the two surfaces composing a fracture are often matched at large scales, in the sense that they display identical large scale fluctuations, while their small scales fluctuations are uncorrelated with each other. As a consequence, and although the fracture surfaces display a Fourier spectrum with a power-law behavior over the full range of length scales extending to the fracture size L, local apertures have a Fourier spectrum that is a power law (i.e., self-affine) up to a critical length scale, $L_c < L$, and that flattens out at large scales. This mismatch scale, L_c, is the only characteristic scale smaller than L available to describe the aperture geometry. We expect it to have a strong impact on the flow.

5.1. *Introducing Surface Matching into the Model*

Brown (1995) recently proposed a simple model for describing rough fractures with matched topographies: the Fourier spectrum of the local apertures is self-affine at scales smaller than a mismatch crossover scale, L_c, and constant at larger scales. Glover *et al.* (1998a) proposed a model with a continuous change from unmatched rough topographies at scales smaller than $L_c/2$ to matched topographies at scales larger than L_c.

In our model, the topographies h_i are uncorrelated at wavelengths smaller than L_c, and display identical fluctuations for wavelengths larger than L_c. Suitable partly-matched topographies are generated in the following way. Two independent rough surfaces of identical vertical extent, h_u and h_l, are first generated (see Section 2.1.1). Distribution h_u is kept unchanged. All Fourier modes of h_l that correspond to wave lengths larger than L_c are then equaled to the corresponding modes of h_u. The resulting local aperture field exhibits a Fourier spectrum similar to that presented by Brown (1995), as shown in Figure 3 for a fracture with a mismatch scale $L_c = L/4$. In Figure 7, we show an example of two sets of profiles extracted from the distributions $h_i(x, y)$ and from the corresponding local apertures $d(x, y)$. In Figure 7a, the surfaces are unmatched up to the fracture size, L, while in Figure 7b a mismatch scale exists: $L_c = L/2$. In the former case, the profiles extracted from the surfaces display fluctuations uncorrelated with each other at all scales. In the latter case, the large scale fluctuations of the profiles are clearly similar.

5.2. *Surface Matching and Lubrication Approximation*

Since the upper limit for the self-affine behavior of the local apertures is L_c, the typical slope along local apertures computed using the smallest scale in the model, l_c, scales as

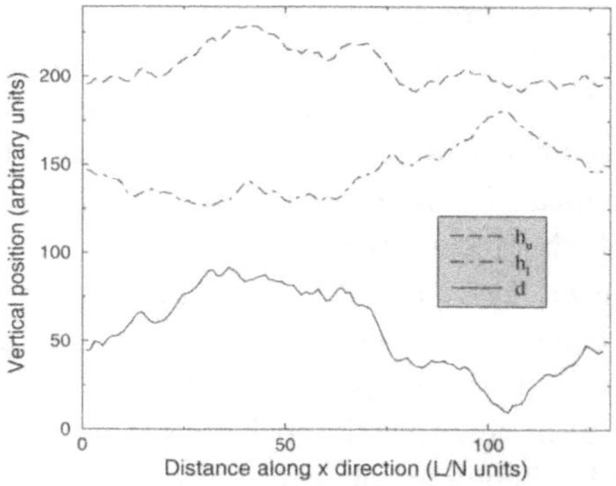

(a)

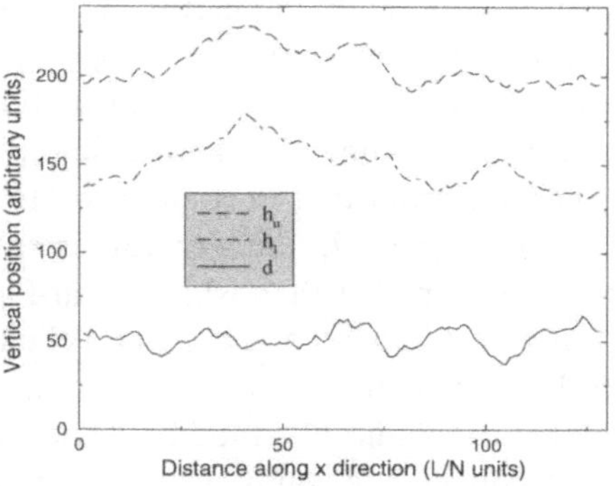

(b)

Figure 7

(a) Profiles extracted from two unmatched topographies h_u and h_l ($N = 128$), and from the resulting local aperture distribution $d = h_u - h_l$; the vertical extent of the latter field is significantly larger than those of h_u and h_l — **(b)** The rough surfaces are matched for wave lengths larger than $L/2$; the profiles extracted from them have identical large-scale fluctuations. Consequently, d has minor fluctuations at the fracture scale, and its vertical extent is significantly reduced in comparison to (a).

$$s(l_c) \sim l_c^{\zeta-1} \frac{\Gamma}{L_c^{\zeta}} , \tag{8}$$

where ζ is the roughness (Hurst) exponent, L_c is the mismatch scale, and Γ is the root-mean-square of the local apertures. Relation (8) is identical to relation (6), with length scale L_c replacing fracture size L.

Surface matching also implies that Γ scales as $\Gamma \sim L_c^{\zeta}$. The root-mean-square of the surface topographies scales as $\sigma \sim L^{\zeta}$. They are related through the scaling relation

$$\Gamma \sim \sigma \left(\frac{L_c}{L} \right)^{\zeta} . \tag{9}$$

Combining Equations (8) and (9) yields Equation (7). For a given fracture, i.e., fixed σ, L and L_c, the lubrication approximation imposes an upper limit value to the grid size N. This limit value is independent of possible matching between the facing surfaces.

5.3. *Influence of Surface Matching on the Hydraulic Behavior of a Fracture*

In this part, we study how surface matching influences the hydraulic behavior of the fracture. For this, we vary the ratio L/L_c, while keeping the same numerical seed for the generation of the rough topographies, and the same root-mean-square for their height distributions. We call family of fractures a set of fractures corresponding to the same numerical seed, and that differ from each other only by the value for the mismatch scale, L_c.

At fixed normalized aperture d_m/σ, as L_c is lowered, two effects are expected to affect the flow between the rough surfaces. First, a systematic decrease in the amplitude of the local aperture fluctuations, Γ, is expected, following scaling law (9). Thus, the protruding of roughness into the void between the rough topographies is to become less important in comparison to the mean fracture separation. This effect is visible in Figure 7: the vertical extent of the local apertures in Figure 7b is significantly smaller than what it is in Figure 7a. This "Γ effect" should always tend to bring the fracture back to a parallel plate behavior with decreasing L_c. Second, the vanishing of spatially correlated fluctuations of the local apertures at large scales is expected to significantly modify the network of preferential channels for the liquid to flow through. We study here the hydraulic behavior of two fracture families generated on a 2048×2048 grid, with respectively a strong flow-inhibiting (family A) and a slight flow-enhancing (family B) behavior observed when $L = L_c$.

The hydraulic aperture of the fractures in family A is investigated in Figure 8. The evolution of the ratio d_h/d_m is shown as a function of d_m/σ for various values of the ratio L/L_c (1/16, 1/8, 1/4, 3/8, 1/2, 3/4, 1), in Figure 8a, and as a function of L/L_c for various normalized apertures (from 3.0 up to 9.0 by steps of 1.0) in Figure 8b. The hydraulic behavior of a given fracture (i.e., at fixed mismatch scale) when varying fracture separation is similar to those shown in Figure 4. At fixed mechanical aperture, the deviation from the cubic law (represented by the horizontal dot-dashed line $d_h/d_m = 1$) decreases monotonously as L_c is decreased. This drop is dramatic between $L_c = L$ and $L_c = 3L/4$, indicating that the loss of correlated fluctuations of $d(x, y)$ at the fracture size has significantly diminished the strong flow-inhibiting

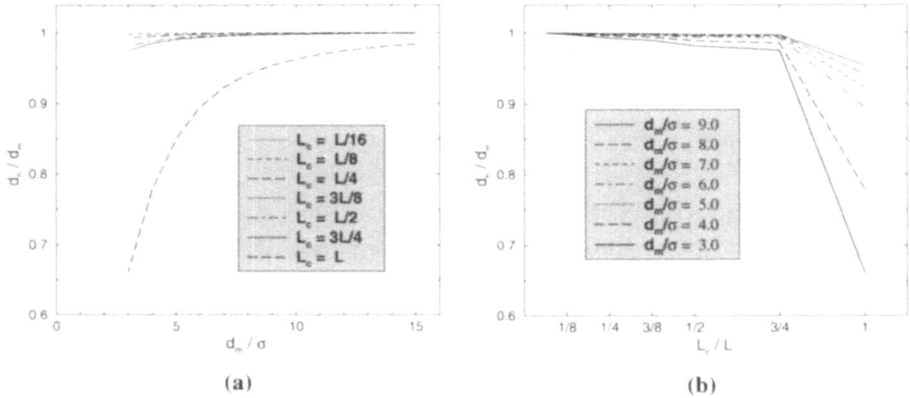

Figure 8

Evolution of the ratio of the hydraulic aperture to the mechanical aperture for fracture family A, **(a)** as a function of the normalized mechanical aperture d_m/σ and for various values of the mismatch scale, L_c, and **(b)** as a function of the ratio L_c/L, for various fractures' separations. The deviation from cubic law drops dramatically when local aperture fluctuations do not exist at the fracture scale. The fracture family exhibits a strong flow-inhibiting behavior for $L_c = L$.

behavior of the fracture. Here, the modifications of the preferential channels inside the fracture, when decreasing the mismatch scale, goes along the systematic "Γ effect," and results in a rapid annihilation of the deviation from the parallel plate behavior.

Plots respectively identical to Figures 8a and 8b are shown in Figures 9a and 9b, for fracture family B. For $L_c = L$, the fracture exhibits a flow-enhancing behavior. This type of peculiar configuration that allows the permeability to be larger than that of a parallel plate with identical opening has been observed in Figure 2b (see also Fig. 4). It was observed experimentally by MÉHEUST and SCHMITTBUHL (2000), and explained in terms of favorable flow channeling in MÉHEUST and SCHMITTBUHL (2001). The latter work demonstrated that it is not the most frequent configuration when considering a statistics of independent fractures with unmatched facing surfaces, but the proportion of such configurations is far from being negligible. We see here that the flow-enhancing configuration holds no more for $L_c = 3L/4$, i.e., when large-scale fluctuations of the local apertures are set to 0: the fracture is flow-inhibiting for $L_c \leq 3L/4$. Though the "Γ effect" tends to make d_h/d_m evolve monotonously to 1 with decreasing mismatch scale, the change in the flow preferential paths when moving from $L_c = L$ to $L_c = 3L/4$ makes the deviation from the parallel plate model change sign and increase in absolute value. This fracture family exhibits a competition between the influence of the decreasing Γ and the more subtle effect related to the change in the main flow channels through the fracture as L_c is decreased.

Families A and B exhibit two very different dependences of their permeability on the mismatch scale. We will now investigate what is the general behavior of a fracture family from the statistical point of view, considering a large number of independent fracture families.

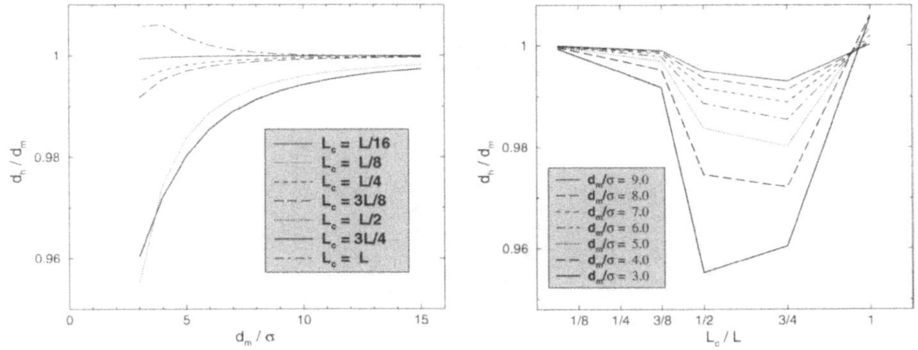

(a) (b)

Figure 9

Same as Figure 8, but for a fracture family with a flow-enhancing behavior for $L_c = L$. Here, the configuration favorable to flow is destroyed when fluctuations at scale L disappear from the local aperture distribution. For $L_c \leq 3L/4$, the fracture is flow-inhibiting.

5.4. *Influence of Surface Matching on the Statistical Hydraulic Behavior*

We have computed the hydraulic aperture of 500 independent fractures with the same mismatch scale, generated on 256 × 256 grids. This has been done for various normalized fracture separations ($d_m/\sigma = 4.0, 5.0, 6.0, 8.0, 10.0$), and for various values of the ratio L_c/L: 1/4, 3/8, 1/2, 3/4, 1. Figure 10 shows the histograms of the values for the ratio of the hydraulic aperture to the mechanical aperture, d_h/d_m, at normalized fracture separation 4.0. Each histogram corresponds to a value for the mismatch scale. The histograms narrow as the mismatch scale is decreased from its maximum value ($L_c = L$) to $L_c = 3L/8$. At the same time, their mean value converges to 1 (parallel plate estimate). This is a confirmation of the role played by long-range correlations of the fracture apertures in the existence of flow channeling and of the subsequent dispersion of the hydraulic behavior over a large statistics of independent fractures.

For each of the histograms computed at a given value of the fracture separation and for a given mismatch length scale, the mean behavior and the fluctuations of the statistics (i.e., the root-mean-square of the histogram) have been computed. They are shown respectively in Figures 11a and 11b, as functions of the ratio L_c/L and for different values of the fracture mean aperture.

At fixed mechanical aperture, the mean behavior converges to the parallel plate estimate when the mismatch scale is decreased. On average, a rough fracture can exhibit a deviation of its hydraulic aperture from its mechanical aperture as large as

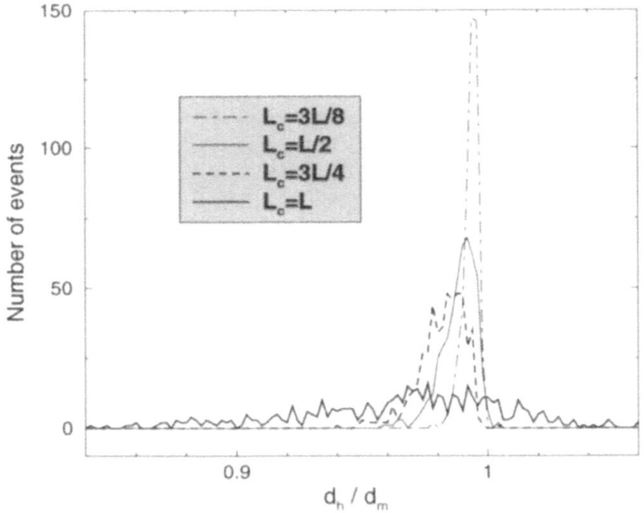

Figure 10
Histograms of the ratio d_h/d_m obtained for 500 independent fractures generated on 256 × 256 grids. Each histogram corresponds to a different value of the fracture mismatch scale, L_c. The histograms broaden as L_c is increased up to the fracture size, L.

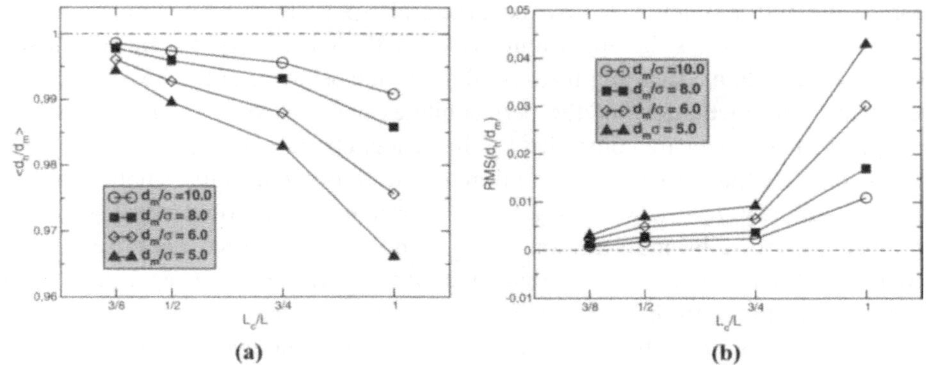

Figure 11
Evolution **(a)** of the mean ratio of the hydraulic aperture to the mechanical aperture and **(b)** of the fluctuations of the statistics for 500 independent fractures generated on a grid 256 × 256, as functions of the ratio of the mismatch scale to the fracture size. The computation has been carried out for different values of the normalized fracture separation: $d_h/d_m = 10.0, 8.0, 6.0, 5.0$. The mean behavior converges to the parallel plate estimate (corresponding to the dot-dashed horizontal line in (a)) when L_c decreases. Fluctuations of the statistics on the scale L_c drop dramatically for $L_c < L$. This is due to the disappearance of large aperture correlated channels at the fracture scale.

4% at $d_m/\sigma = 4.0$ (which in terms of volumetric flow means a deviation larger than 10%), if the rough surfaces are unmatched. If the mismatch scale is $3L/4$, the deviation drops by a half. At the same time, the fluctuations of the statistics decrease. The drop in the root-mean-square of the results over the statistics is especially dramatic when L_c is changed from L to $3L/4$: it represents 80% of the value at $L_c = L$. Thus, the impact of correlated fluctuations of the local apertures at the fracture scale is very important. They are responsible for the existence of prefential large aperture channels, or similarly the existence of small aperture barriers across the whole fracture. They therefore allow a broad variety of hydraulic behaviors over a large population.

For a given mismatch scale, the deviation of the mean behavior from the parallel plate estimate is negative and converges to 1 as fracture separation increases (see Fig. 11a). This means that a majority of rough fractures are flow-inhibiting, and that flow-inhibition vanishes when the roughness amplitude becomes negligible in comparison to the vertical extent of the void between the rough surfaces. At the same time, the dispersion of the behaviors observed over the statistics decreases to 0 for decreasing fracture separations. This result was previously observed by MÉHEUST and SCHMITTBUHL (2000) in the case $L_c = L$.

5.5. *Discussion*

The Stokes flow (low Reynolds number) through a fracture with matched surfaces appears to be highly dependent on the mismatch scale, L_c. Flow channeling inside the fracture is controlled by the large-scale spatially correlated fluctuations of the local apertures. As the upper limit scale for the self-affinity, the mismatch scale controls channeling effects and the related hydraulic behavior at the fracture scale. In particular, the fluctuations of the permeabilities observed over a large statistics of independent fractures are controlled by the mismatch scale. An important point lies within the fact that as soon as the mismatch scale is significantly smaller than the fracture size, say, $L_c < L/8$, the fracture's hydraulic aperture does not deviate from the mean aperture by more than a few percent, for normalized apertures d_m/σ as small as 4.0. However, surface matching entails that the fracture separation can be decreased all the more before surface contact as the mismatch scale is smaller. By going close to surface contact, significant deviations of the hydraulic behavior from the cubic law are observed, as for fractures with no surface matching. Exclusively, these situations occur for mechanical apertures that are smaller than those required for surface contact when surfaces are not matched. In other words, the relevant parameter for studying fracture closure and its hydraulic consequences is the vertical extent of the local apertures, Γ, rather than that of the rough topographies, though the latter are easier to determine from an experimental point of view.

6. *Influence of an Arbitrary Pressure Drop Orientation*

6.1. *Introduction to Hydraulic Anisotropy*

In a previous paper (MÉHEUST and SCHMITTBUHL, 2001), we had shown how it is possible to rotate by any angle α the macroscopic pressure gradient applied to the fracture, considering a fracture with periodic boundary conditions. We computed a hydraulic aperture for the fracture from the volumetric flow through a given section of horizontal length L and perpendicular to the macroscopic flow. We had chosen a particular section $\mathscr{S}^{(\alpha)}$, the horizontal trace of which is centered on the center of the grid (see Fig. 12). We obtained a value for the hydraulic aperture of the fracture, $d_h(\alpha)$, that was dependent on the pressure head orientation, thus displaying an anisotropy of the fracture permeability. An example of such a dependence, borrowed from (MÉHEUST and SCHMITTBUHL, 2001) and addressing a fracture with fully-unmatched topographies, is given in Figure 13 (plain line). Due to the periodicity of the model along the directions x and y, and not along the direction defined by α (for $\alpha \neq 0$ and $\alpha \neq 90°$), the resulting dependence of the hydraulic aperture on the orientation of the pressure head with respect to the fracture was related to the particular section chosen for the calculation. Choosing another section of length L and transverse to flow, like that denoted $\mathscr{S}^{(\alpha)'}$ in Figure 12, would have provided a different evolution of d_h with α (MÉHEUST and SCHMITTBUHL, 2001).

Defining a hydraulic aperture in the direction α from the computation of the flow through a section transverse to macroscopic flow and of infinite horizontal length,

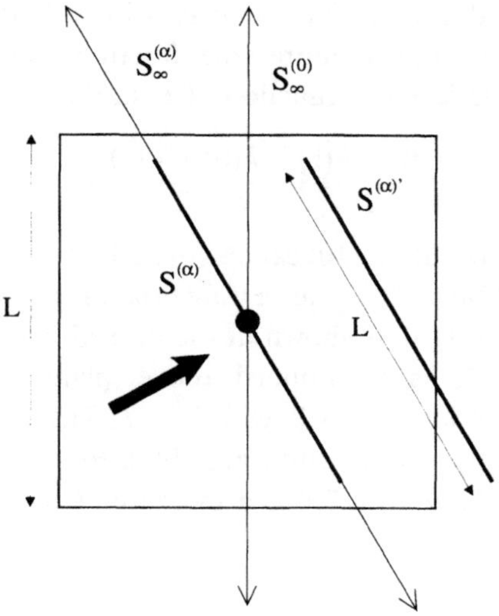

Figure 12
The hydraulic aperture of the fracture in the direction α can be defined from the flux through a section transverse to macroscopic flow, of horizontal length L, like $\mathscr{S}^{(\alpha)}$ or $\mathscr{S}^{(\alpha)'}$, or from that through a section transverse to macroscopic flow and of infinite horizontal length, like $\mathscr{S}_\infty^{(\alpha)}$. The former definition provides results that are dependent on the section chosen, the latter yields results that are independent of the section. The large arrows denotes the direction of macroscopic flow.

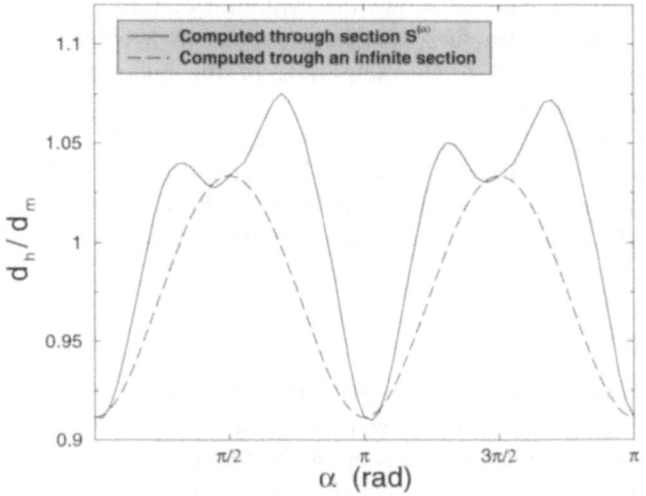

Figure 13
Evolution of the ratio d_h/d_m for a fracture with unmatched surfaces, and for $d_m/\sigma = 3.5$. The plain line curve is issued from (MÉHEUST and SCHMITTBUHL, 2001), and corresponds to a hydraulic aperture computed from the flux through section $\mathscr{S}^{(\alpha)}$ (see Fig. 12). The long-dashed line is the curve for a hydraulic aperture computed from the volumetric flow through the section of infinite horizontal trace, $\mathscr{S}_\infty^{(\alpha)}$, according to Equation (10).

$\mathscr{S}_\infty^{(\alpha)}$ (see Fig. 12), provides an evolution of d_h with α that is independent of the particular section chosen. Furthermore, the fracture transmittivity, $k = d_h^3/12\eta$, exhibits a sinusoidal evolution as a function of α, in the form

$$k(\alpha) = \frac{k(0) + k(\frac{\pi}{2})}{2} + \frac{k(0) - k(\frac{\pi}{2})}{2} \cos(2\alpha) \ . \tag{10}$$

A mathematical proof of equation (10) can be found in Appendix B. In Figure 13, the evolution of $d_h/d_m = (12\eta k)^{(1/3)}/d_m$ as a function of α, with $k(\alpha)$ evolving as a function of α according to (10), is shown in the dashed line. It can be compared to that computed using $\mathscr{S}^{(\alpha)}$, as mentionned above (plain line). The corresponding surface is generated on a grid 512×512, with $L_c = L$. The normalized mean aperture is $d_m/\sigma = 3.5$. Extreme values of the curve are obtained for $\alpha = 0$ and $\alpha = \pi/2$. Their difference, $|(d_h/d_m)(0) - (d_h/d_m)(\pi/2)|$, is a measure of the hydraulic anisotropy of the fracture.

6.2. *Effect of Surface Matching*

The hydraulic anisotropy of a rough fracture with unmatched topographies results from the correlations in the aperture distribution, as does the dispersion of the hydraulic behavior over a large statistics of independent fractures (MÉHEUST and SCHMITTBUHL, 2001). For a fracture with partly matched surfaces, we expect the amplitude of the sinusoidal fluctuations in (10) to be strongly dependent on the mismatch scale. Figure 14 shows the evolution of $|(d_h/d_m)(0) - (d_h/d_m)(\pi/2)|$ with L_c/L, for the two fracture families presented in Section 5.3, at normalized fracture opening $d_m/\sigma = 4.0$. The amplitude of the fluctuations with varying α decreases with decreasing mismatch scale L_c. They are significantly smaller if fluctuations of the local apertures do not exist at the fracture scale, i.e., as soon as $L_c < L$, for both fractures. Since it is related to channeling too, fracture anisotropy is dependent on the mismatch scale in the same way as the root-mean-square of d_h/d_m computed over a large number of independent fractures (see Section 5.4).

6.3. *Discussion*

The dependence of the permeability of a fracture on the orientation of the applied pressure drop is a consequence of flow channeling. It results from the orientation of the large aperture correlated channels inside the fracture with respect to the imposed macroscopic pressure gradient. Consequently, it is controlled by the upper limit scale for the aperture field self-affinity, L_c. A fracture with partly matched surfaces and a mismatch scale that is significantly smaller than the fracture size ($L_c < L/4$) displays little anisotropy: $|(d_h/d_m)(0) - (d_h/d_m)(\pi/2)| < 1$. This is consistent with the fact that under these conditions, the variety of possible behaviors for a fracture is quite narrow (see Section 5).

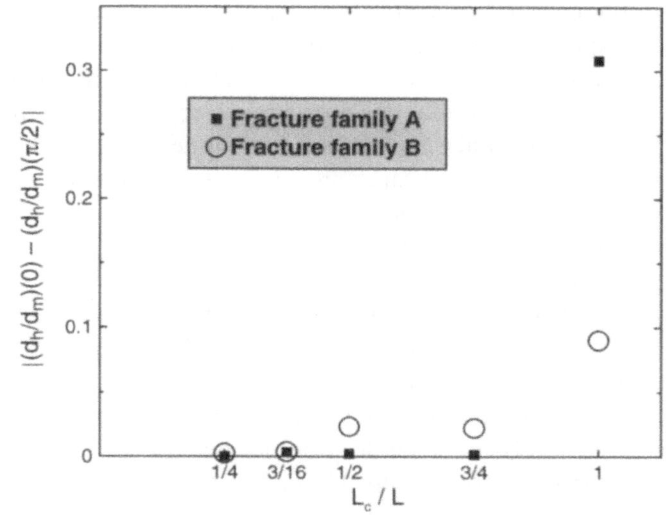

Figure 14
Amplitudes of the theoretical fluctuations of the ratio of the hydraulic aperture to the mechanical aperture when rotating the pressure gradient by 2π, for the two fracture families presented in Section 5.3, labeled A and B as previously, and for $d_m/\sigma = 4.0$. In both cases, the anisotropy is greatly reduced as soon as the mismatch length scale is smaller than the fracture size.

7. Conclusions

In this study we have simulated the flow through a rough fracture, using the Reynolds equation. The numerical model for the rough fracture consists of two rough isotropic self-affine topographies of identical vertical extent. It allows moderate plastic contact between them and possible surface matching at large scales.

We show that the self-affinity of the rough surfaces results in a self-affine behavior of the local apertures, with an identical roughness exponent, up to the mismatch scale only. The latter length scale is the only characteristic scale other than the fracture size which describes the geometry of the fracture. It also appears to be the only scale that controls the behavior of the flow inside the fracture. Indeed, long-range correlations in the local apertures result in long-range correlations in the local fluxes. Flow channeling is the consequence of this in terms of spatial distribution of the flow in the fracture plane. It is controlled by the correlated fluctuations of the local aperture field, $d(x, y)$, at the largest scale of the self-affine description for d, that is, at mismatch scale L_c. The consequences of channeling in terms of hydraulic behavior at the fractures scale, such as the anisotropy of a given fracture toward pressure head orientation, or the dispersion of the behaviors of a large number of independent fractures with identical statistical description (same mismatch scale, same vertical extent for roughness, same roughness exponent), are therefore controlled by the mismatch length scale. When trying to estimate the deviation of the fracture permeability from the parallel plate estimate, the mean opening has to be compared to the vertical extent of roughness at the scale L_c, and not at the fracture scale. In other words, the vertical extent of the local apertures at the fracture scale is the right parameter to compare to the fracture separation, not the vertical extent of fracture walls' roughness at the fracture scale.

The latter conclusions hold for a slow viscous flow inside the fracture, i.e., a flow that is only dependent on the local apertures. In particular, for such a flow, two rough topographies that are perfectly matched at all scales are perfectly equivalent to a configuration with two flat planes and the same surface separation. This is no longer true for flows at higher Reynolds number. For such flows, the role of the mismatch scale is probably more subtle. Another hypothesis of our model is that moderate contact occurs through pure plasticity. A study with a realistic elastic redistribution of contacts is in preparation.

Acknowledgements

Yves Méheust acknowledges fruitful discussions with Y. Bernabé, K. Evans, F. Lanaro, S.-D. Lee, O. Stephansson, T.-F. Wong, and R. W. Zimmerman. Yves Guéguen is acknowledged for his advice during manuscript preparation. The authors thank H.-J. Kümpel for organizing the third Euroconference on Rock Physics and Rock Mechanics. G. Zimmermann and the other (anonymous) reviewer are acknowledged for enhancing the manuscript.

Appendix A

Self-affinity of the Local Apertures Between Two Self-affine Topographies

The fluctuations of the height distributions for the top and bottom surfaces of the fracture around their mean altitude, $h_u(x, y)$ and $h_l(x, y)$ respectively, are two self-affine fractals with identical Hurst exponents. They are uncorrelated with each other. This means that the modula of their respective Fourier transforms $\tilde{h}_u(k_x, k_y)$ and $\tilde{h}_l(k_x, k_y)$ display the same stochastic power law behavior, while their respective phases, θ_u and θ_l, are random and mutually uncorrelated:

$$\tilde{h}_u(k_x, k_y) = \left(k_x^2 + k_y^2\right)^{-\frac{1+\zeta}{2}} M_u(k_x, k_y)\, e^{i\theta_u(k_x, k_y)} \tag{A1}$$

and

$$\tilde{h}_l(k_x, k_y) = \left(k_x^2 + k_y^2\right)^{-\frac{1+\zeta}{2}} M_l(k_x, k_y)\, e^{i\theta_l(k_x, k_y)} \quad , \tag{A2}$$

where $M_u(k_x, k_y)$ and $M_l(k_x, k_y)$ are stochastic real functions following the same Gaussian distribution of mean 1. In what follows, we omit the dependence of M_u, M_l, θ_u, and θ_l on k_x and k_y. The local aperture field $d(x, y)$ is defined by Equation (3). Its mode at $\boldsymbol{k} = \boldsymbol{0}$ is the fracture mean aperture d_m. Due to the linearity of the Fourier transform, its spectral modes for $k_x > 0$ and $k_y > 0$ are

$$\tilde{d}(k_x, k_y) = \left(k_x^2 + k_y^2\right)^{-\frac{1+\zeta}{2}} \left[M_u\, e^{i\theta_u} - M_l\, e^{i\theta_l}\right] \ . \tag{A3}$$

Consequently, the aperture field Fourier spectrum can be written as

$$|\tilde{d}(k_x, k_y)|^2 = \left(k_x^2 + k_y^2\right)^{-(1+\zeta)} M(k_x, k_y) \ , \tag{A4}$$

where

$$M = M_u^2 + M_l^2 - 2M_u M_l \cos[\theta_u - \theta_l] \tag{A5}$$

is a stochastic function of mean 1.

The resulting stochastic behavior of the local apertures spectrum is a power law of $\|\boldsymbol{k}\|$ with an exponent $-2(1+\zeta)$, characteristic of a self-affine behavior with the same Hurst exponent ζ as that displayed by the height distributions of the surfaces.

Appendix B

Sinusoidal Dependence of d_h^3 on the Orientation of $\overline{\nabla P}$

We call $\boldsymbol{q}_\alpha$ the local flux field for an orientation α of the pressure head, and $d_h(\alpha)$ the corresponding hydraulic aperture. Hydraulic aperture $d_h(\alpha)$ is defined from the flux per unit lateral length of $\boldsymbol{q}_\alpha$ through an infinite section transverse to macroscopic flow. Due to mass conservation, the latter flux is independent of the particular section chosen. We use the section that contains the center of the grid, $\mathscr{S}_\infty^{(\alpha)}$ (see Fig. 12). We denote $k(\alpha)$ the fracture transmittivity in the direction α: $k(\alpha) = d_h^3(\alpha)/12\eta$, and $\left.\Phi_\alpha\right]_{\mathscr{S}}$ the flux per unit length of local fluxes $\boldsymbol{q}_\alpha$ through a given section, $\mathscr{S}$. As a consequence of the linearity of Equation (5), the pressure and 2-D velocity fields obtained for an orientation α of $\overline{\nabla P}$ from the solution of (5) are a linear combination of those obtained for $\alpha = 0$ and $\alpha = \pi/2$:

$$\boldsymbol{q}_\alpha = \boldsymbol{q}_0 \cos(\alpha) + \boldsymbol{q}_{\frac{\pi}{2}} \sin(\alpha) \ . \tag{B1}$$

Mass conservation imposes that the flux of $\boldsymbol{q}_0$ through $\mathscr{S}_{\infty}^{(\alpha)}$ be the same as that through section $\mathscr{S}_{\infty}^{(0)}$ (see Fig. 12). Thus

$$\Phi_0\big]_{\mathscr{S}_{\infty}^{(\alpha)}} = \Phi_0\big]_{\mathscr{S}_{\infty}^{(0)}} \cos\alpha = \|\nabla P\| k(0) \cos\alpha \ . \tag{B2}$$

The volumetric flow of $\boldsymbol{q}_{\frac{\pi}{2}}$ per unit length through $\mathscr{S}_{\infty}^{(\alpha)}$ can be expressed in the same way as

$$\Phi_{\frac{\pi}{2}}\big]_{\mathscr{S}_{\infty}^{(\alpha)}} = \Phi_{\frac{\pi}{2}}\big]_{\mathscr{S}_{\infty}^{(\frac{\pi}{2})}} \sin\alpha = \|\nabla P\| k\left(\frac{\pi}{2}\right) \sin\alpha \ . \tag{B3}$$

Following (B1), the resulting flux of $\boldsymbol{q}_\alpha$ per unit length trough $\mathscr{S}_{\infty}^{(\alpha)}$ is

$$\Phi_\alpha\big]_{\mathscr{S}_{\infty}^{(\alpha)}} = \Phi_0\big]_{\mathscr{S}_{\infty}^{(\alpha)}} \cos\alpha + \Phi_{\frac{\pi}{2}}\big]_{\mathscr{S}_{\infty}^{(\alpha)}} \sin\alpha \tag{B4}$$

$$= \|\nabla P\| \left(k(0)\cos^2\alpha + k\left(\frac{\pi}{2}\right)\sin^2\alpha\right) \ . \tag{B5}$$

Since $\Phi_\alpha\big]_{\mathscr{S}_{\infty}^{(\alpha)}} = \|\nabla P\|\ k(\alpha)$, (B5) finally yields

$$k(\alpha) = k(0)\cos^2\alpha + k\left(\frac{\pi}{2}\right)\sin^2\alpha \ , \tag{B6}$$

which is equivalent to (10).

References

AMEDEI, B. and ILLANGASEKARE, T. (1994), *A Mathematical Model for Flow and Solute Transport in Non-homogeneous Rock Fractures*, Int. J. Rock. Mech. Min. Sci. Geomech. *31*, 719–731.

BEAR, J., TSANG, C. F., and DE MARSILY, G., eds., (1993) *Flow and Contaminant Transport in Fractured Rock*.

BERKOWITZ, B. and ADLER, P. M. (1998), *Stereological Analysis of Fracture Network Structure in Geological Formations*, J. Geophys. Res. *103*(B7), 15,339–15,360.

BOFFA, J., ALLAIN, C., and HULIN, J. (1998), *Experimental Analysis of Fracture Rugosity in Granular and Compact Rocks*, Eur. Phys. J. Appl. Phys. *2*(3), 281–289.

BOUCHAUD, E. (1997), *Scaling Properties of Cracks*, J. Phys. Condens. Matter *9*, 4319–4344.

BOUCHAUD, E., LAPASSET, G., and PLANÈS, J. (1990), *Fractal Dimension of Fractured Surfaces: A Universal Value?* Europhys. Lett. *13*, 73–79.

BROWN, S. R. (1986), *Correlation Between the Surfaces of Natural Rock Joints*, Geophys. Res. Lett. *13*, 1430–1433.

BROWN, S. R. (1987), *Fluid Flow Through Rock Joints: The Effect of Surface Roughness*, J. Geophys. Res. *92*, 1337–1347.

BROWN, S. R. (1989), *Transport of Fluid and Electric Current Through a Single Fracture*, J. Geophys. Res. *94*(7), 9429–9438.

BROWN, S. R. (1995), *Simple Mathematical Model of a Rough Fracture*, J. Geophys. Res. *100*(B4), 5941–5952.

BROWN, S. R. and SCHOLZ, C. H. (1985), *Broad Bandwidth Study of the Topography of Natural Rock Surfaces*, J. Geophys. Res. *90*, 12,575–12,582.

BROWN, S. R., STOCKMAN, H. W., and REEVES, S. J. (1995), *Applicability of the Reynolds Equation for Modeling Fluid Flow Between Rough Surfaces*, Geophys. Res. Lett. *22*(18), 2537–2540.

BROWN, S. R., CAPRIHAN, A., and HARDY, R. (1998), *Experimental Observation of Fluid Flow Channels in a Single Fracture*, J. Geophys. Res. *103*(B3), 5125–5132.

COOK, N. G. W. (1992), *Natural Joints in Rock: Mechanical, Hydraulic and Seismic Behavior and Properties under Normal Stress*, Int. J. Rock. Mech. Min. Sci. Geomech. Abstr. *29*(3), 198–223.

DIJK, P. E., BERKOWITZ, B., and BENDEL, P. (1999), *Investigation of Flow in Water-saturated Rock Fractures Using Nuclear Magnetic Resonance Imaging (nmri)*, Water Resour. Res. *35*(2), 347–360.

DURHAM, W. B. (1997), *Laboratory Observations of the Hydraulic Behavior of a Permeable Fracture from 3800 m Depth in the ktb Pilot Hole*, J. Geophys. Res. *102*(B8), 18,405–18,416.

DURHAM, W. B. and BONNER, B. (1994), *Self-propping and Fluid Flow in Slightly Offset Joints at High Effective Pressures*, J. Geophys. Res. *99*(B5), 9391–9399.

GLOVER, P., MATSUKI, K., HIKIMA, R., and HAYASHI, K. (1998a), *Synthetic Rough Fractures in Rocks*, J. Geophys. Res. *103*(B5), 9609–9620.

GLOVER, P. W. J., MATSUKI, K., HIKIMA, R., and HAYASHI, K. (1998b), *Fluid Flow in Synthetic Rough Fractures and Application to the Hachimantai Geothermal Hot Rock Test Site*, J. Geophys. Res. *103*(B5), 9621–9635.

IWAI, K., *Fundamental Studies of Fluid Flow Through a Single Fracture*, Ph.D. Thesis, Univ. of Calif., Berkeley (1976).

MÉHEUST, Y. and SCHMITTBUHL, J. (2000), *Flow Enhancement of a Rough Fracture*, Geophys. Res. Lett. *27*(18), 2989–2992.

MÉHEUST, Y. and SCHMITTBUHL, J. (2001), *Geometrical Heterogeneities and Permeability Anisotropy of Rough Fractures*, J. Geophys. Res. *106*(B2), 2089–2102.

MORENO, L., TSANG, Y. W., TSANG, C., HALE, F., and NERETNIEKS, I. (1988), *Flow and Tracer Transport in a Single Fracture: A Stochastic Model and its Relation to Some Field Observations*, Water Resour. Res. *24*(12), 2033–2048.

MOURZENKO, V. V., THOVERT, J.-F., and ADLER, P. (1995), *Permeability of a Single Fracture; Validity of the Reynolds Equation*, J. Phys. *II*, 5, 465–482.

MOURZENKO, V. V., GALAMAY, O., and THOVERT, J.-F. (1997), *Fracture Deformation and Influence on Permeability*, Phys. Rev. E *56*(3).

NICHOLL, M. J., RAJARAM, H., GLASS, R. J., and DETWILER, R. (1999), *Saturated Flow in a Single Fracture: Evaluation of the Reynolds Equation in Measured Aperture Fields*, Water Resourc. Res. *35*(11), 3361–3373.

PLOURABOUÉ, F., HULIN, J. P., ROUX, S., and KOPLIK, J. (1998), *Numerical Study of Geometrical Dispersion in Self-affine Rough Fractures*, Phys. Rev. E *58*(3), 3334–3346.

SAHIMI, M. (1993), *Flow Phenomena in Rocks: From Continuum Models to Fractals, Percolation, Cellular Automata, and Simulated Annealing*, Rev. Mod. Phys. *65*, 1393–1534.

SCHMITTBUHL, J., GENTIER, S., and ROUX, S. (1993), *Field Measurements of the Roughness of Fault Surfaces*, Geophys. Res. Lett. *20*(8), 639–641.

SCHMITTBUHL, J., SCHMITT, F., and SCHOLZ, C. H. (1995a), *Scaling Invariance of Crack Surfaces*, J. Geophys. Res. *100*, 5953–5973.

SCHMITTBUHL, J., VILOTTE, J. P., and ROUX, S. (1995b), *Reliability of Self-affine Measurements*, Phys. Rev. E *51*, 131–147.

SIMONSEN, I., HANSEN, A., and MAGNAR NES, O. (1998), *Using Wavelet Transforms for Hurst Exponent Determination*, Phys. Rev. E *58*, 2779.

THOMPSON, M. E. (1991), *Numerical Simulation of Solute Transport in Rough Fractures*, J. Geophys. Res. *96*(B3), 4157–4166.

THOMPSON, M. E. and BROWN, S. R. (1991), *The Effect of Anisotropic Surface Roughness on Flow and Transport in Fractures*, J. Geophys. Res. *96*(B13), 21,923–21,932.

TSANG, Y. W. (1984), *The Effect of Tortuosity on Fluid Flow Through a Single Fracture*, Water Resour. Res. *20*(9), 1209–1215.

TSANG, Y. W. and WITHERSPOON, P. A. (1981), *Hydromechanical Behavior of a Deformable Rock Fracture Subject to Normal Stress*, J. Geophys. Res. *86*(B10), 9287–9298.

UNGER, A. J. A. and MASE, C. (1993), *Numerical Study of the Hydromechanical Behavior of Two Rough Fracture Surfaces in Contact*, Water Resour. Res. *29*(7), 2101–2114.

YEO, I., FREITAS, M. D., and ZIMMERMAN, R. (1998), *Effect of Shear Displacement on the Aperture and Permeability of a Rock Fracture*, Int. J. Rock Mech. Min. Sci. *35*(8), 1051–1070.

ZIMMERMAN, R. and BODVARSSON, G. (1996), *Hydraulic Conductivity of Rock Fractures*, Transp. Porous Media *23*, 1–30.

ZIMMERMAN, R. W. and YEO, I.-W., *Fluid flow in rock fractures: from the Navier-Stokes equations to the cubic law*, in *Dynamics of Fluids in Fractured Rock* (2000).

(Received February 6, 2001, revised June 27, 2001, accepted August 1, 2001)

Pure appl. geophys. 160 (2003) 1051–1066
0033–4553/03/061051–16

Pure and Applied Geophysics

Triggering of Seismicity by Pore-pressure Perturbations: Permeability-related Signatures of the Phenomenon

S. A. Shapiro,[1] R. Patzig,[1] E. Rothert[1] and J. Rindschwentner[1]

Abstract—We consider various cases of seismicity, induced by artificial fluid injections in boreholes. Like many other authors, we support the hypothesis that to a large extent the triggering of this seismicity is caused by a diffusive process of the pore pressure relaxation in porous (or fractured), saturated rocks. We show that if this hypothesis is correct, then the spatio-temporal distributions of the seismic events must have several specific features related to the effective permeability of the rock. As a rule the fluid injection-induced seismicity obeys such features. These features can be indications of the diffusive and even hydraulic nature of the seismicity triggering process.

From this point of view we analyze the spatio-temporal distribution of the late aftershocks of the Antofagasta (northern Chile) 1995 earthquake. These aftershocks were concentrated in a plane, an approximately 3 km-thick spatial zone. This thin seismogenic layer is a part of the South American subduction zone. The time-distance distribution of the aftershocks along this layer indicates that they could be triggered by a diffusion-like process. Possibly, such a process is the relaxation of the pressure perturbation caused in the pore fluid by the main Antofagasta event. We estimated the permeability required to explain the spatio-temporal distribution of the aftershocks by such a triggering mechanism. The obtained value, 60 mD, is very large. However, it is realistic for a long-time existing and large-scale fault zone.

Key words: Diffusivity, permeability, aftershocks, seismicity, subduction.

1. Introduction

Fluid injections in boreholes as well as perturbations of the pore pressure of a natural origin like rainfall (see e.g., Ventura and Vilardo, 1999), or water-level fluctuations in artificial lakes (see e.g., Ferreira *et al.*, 1995) can generate a seismicity. Possible hypothetical mechanisms of the induced microseismicity triggering which are under discussion now in seismological literature (see e.g., Maillot *et al.*, 1999 and references therein) are for example the hydraulic fracturing, the redistribution of elastic stresses, the pore pressure relaxation or any combination of these mechanisms.

[1] Freie Universität Berlin, Fachrichtung Geophysik, Malteserstr. 74-100, 12249 Berlin, Germany. E-mail: shapiro@geophysik.fu-berlin.de, rip@geophysik.fu-berlin.de, rothert@geophysik.fu-berlin.de, janr@geophysik.fu-berlin.de

The question arises, which of these mechanisms under which conditions is the dominant one, and what are its specific signatures. We support the hypothesis that typically the diffusive process of the pore pressure relaxation in porous (or fractured) saturated rocks is a dominant mechanism of triggering of the fluid injection induced microseismicity. In some locations in the crust, rocks are close to a critical state of a failure equilibrium so that small perturbations of the pore pressure modifying the effective normal stress can lead to seismic events (see NUR and BOOKER, 1972; FLETCHER and SYKES, 1977; OHTAKE, 1974; TALWANI and ACREE, 1985; ZOBACK and HARJES, 1997). In this paper we discuss specific features of this mechanism.

If the above hypothesis is valid, then, at least in the first approximation, the behavior of the seismicity triggering in space and in time must be controlled by the process of the relaxation of the pore pressure perturbation $p(t, \mathbf{r})$ initially created at the injection source. For such a very simplified description, the differential equation of diffusion can be applied:

$$\frac{\partial p(t, \mathbf{r})}{\partial t} = D\nabla^2 p(t, \mathbf{r}) \ , \tag{1}$$

where D is the hydraulic diffusivity, t is the time and $\mathbf{r}$ is the radius vector of the observation point. This equation corresponds to the second-type P wave, the so-called slow- or Biot-wave (BIOT, 1962; DUTTA and ODE, 1979) in the low-frequency limit. Equation (1) describes linear relaxation of pore-pressure perturbations in a homogeneous, isotropic, poroelastic, fluid-saturated medium. This linear equation was implicitly or explicitly used in many works on hydraulically-induced seismicity (see, for example, all papers cited above).

Recently, SHAPIRO *et al.* (1997–2000) and AUDIGANE (2000) proposed describing the spatio-temporal evolution of clouds of hydraulically-induced seismic events in terms of pore pressure relaxation in anisotropic and heterogeneous poroelastic media (see also the discussion CORNET, 2000 and SHAPIRO *et al.*, 2000). For this a more general form of equation (1) is applied:

$$\frac{\partial p}{\partial t} = \frac{\partial}{\partial x_i}\left[D_{ij}\frac{\partial}{\partial x_j}p\right] \ , \tag{2}$$

where D_{ij} are components of the tensor of the hydraulic diffusivity heterogeneously distributed in the medium and x_j ($j = 1, 2, 3$) are the components of the radius vector.

The equations above imply several very simple features of the spatio-temporal distributions of the fluid-induced seismicity. These features will be described in the next chapter. Then we will demonstrate these features with examples of microseismicity induced by fluid injections in boreholes. Finally, we will analyze the spatio-temporal distribution of the late aftershocks of the Antofagasta (northern Chile) 1995 earthquake from the point of view of these features. We will then discuss the applicability of the hypothesis of aftershock triggering by pore-pressure perturbations to natural earthquakes.

2. Features of Hydraulically-Induced Seismicity

The permeability-related features of the hydraulically-induced seismicity can be identified in a very natural way from the notion of triggering fronts (SHAPIRO *et al.*, 1997–2000). To recall the notion of a triggering front let us approximate a real configuration of a fluid injection in a borehole by a point source of pore pressure perturbation in an infinite, heterogeneous, anisotropic, poroelastic, fluid-saturated medium. The time evolution of the pore pressure at the injection point is assumed to be a step function. It is natural to assume that the probability of the triggering of microseismic events is an increasing function of the power of the pore pressure perturbation. Thus, at a given time t_0 it is probable that events will occur at distances which are smaller or equal to the size of the relaxation zone of the pore pressure. The events are characterized by a significantly lower occurrence probability for larger distances. The spatial surface which separates these two spatial domains is called the 'triggering front'. Because the size of the relaxation zone is to some extent a heuristic parameter, we introduce a more formal quantity which is directly proportional to this size. The triggering front at the time t_0 is the front of the zero phase at the time t_0 of a harmonic diffusion wave of pore pressure relaxation radiated with zero phase at the time 0 at the injection source with the frequency $2\pi/t_0$.

Let us first assume that the medium is homogeneous and isotropic. Then the following equation can be obtained for the triggering front (see SHAPIRO *et al.*, 1997, 1999):

$$r = \sqrt{4\pi D t} \ , \tag{3}$$

where t is the time of the injection start, D is a scalar hydraulic diffusivity and r is the radius of the triggering front (which is a sphere in a homogeneous isotropic medium).

It is clear that geological media are usually hydraulically heterogeneous. Equation (3) can then be considered as an equation for the triggering front in an effective isotropic homogeneous poroelastic medium with the scalar hydraulic diffusivity D.

Because seismicity is considerably more probable in the relaxation zone than at larger distances, equation (3) corresponds to the upper bound of the cloud of events in the plot of their spatio-temporal distribution (i.e., the plot of r versus t). In Figure 1a such a spatio-temporal distribution of the microseismicity is shown for the microseismic data collected in December 1983 during the hydraulic injection into crystalline rock at a depth of 3463 meters at Fenton Hill, a HDR site in the USA (for details and further references see FEHLER *et al.*, 1998). There is good agreement between the theoretical curve with $D = 0.17\,\mathrm{m}^2/\mathrm{s}$ and the data.

Such a good agreement supporting the above concept of the triggering of microseismicity can be observed in many other cases. For example, Figure 1b shows a similar plot for the Soultz-sous-Forêts experiment in 1993 in France, where about

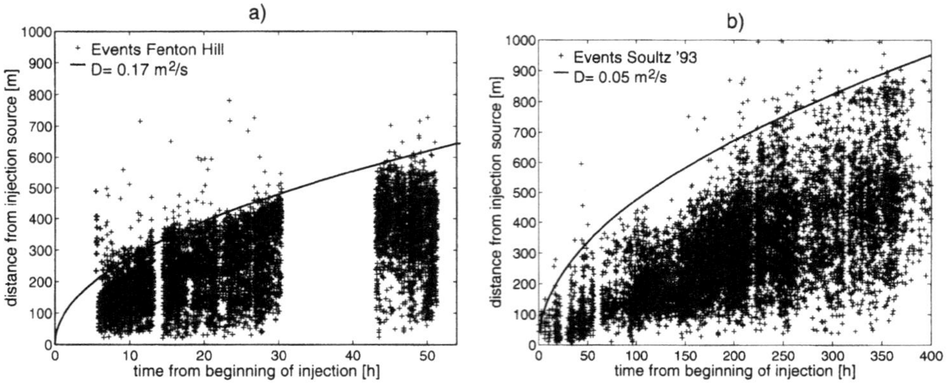

Figure 1
Distances of events from the injection source versus their occurrence time for a) the Fenton Hill experiment, 1983 and b) the Soultz-sous-Forêts experiment, 1993.

9000 events were localized during an injection (see DYER *et al.*, 1994). The diffusivity $D = 0.05\,\mathrm{m}^2/\mathrm{s}$ was observed for the seismically-active volume of the crystalline rock at the depth of 2500–3500 m.

We consider the existence of a parabolic-like envelope of the cloud of events in the plot of r versus t as an important signature of the hydraulically triggered seismicity.

Let us now assume that the components of the tensor of hydraulic diffusivity D_{ij} are homogeneously distributed in the medium. Under such an assumption we replace the complete heterogeneous, seismically-active rock volume by an effectively homogeneous, anisotropic, poroelastic, fluid-saturated medium. The permeability tensor of this effective medium is the permeability tensor of the heterogeneous rock upscaled to the characteristic size of the seismically-active region.

Performing very similar consideration as in SHAPIRO *et al.* (1997), but now using equation (2) for homogeneous but anisotropic media, the following equation for the triggering front can be obtained for anisotropic media (SHAPIRO *et al.*, 1999):

$$r = \sqrt{\frac{4\pi t}{\mathbf{n}^T \mathbf{D}^{-1} \mathbf{n}}} \ . \tag{4}$$

T denotes that the matrix (vector) is transposed, $n = \mathbf{r}/|\bar{\mathbf{r}}|$ and $\mathbf{D}^{-1}$ is the inverse of $\mathbf{D}$. Let us consider this equation in the principal coordinate system of the diffusivity tensor. Then, the matrix D_{ij} becomes diagonal (D_{11}, D_{22}, D_{33}), and we obtain:

$$x_1^2/D_{11} + x_2^2/D_{22} + x_3^2/D_{33} = 4\pi t \ . \tag{5}$$

If we scale the principal coordinate system in the following way

$$x_{sj} = x_j/\sqrt{4\pi t} \ , \tag{6}$$

then the equation of the triggering front will be an equation of an ellipsoid:

$$x_{s1}^2/D_{11} + x_{s2}^2/D_{22} + x_{s3}^2/D_{33} = 1 \tag{7}$$

with the half-axes equal to the square roots of the principal diffusivities. It is clear that the ellipsoid will keep its shape in an arbitrary rotated coordinate system. Therefore, if we scale the coordinates of all events by the square-root of their occurrence time using equation (6), then — by analogy with equation (3) and Figure 1 — the ellipsoid (equation (7)) will be an envelope of the cloud of events, although now in a normalized 3-D space.

Let us now return to the examples of the Fenton Hill and Soultz HDR experiments to demonstrate this signature of the diffusive nature of the triggering of events. Figure 2 on the left shows a view of the Soultz microseismicity cloud from the south. On the right, this Figure shows the same view of the cloud in the coordinate system scaled in accordance with equation (6). The same is shown in Figure 3 for the Fenton Hill data set. In both cases we observe that in the scaled coordinate systems the shapes of the microseismic clouds become more similar to ellipsoids than in the original coordinate system. This happens because in reality, locations where rocks are close to a failure equilibrium are randomly distributed in space. Therefore, even if the medium is hydraulically homogeneous, the seismicity cloud can be very irregular. By normalization according to equation (6), some irregularities will be eliminated, and the seismicity cloud will tend to an ellipsoid closely related to the diffusivity tensor, as equation (7) shows. Of course, all features considered above can be significantly disturbed in strongly heterogeneous hydraulic media. However, as the examples above demonstrate, they can be

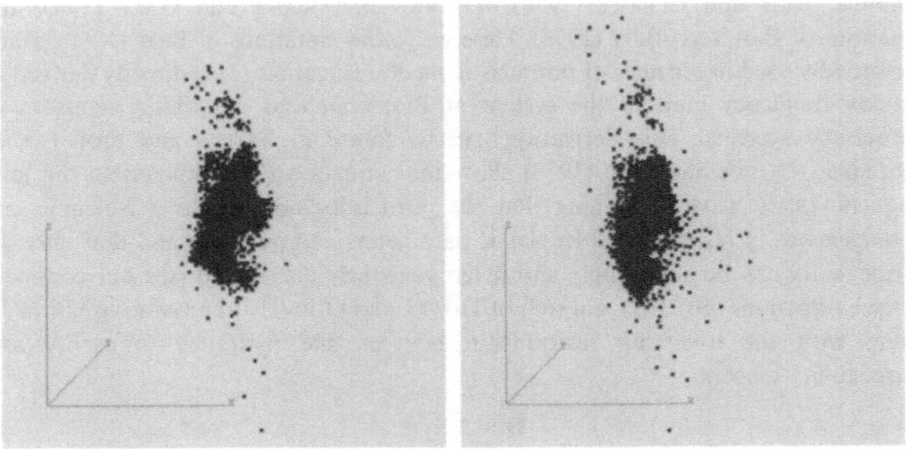

Figure 2
On the left: A view of the Soultz microseismicity cloud from the south; on the right: the same view of the cloud in the coordinate system scaled in accordance with equation (6).

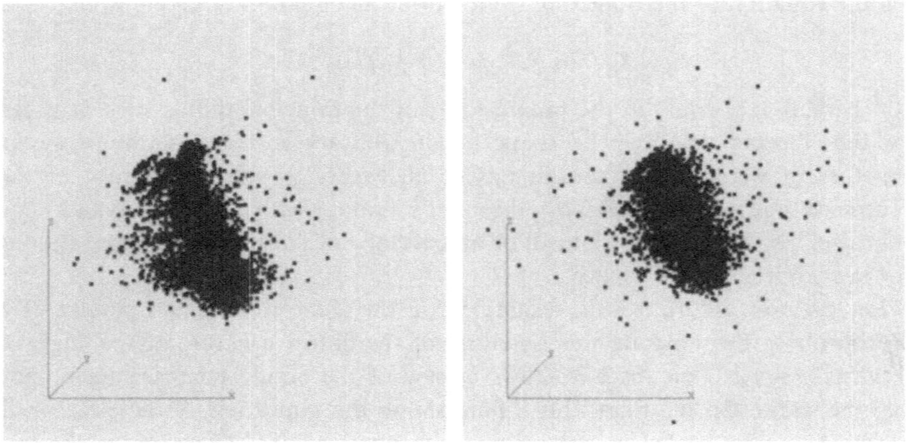

Figure 3
On the left: A view of the Fenton Hill microseismicity cloud from the south; on the right: the same view of the cloud in the coordinate system scaled in accordance with equation (6).

observed in reality. Features of hydraulically-induced seismicity considered above are controlled by the diffusivity tensor. This tensor is directly proportional to the tensor of permeability.

The relationship between the hydraulic diffusivity and the permeability is derived in the same way as equation (2). For this Darcy's law must be combined with the equation of mass conservation (see e.g., MAVKO *et al.*, 1996, Section 8.1). In such derivations different authors use different notations for analogous relationships. For example, RICE and CLEARY (1976) and VAN DER KAMP and GALE (1983) use notations of Biot, e.g., BIOT (1956). Here we follow notations of BIOT (1962) which are broadly used in seismics of poroelastic media. Equation (2) is directly derived as the low-frequency limit of the system of Biot equations describing acoustics of poroelastic systems. This derivation can be found in DUTTA and ODE (1979). Moreover, DUTTA and ODE (1979) show that equation (2) characterizes the low-frequency slow wave. Assuming that the poroelastic continuum is isotropic and homogeneous in respect to all its elastic parameters and porosity, and that only the permeability can be anisotropic and heterogeneously distributed (see our comment below), equations (60), (62) and (63) of DUTTA and ODE (1979) provide equation (2) along with the following relationship between the hydraulic diffusivity and permeability tensors:

$$\mathbf{D} = N\mathbf{K}/\eta \ , \tag{8}$$

where $\mathbf{K}$ is the permeability tensor, η is the pore-fluid dynamic viscosity and N is a poroelastic modulus defined as follows: $N = MP_d/H; M = (\phi/K_f + (\alpha - \phi)/K_g)^{-1}$; $\alpha = 1 - K_d/K_g$; $H = P_d + \alpha^2 M$; $P_d = K_d + 4/3\mu_d$. Here $K_{f,d,g}$ are bulk moduli of the fluid, dry frame and grain material respectively; μ_d is the shear modulus of the frame and ϕ is the porosity.

Note also, that we neglected here the elastic anisotropy in comparison with the anisotropy of the permeability. It is clear that in reality an anisotropy and heterogeneity of the permeability is related to an anisotropy and a heterogeneity of elastic properties of rocks, at least on a micro-scale. However, usually the elastic anisotropy has the order of 5 to 30 percent. On the other hand, the permeability has the anisotropy of the order of several hundreds percent or more. Thus, in equation (8) the elastic anisotropy can be neglected for simplicity. Further, for the case of highly porous rocks, an approximation $N \approx K_f/\phi$ can be used. For the case of low-porosity crystalline rocks N can be approximated as follows: $N = [\phi/K_f + \alpha/K_g]^{-1}$.

For the Fenton Hill experiment, accepting the following estimates used in the literature for the crystalline rocks at the depth of 3500 m: $\phi = 0.003$, $\eta = 1.9 \cdot 10^{-4}$ Pa·sec. (dynamic viscosity of salt water at 150 °C), $K_d = 49$ GPa, $K_g = 75$ GPa and $K_f = 2.2$ GPa, we obtain $N \approx 1.68 \cdot 10^{11}$ Pa, and the permeability tensor in the principle coordinate system is:

$$\mathbf{K} = \mathrm{diag}(0.2; 0.8; 1.8) \cdot 10^{-16}\,\mathrm{m}^2 \ . \tag{9}$$

Hydraulic tests in Fenton Hill show estimates of the large-scale permeability of the order of $10^{-17} - 10^{-15}$ m^2 (M. Fehler, personal communication). The microseismic experiment in Soultz mentioned above provided the following estimates (SHAPIRO *et al.*, 1999):

$$\mathbf{K} = \mathrm{diag}(0.7; 1.9; 5.2) \cdot 10^{-17}\,\mathrm{m}^2 \ . \tag{10}$$

Determinations of the apparent permeability from independent hydraulic data (see JUNG *et al.*, 1996) provided the estimate of 2.5×10^{-17} m^2.

3. Antofagasta Earthquake Aftershocks

An aftershock sequence of a large earthquake is a much more complex phenomenon than the fluid injection induced microseismicity. However, it can be of importance to analyze such sequences from the point of view of the possible diffusive nature of their triggering. For such an analysis a high precision of the hypocenter localization is required. Unfortunately, data sets including a large number of events localized with a high precision are practically nonexistent. In this respect the aftershock series of the Antofagasta 1995 earthquake is of unique character. This large earthquake ($M_w = 8.0$) occurred at 05:11 h, July 30, 1995 at the Antofagasta region in the North of Chile. The body-wave motion analysis of DELOUIS *et al.* (1997) shows that the main rupture has a character of a double event thrust mechanism. The depth of the main shock was estimated as 36 km.

Approximately 74 percent of the seismic moment was released in two main events located approximately 60 km from each other.

The aftershocks series of the Antofagasta earthquake constitutes a good database to study the spatial structure of the shallow part of the Central-Andean Wadati-Benioff zone in detail. Several temporary seismological networks were run from the Chilean coast to the Puna region in Argentina within the framework of the Collaborative Research Center 267 (SFB 267 of the DFG). The westernmost of those networks was installed within the CINCA (*Crustal Investigations On- and Offshore Nazca Plate and Central Andes*) project 1995. Some of the stations were separately provided from the German Task Force for Earthquakes. During three months (Aug. 10 to Oct. 11, 1995) these networks recorded the aftershocks of the Antofagasta earthquake. The networks consisted of onshore and offshore stations (Fig. 4). Onshore, the stations were maintained by the GeoForschungsZentrum Potsdam (GFZ) and the Freie Universität Berlin. Each station was equipped with a 3-component geophone (Mark L-4 3D, 1 Hz). The geophone signals were digitized with 100 Hz frequency. GPS was used for the determination of the position and for updating the internal clock every two hours. Offshore, the Ocean Bottom Hydrophones (OBH) were maintained by Geomar (Kiel). The position of each OBH was determined with GPS during expose and salvage. Like the onshore data, the OBH data was digitized with 100 Hz. The onset times of the *P* and *S* phases of 1878 shallow (down to 75 km depth) events were determined interactively by groups at Geomar, GFZ and FU-Berlin with GIANT/PITSA software (RIETBROCK and SCHERBAUM, 1998). All events with magnitudes above 2.4 were detected (SOBESIAK, 2000).

A subset of the aftershocks was used to gain a 1-D velocity model (HUSEN *et al.*, 1999). 3-D tomographic inversions of the onset times were performed by PATZIG (1999) and HUSEN (1999). For the inversion, a subset of 883 events was used. These events were most suitable for the tomographic inversion due to their epicenter location relative to the set of stations. The optimization criterion for the inversion was a minimization of the time misfit on the entire set of stations and events. The inversion procedure led to a detailed 3-D model of the *P* and *S* velocity in the coastal area of Antofagasta as well as to a localization of the hypocenters. For a single model the standard deviation of the event locations was formally computed to be less than 125 m at the depth range from zero to 75 km. Apart from the standard deviation, the difference of the final locations due to the different initial models (velocity, hypocenters and node positions) must be taken into account, because results of a tomographic inversion always depend on the initial model. The location differences are smaller than 1 km for the depth range 23 to 48 km. For the complete depth range of 15 to 75 km the location differences are smaller than 3 km.

A Cartesian coordinate system with two horizontal axes for longitude and latitude, respectively, and a vertical axis for the depth is a good approximation for the spherical earth within a roughly $300 \times 200 \times 100$ km region. In this Cartesian

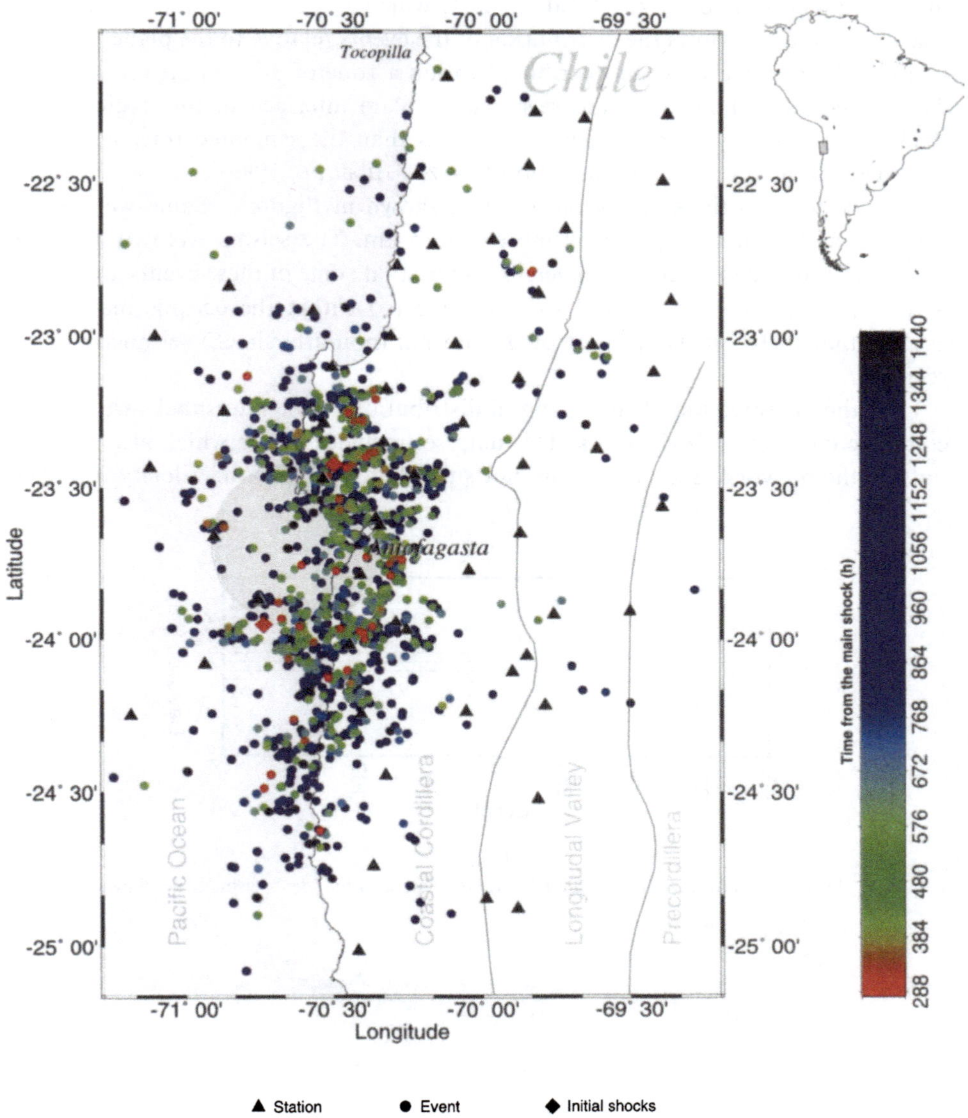

Figure 4

Plane view of the Antofagasta region. Triangles mark the seismological stations of the CINCA network (onshore and offshore). Circles mark the 1878 epicenters of the aftershocks. The color denotes the occurrence times of the events relative to the occurrence time of the initial shock.

coordinate system the spatial distribution of the 883 aftershocks, which were used for the tomographic inversion, shows a strong clustering of the events within a thin spatial domain which is very close to an ideal plane. This plane dips with $20.9° \pm 0.1°$

in E-W direction and strikes N6.0° ± 0.2°E which is parallel to the strike of the coastline. Figure 5 shows the distribution of the events relative to the plane. Seventy-two percent of the events are distributed within a zone of ±1.5 km from the plane. The distribution shows that the seismogenic plate interface in this region has a thickness of roughly 3 km, which is 4 km less than the estimated thickness of the oceanic crust at the Antofagasta region (PATZWAHL *et al.*, 1999).

A side view to the seismogenic plane is shown in Figure 6. Some well-located events deviate from the plane by more than 5 km. The seismic velocity structure obtained from the tomographic inversion shows that some of these events are located within the continental crust and some are located within the oceanic mantle. We assume that these events are part of the normal (non-aftershock) seismicity in this region.

For the investigation of the temporal distribution of the aftershocks we used the complete data set of 1878 events. This data set includes events which also occurred outside the network. Such events are not suitable for the task of velocity inversion,

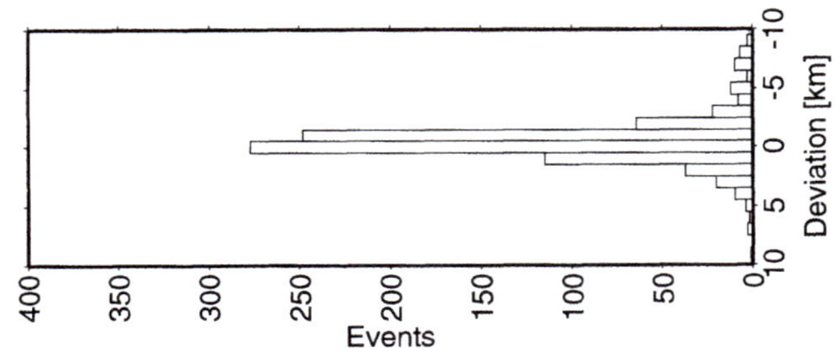

Figure 5
Distribution of deviation of the 883 events relative to the ideal seismogenic plane.

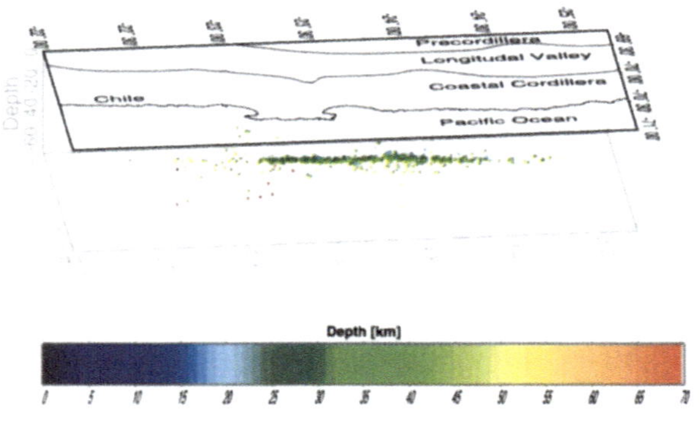

Figure 6
Above: A side view to 883 seismic events which were located by the tomographic inversion. In this view the ideal seismogenic plane would be seen as a straight line. The color of the events denotes their depth.

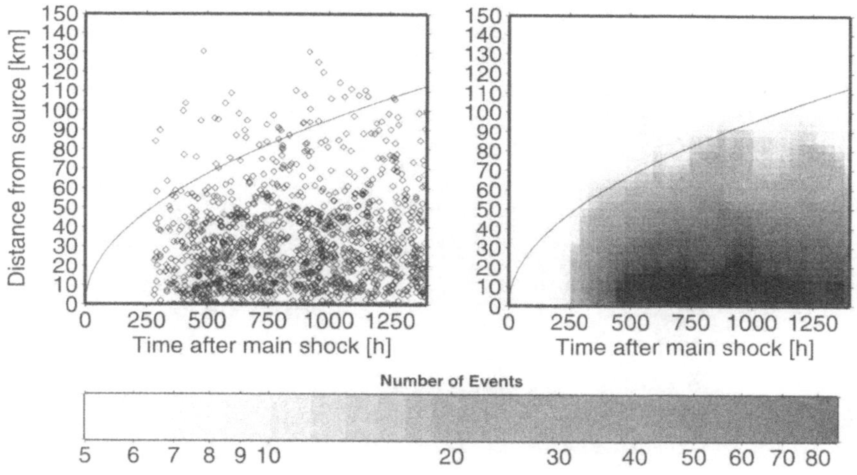

Figure 7
Distances of the aftershocks from the hypothetical perturbation source versus their occurrence time. The right-hand part plot includes a grey coded representation of the number of events within a moving window of 10 h × 10 km. The window was moved within increments of 2.5 h and 2.5 km, respectively.

but are valuable in the investigation of the spatio-temporal distribution. This data set is shown in Figure 4. The color indicates the aftershock occurrence times relative to the occurrence time of the main event. We note a dense cloud of relatively early events composing boundaries of a circle-like domain of approximately 30 km radius with the center between the two main events of the Antofagasta earthquake. This circle approximately coincides with the hypocenter domain of the main event found by DELOUIS (1997) on the base of aftershock observations during the first 20 h. During the first two weeks after the main shock an increase in the aftershock domain was observed. Our aftershock data, which were collected within the subsequent two months, confirms this observation.

Taking this into account, we propose a hypothesis that the aftershocks are triggered by a physical process requiring a finite time to propagate from the main rupture to a given spatial point. In order to show that this hypothesis is in agreement with the data, we computed the shortest distances from all aftershocks with deviations from the seismogenic plane smaller than ±3 km to a hypothetical perturbation source. This source was assumed to be a 6 km thick penny-shaped circular cylinder with radius 30 km and centered in the center of the main rupture. Figure 5 (left and right) shows a plot of the distances r between the events and the border of this circular cylinder versus the event times t. Additionally, a curve computed on the basis of equation (3) is shown. Here, $D = 200$ m^2/s describes the process of the aftershocks spreading sufficiently well, at least in the time range of 250 to 1000 h. Also, the plot of the aftershock cloud (see Fig. 8) in the coordinate system with the origin in the center of the main rupture and scaled in accordance with

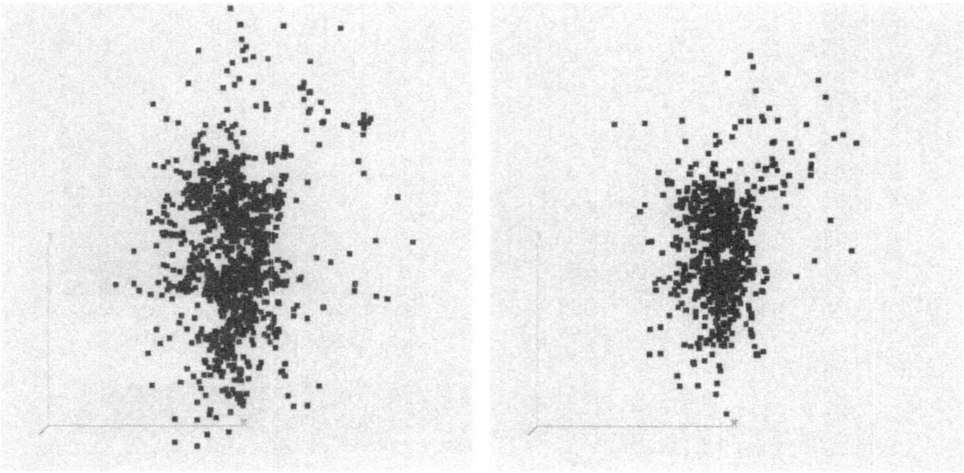

Figure 8
On the left: A view of the cloud of the Antofagasta aftershocks projected to the seismicity plane; on the right: the same view of the cloud in the coordinate system scaled in accordance with equation (6).

equation (6) indicates an ellipsoid-like, relatively dense set of points similar to Figures 2 and 3.

It is obvious that a direct comparison of the Antofagasta aftershock series with a fluid-injection induced microseismicity requires an additional discussion. First, even if we postulate that the Antofagasta aftershocks were also triggered by a pore pressure relaxation we have to point out the difference in boundary conditions in these two situations. The fluid injections in Soultz and Fenton Hill were continuous during the entire recording. The pore pressure signal at the injection source can be approximated as a step function of time. A hypothetical pore pressure perturbation at the hypocenter of the Antofagasta earthquake could be created by a rupture deformation of fluid-saturated rocks. The time dependence of this deformation is also approximately a step function. Therefore, in the first approximation one can expect a step function like perturbation of the pore pressure due to the earthquake. A similar scenario was also described by NUR and BOOKER (1972). However, especially at long time delays after the main event the pore pressure relaxation process as well as creep processes on the rupture will reduce the pore pressure perturbation. Thus, in the case of an earthquake the pore pressure signal can deviate from a step function significantly, especially for long time delays. This could be one of the reasons of deviation of the Antofagasta aftershocks from a parabolic envelope for times larger than 1000 h on Figure 5.

Second, the triggering mechanism of the Antofagasta aftershocks can be much more complex than a diffusion process in a homogeneous medium. Here, a combination of a slow diffusion process with quick stress redistributions is possible.

Also in this case, the envelope of events might differ from an ideal parabola. Moreover, the estimate of the diffusivity $D = 200$ m^2/s would provide then an upper possible limit of this coefficient.

In order to find an additional indication whether the pore pressure relaxation could be responsible for the triggering of the aftershocks, it is reasonable to estimate the permeability required for a corresponding triggering front and then judge if the estimate is realistic. For estimating the order of the permeability, we use the following elastic and viscoelastic properties: for the grain material we accepted the density, the *P*-wave and the *S*-wave velocities of a plagioclase-pyroxene-olivine composite: 3000 kg/m^3, 7500 m/s, and 4200 m/s, respectively. Additionally we used the velocity values obtained *in situ* by the velocity tomography described above: $V_p = 7000$ m/s and $V_s = 3900$ m/s. Finally, using for the porosity and the fluid bulk modulus the estimations $\phi \leq 0.001$ and $K_f = 8.4 \times 10^9$ Pa (an estimate for the water under very high pressure, about 10 kbar and $T = 200\,^\circ$C) we obtain $K_g = 9.8 \times 10^{10}$ Pa, $K_d = 8.5 \times 10^{10}$ Pa. Thus, we arrive at the following value: $N \approx 6.4 \times 10^{11}$ Pa. Finally, accepting the value $\eta = 2 \times 10^{-4}$ Pa·s (an approximate value of the dynamic viscosity of the water under 10 kbar, $T = 200\,^\circ$C HAAR *et al.* (1984)) we obtain the following estimation: $K = 6 \times 10^{-14}\,\mathrm{m}^2 = 60\,\mathrm{mD}$.

For our estimates we used the temperature of 200 °C. Such a temperature coincides with values found by SPRINGER (1999) for the upper crust of the cold subducting oceanic plate at the depth of 20 to 50 km. An increase of the temperature value can decrease the viscosity and correspondingly decrease the permeability estimate. Note, that the permeability of well-consolidated sedimentary reservoir rocks is of the order of 0.1–1000 mD. Fractured igneous and metamorphic rocks and basaltic oceanic crust can have permeabilities in the range of 0.001–10000 mD (see e.g., SCHÖN, 1996 and FISHER, 1998). Generally, 60 mD is a rather high value for a deep domain of the lithosphere. However, taking into account that the seismogenic contact zone must be a long-existing, active and large-scale fault zone (possibly including even some rests of sedimentary laminated structures involved in the subduction process), such a value seems to be realistic. Moreover, in the case of a mixed pore pressure relaxation and stress redistribution triggering mechanism (see the discussion above) the value of 60 mD must be interpreted as an upper bound estimate of the permeability along the contact zone.

Therefore, we conclude that it is possible that the main Antofagasta earthquake caused a pore-fluid pressure perturbation in its hypocenter. This pressure perturbation can be a result of deformations of the saturated porous space of rocks due to the rupture process in the hypocenter. Our observation indicates that the diffusion-like relaxation of this pore-pressure perturbation in the fractured rocks along the subduction plane can be an important component of the triggering mechanism of the Antofagasta aftershocks.

4. Conclusions

We show that a spatio-temporal distribution of a fluid-injection induced seismicity obeys several features directly related to the large-scale permeability tensor of the seismogenic rock volume. For example, in the crossplot of hypocentral distances versus occurrence times, the seismicity cloud usually has a parabolic envelope. In a homogeneous medium, the surface of the seismicity triggering front has an ellipsoidal form. These features can also be considered as indications of the diffusive and even hydraulic nature of the seismicity triggering process.

The spatio-temporal distribution of aftershocks of the Antofagasta (northern Chile) 1995 earthquake seems to obey such features. This indicates the possibility that the main Antofagasta earthquake caused a pore-fluid pressure perturbation in its hypocenter. At least a significant proportion of the aftershocks could have been triggered by the relaxation of this perturbation of the pore pressure.

Acknowledgements

Some of the graphics have been generated with the use of the Generic Mapping Tool (GMT) by Wessel and Smith. The CINCA project was financed by the German Science Foundation (DFG) through the Collaborative Research Centre (SFB) 267 "Deformation Processes in the Andes", by the GeoForschungsZentrum Potsdam and the Freie Universität Berlin. Additionally, financial support was given by the Bundesministerium für Forschung und Technologie (BMFT), the Bundesanstalt für Geowissenschaften und Rohstoffe (BGR) Hannover and Geomar (Kiel). Scientists from the mentioned Institutions, the Geophysical Institute of the Ludwig-Maximilians-Universität of München, and the Universidad del Norte, Antofagasta, elaborated the project in the field campaign. Support for this work was provided in part from the sponsors of the WIT university-consortium project and in part from SFB 267. We thank Alexandre Schubnel and one anonymous reviewer for constructive comments improving the manuscript.

References

Audigane, P. *Caractérisation microsismique des massif rocheux fracturés. Modélisation thermo-hydraulique. Application au concept géothermique de Soultz.* Dissertation, 219pp, ENSG, INPL, Nancy, France, February 2000.

Biot, M. A. (1956), *Theory of Propagation of Elastic Waves in a Fluid-saturated Porous Solid. I. Low-frequency Range*, J. Acoust. Soc. Am. *28*, 168–178.

Biot, M. A. (1962), *Mechanics of Deformation and Acoustic Propagation in Porous Media*, J. Appl. Phys. *33*, 1482–1498.

DELOUIS, B. T. (1997), *The $M_w = 8.0$ Antofagasta (Northern Chile) Earthquake of July 30, 1995: A Precursor to the End of the Large 1877 Gap*, Bull. Seismol. Soc. Am. *87*, 427.

DUTTA, N. C. and ODE, H. (1979), *Attenuation and Dispersion of Compressional Waves in Fluid-filled Porous Rocks with Partial Gas Saturation (White Model)-Part i: Biot Theory*, Geophys. *44*, 1777–1788.

DYER, B., JUPPE, A., JONES, R. H., THOMAS, T., WILLIS-RICHARDS, J., and JAQUES, P. (1994), *Microseismic Results from the European HDR Geothermal Project at Soultz-sous-Fôrets, Alsace, France*, CSM Associated Ltd, IR03/24.

FEHLER, M., HOUSE, L., PHILLIPS, W. S., and POTTER, R. (1998), *A Method to Allow Temporal Variation of Velocity in Travel-time Tomography Using Microearthquakes Induced During Hydraulic Fracturing*, Tectonophys. *289*, 189–202.

FERREIRA, J. M., DE OLIVEIRA, R. T., ASSUMPCAO, M., MOREIRA, J. A. M., PEARCE, R., and TAKEYA, M. K. (1995), *Correlation of Seismicity and Water Level in the acu Reservoir – An Example from Northeast Brazil*, Bull. Seismol. Soc. Am. *85*(5), 1483–1489.

FISHER, A. T. (1998), *Permeability within Basaltic Oceanic Crust*, Rev. Geophys. *36*, 143–182.

FLETCHER, J. B. and SYKES, L. R. (1977), *Earthquakes Related to Hydraulic Mining and Natural Seismic Activity in Western New York State*, J. Geophys. Res. *82*, 3767–3780.

HAAR, L., GALLAGHER, J. S., and KELL, G. S., *NBS/NRC Steam Tables; Thermodynamic and Transport Properties and Computer Programs for Vapor and Liquid States of Water in SI Units* (Hemisphere Publishing Corporation, New York, 1984) 320 pp, New York, 1 edition.

HUSEN, E., KISSLING, S., FLUEH, E., and ASCH, G. (1999), *Accurate Hypocentre Determination in the Seismogenic Zone of the Subducting Nazca Plate in Northern Chile Using a Combined on-offshore Network*, Geophys. J. Int. *138*, 687–701.

HUSEN, S. (1999), *Local Earthquake Tomography of a Convergent Margin, North Chile*. Ph.D. Thesis, Universität zu Kiel, Germany.

JUNG, R., CORNET, F., RUMMEL, F., and WILLIS-RICHARD, J. *Hydraulic stimulation results 1992/1993*. In *European Hot Dry Rock Programme, 1992–1995*. Extended Summary of the Final Report to European Community (DG XII), Contract N JOU2-CT92-0115 (Date of issue: 28 March 1996, Version 1). (Baria, R., Baumgärtner, J., and Gérard, A. eds.) pages 31–41, 1996.

MAILLOT, B., NIELSEN, S., and MAIN, I. (1999), *Numerical Simulation of Seismicity Due to Fluid Injection in a Brittle Poroelastic Medium*, Geophys. J. Int. *139*, 263–272.

MAVKO, G., MUKERJI, T., and DVORKIN, J. *Rock Physics Handbook* (Stanford University, Stanford, 1996).

NUR, A. and BOOKER, J. (1972), *Aftershocks Caused by Pore Fluid Flow*? Science *175*, 885–887.

OHTAKE, M. (1974), *Seismic Activity Induced by Water Injection at Matsushiro, Japan*, J. Phys. Earth *22*, 163–176.

PATZIG, R. *Lokalbeben – Tomographie der Umgebung von Antofagasta (Nordchile)*, Ph.D. Thesis, Freie Universität Berlin, 1999.

PATZWAHL, R., MECHIE, J., SCHULZE, A., and GIESE, P. (1999), *Two-dimensional Velocity Models of the Nazca Plate Subduction Zone between 19.5 degs and 25 degs from Wide-angle Seismic Measurements during the cinca 95 Project*, J. Geophys. Res. *104*, 7293–7317.

RICE, J. R. and CLEARY, M. P. (1976), *Some Basic Stress Diffusion Solutions for Fluid-saturated Porous Media with Compressible Constituents*, Rev. Geophys. Space Phys. *14*, 227–241.

RIETBROCK, A. and SCHERBAUM, F. (1998), *The Giant Analysis System*, Seismol. Res. Lett. *69*, 40–45.

SCHÖN, J. *Physical Properties of Rocks* (Pergamon, Elsevier, 1996).

SHAPIRO, S. A., AUDIGANE, P., and ROYER, J.-J. (1999), *Large-scale in situ permeability tensor of rocks from induced microseismicity*, Geophys. J. Internat. *137*, 207–213.

SHAPIRO, S. A., HUENGES, E., and BORM, G. (1997), *Estimating the Crust Permeability from Fluid-injection-induced Seismic Emission at the KTB Site*, Geophys. J. Internat. *131*, F15–F18. (See also Corrigendum, Geophys. J. Internat., 1998, v. 134, p. 913).

SHAPIRO, S. A., ROYER, J.-J., and AUDIGANE, P. *Estimating the permeability from fluid-injection induced seismic emission*. In *Poromechanics* (Thimus, J.-F., Abousleiman, Y., Cheng, A.H.-D., Coussy, O., and Detournay, E. eds.) pages 301–305 (A. A. Balkema, Rotterdam, Brookfield, 1998).

SHAPIRO, S. A., ROYER, J.-J., and AUDIGANE, P. (2000), *Reply to Comment by F.H. Cornet on "Large-scale in situ Permeability Tensor of Rocks from Induced Microseismicity."* Geophys. J. Internat. *140*, 470–473.

SOBESIAK, M. (2000), *Fault Plane Structure of the Antofagasta, Chile, Earthqake of 1995*, Geophys. Res. Lett. *27*, 577–600.

SPRINGER, M. (1999), *Interpretation of Heat Flow Density in the Central Andes*, Tectonophys. *306*, 377–395.

TALWANI, P. and ACREE, S. (1985), *Pore Pressure Diffusion and the Mechanism of Reservoir-induced Seismicity*, Pure Appl. Geophys. *122*, 947–965.

VAN DER KAMP, G. and GALE, J. E. (1983), *Theory of Earth Tide and Barometric Effects in Porous Formations with Compressible Grains*, Water Resources Res. *19*, 538–544.

VENTURA, G. and VILARDO, G. (1999), *Seismic-based Estimate of Hydraulic Parameters at Vesuvius Volcano*, Geophys. Res. Lett. *26*, 887–890.

ZOBACK, M. and HARJES, H.-P. (1997), *Injection Induced Earthquakes and the Crustal Stress at 9 km Depth at the ktb Deep Drilling Site*, J. Geophys. Res. *102*, 18, 477–492.

(Received February 21, 2001, revised June 25, 2001, accepted August 1, 2001)

Pure appl. geophys. 160 (2003) 1067–1085
0033 – 4553/03/061067–19

Pure and Applied Geophysics

Scale Dependence of Hydraulic and Structural Parameters in the Crystalline Rock of the KTB

GÜNTER ZIMMERMANN,[1] HANS BURKHARDT,[1] and LUDWIG ENGELHARD[2]

Abstract—Knowledge of rock properties controlling the fluid movement is a basic prerequisite to understand the dynamical processes and the temperature and stress regime of the upper crust. Fracture networks were investigated on different scales to obtain quantitative results of fracture geometry like fracture length, orientation and fracture frequencies. Due to the scale effect, these parameters differ in several orders of magnitude in dependence of the scale of investigation. On the microscopic scale, fluorescent thin sections from cores were analysed and permeability was estimated for 2-D hydraulic networks. On the macroscopic scale, fracture parameters were determined from images of structural borehole measurements. The vicinity of the drill site represents the megascopic scale, where seismic reflectors were assumed as active hydraulic structures for construction of a fracture network. Compiling the fracture densities from all investigated scales and taking into consideration only the networks above the percolation threshold, the fracture length distribution follows a power law with an exponent of -1.9 ± 0.05. Besides the scale differences of the geometric parameters like fracture density and length and the hydraulic parameters like permeability, the connectivity of the networks seems to be a confining characteristic. This is quantitatively described by the percolation parameter and the mean number of intersections per fracture. When assuming a macroscopic hydraulic system at the percolation threshold for the KTB site, the macroscopic mean fracture length can be estimated to approximately 30 m. This stands in agreement with the hydraulic experiments on site.

Key words: Permeability, crystalline rock, scale dependence, KTB.

Introduction

Permeability is one of the essential parameters for the description of the hydraulic properties of rocks and a prerequisite to describe the hydraulic pathways for fluids in the upper crust. Therefore, the investigation of permeability was one of the main objectives for the two boreholes in the Continental Deep Drilling Program (KTB). Permeability is a quantity that varies depending on the scale of the investigation (i.e., it is not scale-invariant) (BRACE, 1984; CLAUSER, 1992). Permeability on a

[1] Department of Applied Geophysics, Technical University Berlin, Ackerstr. 76, 13355 Berlin, Germany.
[2] Institute of Geophysics and Meteorology, Technical University Braunschweig, Mendelssohnstr. 3, 38106 Braunschweig, Germany. E-mail: zimmermann@geophysik.tu-berlin.de

macroscopic scale representing an integral value for the borehole vicinity, depends on relatively large, irregular distributed fracture systems which cannot be verified by core measurements. This permeability can be estimated by hydraulic experiments in drill holes yielding integral values (results in context with KTB: KESSELS and KÜCK, 1995; HUENGES *et al.*, 1997; HUENGES and ZIMMERMANN, 1999; ZIMMERMANN *et al.*, 2000). These are normally limited to a few depth intervals. The core measurements represent a permeability with hydraulic pathways on the microscopic scale. Taking the *in situ* confining pressure into account the measurements lead to an estimate of the matrix permeability. This matrix permeability cannot generally be up-scaled.

Beside various methods to measure permeability directly or indirectly, diverse concepts exist to model the hydraulic flow in fractured media. To transfer the complex structures of natural rocks into adequate equivalent models is a non-trivial problem. One can distinguish between deterministic fracture networks (KOLDITZ, 1992; DAVID, 1993), fractal fracture networks (KOSAKOWSKI, 1994; 1996; ACUNA and YORTSOS, 1995) and stochastic fracture networks (WOLLRATH, 1990; ZIMMERMANN *et al.*, 2000). Another technique to estimate the hydraulic properties uses the micro-seismic activity induced during hydraulic experiments (SHAPIRO *et al.*, 1997, 1999). The permeability was calculated from hydraulic diffusivity, which can be estimated from the spatio-temporal distribution of all located events.

The scale dependence of the hydraulic properties is connected to the existence and the size of a potential representative elementary volume (REV). This represents an unsolved problem, which has been investigated with various approaches and model concepts. Below the REV a medium appears as heterogeneous with a high variability of its properties. Above the REV it can be considered as a statistically homogeneous and ergodic medium. Hence it can be modelled as an "equivalent homogeneous" medium. GUÉGUEN *et al.* (1996) give an overview of the scale effect and the corresponding concepts to model the hydraulic flow. The description of fracture models and its characteristic parameters is achieved with various theoretical approaches. The connectivity of discrete fracture structures, i.e., planar discontinuities, can be described for instance by percolation theory (e.g., STAUFFER and AHARONY, 1992). Percolation theory describes the fluid flow in a hypothetical medium or network and was first introduced by BROADBENT and HAMMERSLEY (1957). Generally, this process can be described as a diffusion of fluids with corresponding randomness of its pathways in the medium or with stochastic properties of the medium itself. An introduction to percolation theory and to the fluid transport in fractured media can be found in BERKOWITZ and BALBERG (1993), SAHIMI (1995), SELYAKOV and KADET (1996) and BERKOWITZ and EWING (1998).

"Effective medium theory" (EMT) describes the transfer of heterogeneous properties into a statistically homogeneous medium (DAVID *et al.*, 1990). This approach is only valid for small heterogeneities and requires a REV (GUÉGUEN *et al.*, 1997). GAVRILENKO and GUÉGUEN (1998) used a modified "renormalization-group" (RG)-theory in conjunction with a percolation theory approach to describe the scale effect of permeability. Their results are in agreement with the observed measured effective permeability on the regional scale, based on the compilation of permeability measurements as a function of scale by CLAUSER (1992).

One of the most important controlling influences of percolation theory is the scaling laws near the percolation thresholds. For fracture networks close to the percolation threshold, power laws describe the hydraulic and structural properties of the networks in a very general and in some cases universal way (ROBINSON, 1983, 1984; BALBERG *et al.*, 1991; DAVY *et al.*, 1992, 1993). A frequently analysed quantity affecting the connectivity of the fracture networks (therefore its permeability) is the fracture length distribution function. This variable also follows a power law (BALBERG *et al.*, 1991; BOUR and DAVY, 1997, 1998) and does not permit an *a priori* definition of a REV.

Modelling Fracture Permeability

The aim of this work is to determine the hydraulic transport parameters of fractured rock on different scales. Comparisons were made on samples from the KTB. To simplify the analysis we assume the fractures to be smooth planes without roughness. Generally, individual fracture planes can be described by the fracture length and width (surface area), the aperture and orientation (dip, azimuth). Projection to a 2-D surface reduces the geometry description of the fracture surface area to a fracture length.

On the microscopic scale, we analysed fluorescent thin sections with 2-D networks, which represent microfractures with a length scale reaching 100 μm. The macroscopic scale is represented by structural borehole measurements with a formation microscanner (FMS), which creates an image of resistivities of borehole wall structures on the metre scale. This provides 3-D information regarding the orientation of the fractures and represents the macroscopic scale but not fracture length and width. The 3-D networks derived from these data are reduced to a 2-D projection area with a preferred orientation (ZIMMERMANN *et al.*, 2000) to assure a remaining connected 2-D network. Seismic structures in a 2-D projection of up to a few kilometres can be interpreted as active hydraulic pathways and represent the megascopic scale. From these basic geometrical data discrete deterministic as well as stochastic fracture networks can be obtained. Subsequent modelling of the hydraulic flow with the finite element program *ROCKFLOW* (LEGE *et al.*, 1996) leads to the calculation of permeability, which can be compared to measured permeability on the different scales. Fluid flow was simulated applying Darcy's law (GUÉGUEN and PALCIAUSKAS, 1994) and assuming fluid velocities with laminar flow. This leads to the well known "cubic law" relationship between permeability and fracture aperture. The problems to solve are highly underdetermined. Therefore, besides the known basic information concerning the fractures, additional data and calibrations are needed to provide a well constrained estimate of permeability. Furthermore, this allows a discussion of the limitations of unknown quantities and their influence on permeability of the geometric networks.

Calculation of the Percolation Parameter

The parameter describing the transition from a non-connected medium to a connected medium is called percolation threshold (STAUFFER and AHARONY, 1992). In the case of continuously distributed structures, a quantitative definition of the percolation threshold is given by the percolation parameter. For 2-D networks with constant fracture length l and randomly distributed position and orientation, the percolation parameter can be defined as (BOUR and DAVY, 1997):

$$p = N l^2 / A = \lambda_A l^2$$

with N = number of fractures, l = fracture length, A = surface of the network, $\lambda_A = N/A$ = density of fractures (2-D). Introducing a fracture length distribution and replacing the constant fracture length by the mean value (expectation value) μ_l leads to a generalisation of the formula. According to PRIEST (1993), a general relationship between the linear frequency of fractures λ_L and the 2-D fracture density λ_A is given by:

$$2/\pi \; \lambda_A \; \mu_l \leq \lambda_L \leq \lambda_A \mu_l$$

with the lower bound for spatially random distributed fractures and the upper bound for parallel fractures. Merging both relations leads to the generalisation of the percolation parameter:

$$p = N \mu_l^2 / A = \lambda_A \mu_l^2 \geq \lambda_L \mu_l \; .$$

Results

Microscopic Scale

On the microscopic scale fluorescent thin sections (Fig. 1) were investigated to obtain geometrical parameters such as mean fracture lengths, orientation, fracture density and the corresponding distribution functions. The images of the hydraulic pathways were illuminated by means of a fluorescent epoxy resin, which has been forced into the micro-fractures using a centrifuge. These visible micro-fractures were represented by a network of straight lines and transformed to a finite element mesh to simulate the hydraulic flow. Figure 1 shows different images of a thin section with two different resolutions (0.25 μm and 1 μm per pixel) of a core taken from the KTB main borehole (KTB-HB007; 4592 m). On the bottom image of Figure 1 the

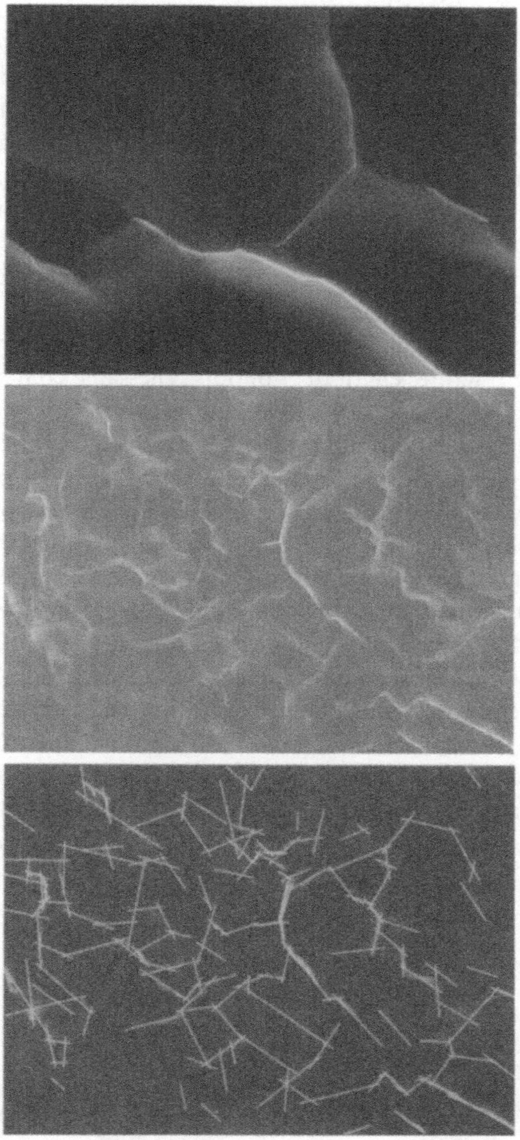

Figure 1

Fluorescent thin sections (KTB-HB H007, 4592 m). Top: detailed photo 192 μm × 144 μm; the fluorescent epoxy resin penetrated into the micro-fractures (resolution 0.25 μm). Middle: section to construct fracture networks (768 μm × 576 μm, resolution 1 μm). Bottom: the same section with corresponding marked micro-fractures for descriptive statistics and modelling of the fluid transport. Measured permeability of the core sample is 4×10^{-20} m^2 (40 nDarcy). Modelling of the network leads to a permeability of approx. 1.5×10^{-19} m^2 (150 nDarcy) assuming a mean fracture aperture of 0.5 μm and an exponential distribution function. *Vice versa* a permeability of 40 nDarcy corresponds to a mean fracture aperture of 0.322 μm.

resulting network of straight lines connecting the pathways to model the hydraulic flow is displayed. The fracture aperture distribution plays a crucial role in fracture permeability, therefore this parameter deserves additional discussion. Aperture distributions can be roughly estimated from visual inspection, however this procedure is quite limited. Fluorescent micro-structures with apertures below the resolution of the microscope are nonetheless visible. Therefore, only an upper bound can be estimated in case of very small structures. Nonetheless the aperture distribution can be obtained indirectly from the aspect ratio. According to HADLEY (1976) the aspect ratio for microcracks is a constant of approx. 10^{-3}. This result implies that the distribution function of the fracture aperture and length are similar. Accordingly, the distribution function of the fracture aperture can be obtained indirectly from the fracture length distribution. The same context is also supported by the work of BROWN and BRUHN (1998) and GAVRILENKO and GUÉGUEN (1998), who assume a linear relationship between fracture aperture and length with an asymptote exceeding a critical scale. Figure 2 shows the histogram of the fracture length distribution. It shows a skew distribution with a higher probability for lower values and can be characterised as a log-normal distribution function or as a decreasing exponential type with a cutoff at low values due to the limited resolution of the image. Furthermore, PYRAK-NOLTE *et al.* (1997) estimated fracture aperture distributions directly from X-ray computerised tomographic imaging from cores of opaque bituminous coal. Cores with few long fractures exhibit a log-normal aperture distribution, whereas cores with many filamentary fractures (indicating a normal distribution of the fracture length) tend to have a normal aperture distribution. This again indicates the linear relationship between fracture length and aperture on the microscopic scale and hence a constant aspect ratio. Taking into account the result from the fracture length distribution in Figure 2, we applied the exponential

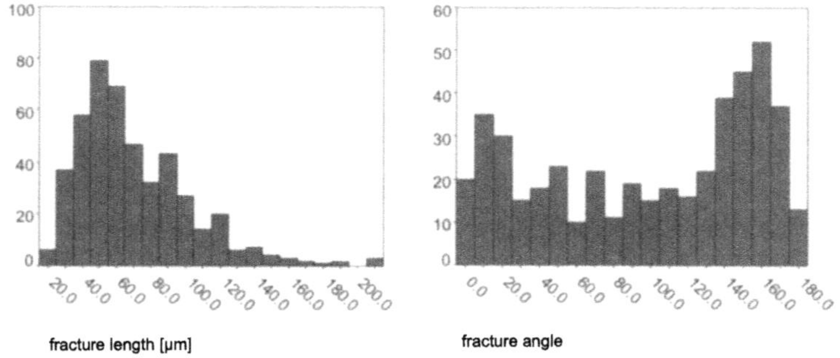

Figure 2

Histograms of the fracture lengths (in µm) and the orientation of the fractures. The distribution function of the fracture lengths is similar to the log-normal distribution. The distribution of the fracture orientation can be divided into two sets with preferred orientations. This is a characteristic which is also valid for the macroscopic and megascopic scale.

distribution for the fracture aperture. The exponential distribution can be described by one characteristic parameter, which is equivalent to the mean as well as the standard deviation of this distribution. An appropriate expectation value of the aperture of the visible microstructures seems to be the resolution of the microscope, therefore the fracture apertures between the nodes of the finite element mesh were varied stochastically with a mean of 0.5 μm and thus a variance of 0.25 $(\mu m)^2$. Modelling of the network with this aperture distribution leads to a mean hydraulic permeability of 1.5×10^{-19} m^2 (150 nDarcy). The corresponding measured gas-permeability of the core sample is 4.0×10^{-20} m^2 (40 nDarcy) and was measured with a confining pressure of 10 MPa and steady-state conditions. This permeability corresponds to a mean fracture aperture of 0.332 μm.

Since the calculation of the fracture orientation is based on the optical alignment of the thin section, only relative values are given. The histogram of the fracture orientation (Fig. 2) shows a slight bimodal distribution. This is also characteristic of other scales.

From the geometrical parameters, the percolation parameter as a qualitative measure of the connectivity of the networks can be calculated according to the above-mentioned formulas. We investigated eight different images with connected and unconnected hydraulic networks and calculated the corresponding percolation parameter in order to compare the results with the actual visual connectivity. The state of the actual 2-D connectivity was determined from visual inspection of a pathway, which connects at least two opposite boundaries of the image. Table 1 shows the results from calculation. The mean fracture length l is calculated from the expectation value of the log-normal distribution according to the result in Figure 2. All networks with a percolation parameter $p > 1.3$ are connected. Bour and Davy (1997) give a theoretical value of ~5.6 in the limit of infinitely large systems for 2-D

Table 1

Geometrical parameters and the percolation parameter of thin sections from KTB drill cores from 4590 m (see Fig. 1) and 7540 m. The percolation parameter was calculated according to Bour and Davy (1997) from the number of fractures and the mean fracture length with regard to the log-normal distribution of the fracture length. Networks which are actually connected (marked with "conn") show a parameter >1.3, whereas the non-connected networks (marked with "not") have lower values

	Number of Fractures N	Mean Fracture length l [μm]	$p = N l^2/A$
H7p-1 not	71	53.7	0.46
H7p-2 not	96	56.2	0.68
H7p-3 conn	143	67.6	1.48
H7p-4 conn	150	70.8	1.70
H7s-1 not	120	69.2	1.30
H7s-2 conn	146	63.1	1.31
H7s-3 not	54	55.0	0.36
7540 m conn	251	74.1	1.74

networks with fractures randomly distributed in position and orientation and a constant fracture length. The thin section data presented here denote fracture lengths, which are log-normal distributed and fractures with a nonrandomly distributed orientation (Fig. 2). Those conditions make connectivity more probable and hence the critical percolation parameter is considerably lower. Applying the mean (expectation value) from the normal (Gaussian) distribution instead of the log-normal distribution leads to percolation parameters which are $\sim$ 20% higher than the results in Figure 2 and the threshold is reached at $p \sim 1.6$.

Macroscopic Scale

In a recent work (ZIMMERMANN *et al.*, 2000) stochastic networks in the depth section of the Franconian Lineament (FL; $\sim$7000 m) were investigated to simulate the hydraulic condition at the two boreholes of the KTB. The KTB is located in the Variscan basement at the northwestern margin of the Bohemian Massif. The FL appears as a NW-SE trending boundary between the Variscan basement and the Permo-Mesozoic sedimentary basin and was formed in the Late Paleozoic and reactivated as a reverse fault in the Mesozoic (WAGNER *et al.*, 1997). This steeply dipping fault system strikes the KTB main borehole at a depth interval from 6900 m to 7140 m and appears as a brittle ductile shear zone and a prominent seismic reflector (HARJES *et al.*, 1997). The steeply inclined reverse faults and fractures can be kept hydraulically conductive over a long period of time under the present stress regime (discussed thoroughly in ZIMMERMANN *et al.*, 2000), which is a strike-slip faulting for this depth interval (BRUDY *et al.*, 1997; ZOBACK and HARJES, 1997). The geometrical parameters of the fractures such as fracture locations (and derived fracture frequency) and orientations were determined from structural borehole measurements, which create a microresistivity image of the borehole wall. The utilised FMS tool consists of four arms with rows of small electrodes to determine the resistivity image. The geometrical orientation of the fractures like dip, azimuth and depth can be determined by interactively picking the images. The selection of potentially hydraulically conductive fractures was carried out according to stress criteria described in detail in ZIMMERMANN *et al.* (2000). As a result of these stress criteria, fracture planes were assumed to be potentially conductive, when possessing an azimuth perpendicular to the maximum horizontal stress field and a steeply dipping angle. The fracture aperture was estimated by comparison with the results of hydraulic tests (HUENGES *et al.*, 1997; HUENGES and ZIMMERMANN, 1999). Based on the FMS data, a fracture network of the FL (depth interval 6900 m to 7140 m) was constructed and permeability calculated for various fracture apertures; a fracture aperture of 25 μm reproduced the *in situ* permeability (ZIMMERMANN *et al.*, 2000). To study the influence of fracture length, four different 2-D stochastic networks based on the same FMS data were constructed with fracture lengths of 100 m, 50 m, 30 m and 10 m, respectively (Fig. 3). The permeability was calculated applying Darcy flow

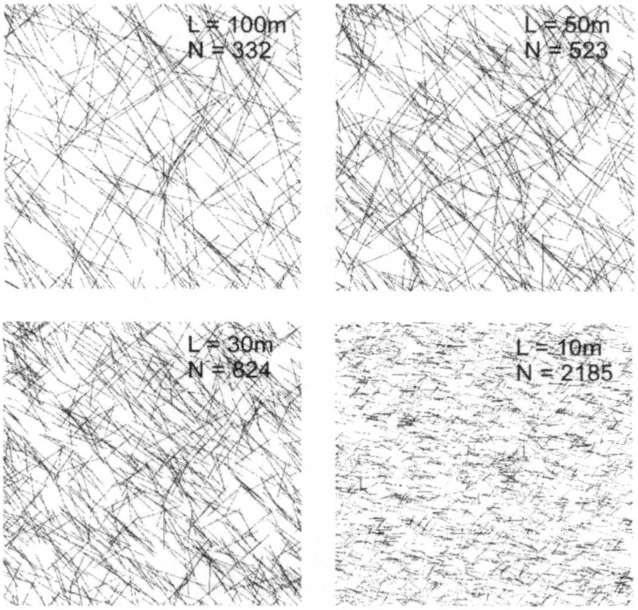

Figure 3
Stochastic fracture networks used for modelling the hydraulic flow (according to ZIMMERMANN *et al.*, 2000). The geometrical parameters of the fractures were obtained by FMS-measurements at the depth section of the Franconian lineament (6900 m to 7140 m KTB-HB, projection in W-E-direction). The number of fractures N per area (2-D fracture density) was calculated from the detected fractures from FMS measurements (1-D fracture frequency), which is the invariant parameter for all networks; the fracture length was set to 100 m, 50 m, 30 m and 10 m. Variation of the mean fracture aperture leads to different permeability estimations (Fig. 4). The networks with fracture length of 10 m is not permeable; that is, there is no hydraulic pathway between the boundaries of the model and hence the fracture system is not connected.

in horizontal and vertical directions. The obtained permeability depends on the fracture length of the networks and decreases with decreasing fracture length (Fig. 4), although the fracture density is increased simultaneously to ensure that all models are equivalent according to the original data from the borehole measurements. The 2-D network with a fracture length of 10 m is not permeable; that is, no connection exists between the boundaries of the model in vertical and horizontal directions.

From the geometrical parameters of the fractures, the percolation parameter can be calculated and compared to the actual connectivity of the fracture models. Figure 5 shows the percolation parameter as a function of the mean fracture length for the networks in Figure 3. The critical value of 5.6 is reached at a mean fracture length of ~15 m in agreement with the results; the network with a mean fracture length of 10 m is not connected.

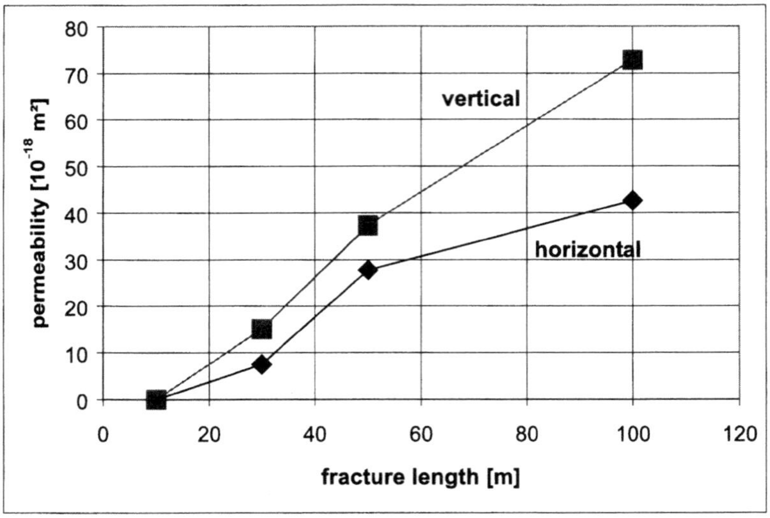

Figure 4
Permeability calculations in horizontal and vertical directions of the stochastic networks in Figure 3 (according to ZIMMERMANN *et al.*, 2000). The mean fracture aperture was set to 25 μm with an exponential distribution function. Permeability decreases for decreasing fracture lengths, although the fracture density is increased simultaneously to ensure that all models are equivalent according to the primary data of the borehole measurements.

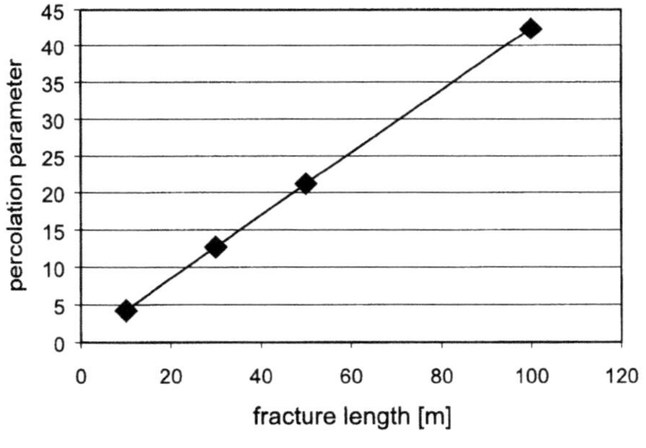

Figure 5
Percolation parameter of the stochastic networks (Fig. 3) according to BOUR and DAVY (1997). The percolation threshold, which is equivalent to the critical percolation parameter of ~5.6, is reached from these calculations at a mean fracture length of approx. 15 m in agreement with the actual network with a mean fracture length of 10 m, which is the only unconnected and hence impermeable one.

The mean number of intersections per fracture describes quantitatively the connectivity of the fracture system. The critical value at the percolation threshold is reached at 3.6 intersections per fracture according to HESTIR and LONG (1990), and

3.1 according to ROBINSON (1984). Figure 6 shows the mean intersections per fracture as a function of the mean fracture length for the stochastic networks in Figure 3. Connectivity and hence permeability decreases with mean fracture length until the percolation threshold is reached. This leads to a similar result as in Figure 5; the corresponding critical mean fracture length is approximately 15 m.

Megascopic Scale

The seismic reflectors in the vicinity of the KTB drill sites represent discontinuities which can be interpreted as active hydraulic systems (Fig. 7) (EMMERMANN and LAUTERJUNG, 1997; HARJES *et al.*, 1997). The seismic program demonstrated that most of the reflections registered originate from fracture zones and faults and hence can be referred as hydraulically significant structures, whereas the lithological boundaries are less important. These discontinuities can be examined stochastically

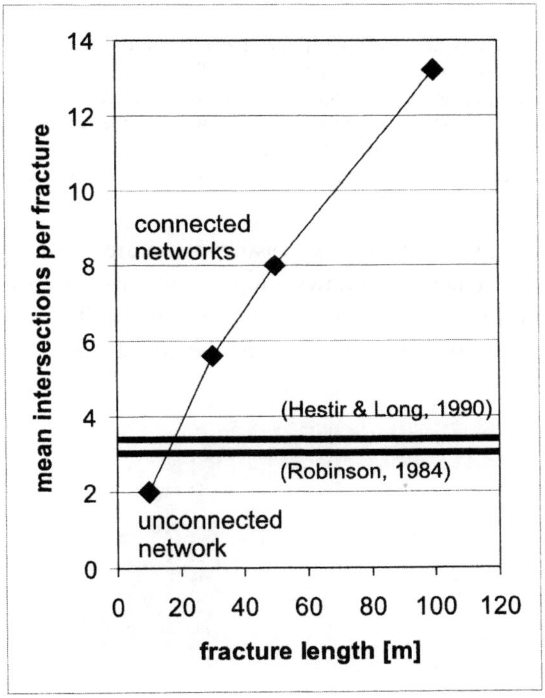

Figure 6

Mean intersections of fractures versus mean fracture length for the stochastic networks (Fig 3). Connectivity and hence permeability decreases with decreasing mean fracture length until the percolation threshold is reached. The critical value is reached at 3.6 intersections per fracture (according to HESTIR and LONG, 1990) respectively at 3.1 (according to ROBINSON, 1984), which corresponds to a mean fracture length of approximately 15 m. This is in agreement with the networks of Figure 3; the network with a mean fracture length of 10 m, which is below the theoretical threshold, is unconnected.

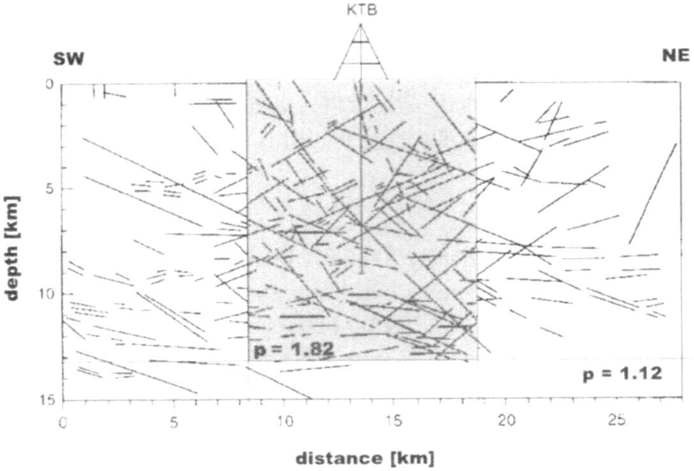

Figure 7
Seismic reflectors in the vicinity of the KTB drill site (according to EMMERMANN and LAUTERJUNG, 1997). These discontinuities can be interpreted as active hydraulic structures. The percolation parameter as a measure of the hydraulic connectivity was calculated from fracture density and mean fracture length. The percolation parameter for the whole section is $p = 1.12$, whereas the restriction to the drill site (central area) leads to a higher percolation parameter of $p = 1.82$. The whole section is not connected, the restriction of the central area leads to a 2-D connected network.

to determine the mean length of the seismic structures and the 2-D density. The length of the seismic structures follows a log-normal distribution consistent with the results from the thin section analysis (Fig. 8). A similar result is obtained for the spatial distribution of the orientation of the structures: as for the thin sections,

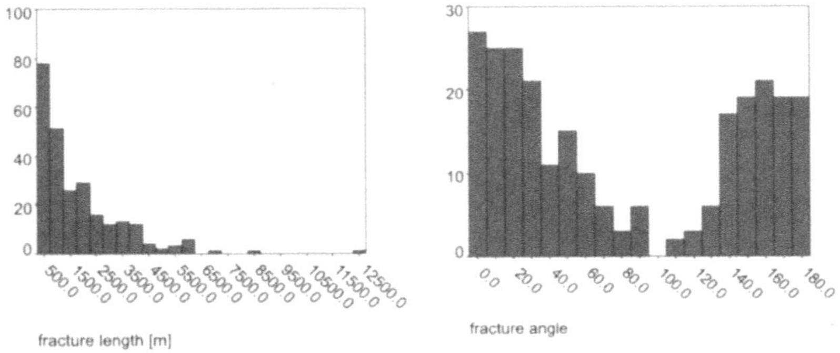

Figure 8
Histograms of the fracture lengths (in m) and the orientation of the fractures. The distribution function of the fracture lengths is similar to the log-normal distribution. The distribution of the fracture orientation can be divided into two sets with preferred orientations. This is a characteristic which is also valid for the other scales.

the structures manifest a bimodal distribution (Fig. 8). From the mean calculated fracture length with regard to the log-normal distribution and the 2-D density of the fractures, the percolation parameter as a measure of the connectivity of the planar structures was calculated. Applying these descriptive statistics to the entire section of Figure 8 leads to a percolation parameter $p = 1.12$ with a corresponding 2-D network of seismic structures which is not connected. The restriction to the central area near the drill site leads to a percolation parameter of 1.82 with a corresponding connected network. These results document that the derived critical value of the percolation parameter from the thin section analysis of approximately 1.3 is consistent with the values of the megascopic scale. A possible reason might be the similar distribution functions for the fracture length and orientation.

Permeability estimation was carried out in a horizontal direction for the central area with the main borehole in the middle (Fig. 7). To the extent the network is connected, i.e., above the percolation threshold, the aperture of the fractures is the dominant parameter for permeability according to the 'cubic law'. There are two special notes which might help to estimate values for permeability and apertures on the regional scale. The first hint is the compilation of permeability measurements from CLAUSER (1992). It is shown that permeability values in the range of borehole measurements (macroscopic scale) are within the same range as permeabilities in the regional (megascopic) scale. That means we can expect permeabilities in the range of 10^{-17} m^2 to 10^{-16} m^2 (10 to 100 μDarcy). The second reference is laboratory observations of a macroscopic hydraulic active fracture from the KTB pilot hole 3800 m deep, which indicate a narrow zone of high permeability and led to a rough estimate of the mean fracture aperture (DURHAM, 1997). Assuming *in situ* confining pressure, the upper bound of the mean fracture aperture should be at approximately 100 μm. Assuming this result with constant fracture apertures for all bonds of 100 μm leads to a calculated permeability in horizontal direction of 2.71×10^{-17} m^2 (27.1 μDarcy), which is in the expected range.

Comparison of the Results

The distribution functions of the fracture lengths on the microscopic and the megascopic scale appear similar to log-normal distributions. This result at first view can be explained by considering that small-scale structures are below or at the resolution threshold for the method applied and therefore do not yield a representative number. Consequently, the probability density function decreases with the decreasing length of the fractures. Therefore, a power-law distribution seems to be more plausible. This is in agreement with previous results from other authors (KOJIMA *et al.*, 1989; BALBERG *et al.*, 1991; NEUMAN, 1994; BOUR and DAVY, 1997; GAVRILENKO and GUÉGUEN, 1998), who suggest power-law distributions for the fracture lengths or distances. Combining the results from all investigated scales in a log-log presentation of fracture density versus mean fracture length for all networks which are above the percolation threshold, yields a clear indication for a power-law (Fig. 9). This graph is equivalent to a histogram and hence the density function is equivalent to a power-law distribution function for the fracture length. Which is in agreement with the above-mentioned citations. The exponent of this power law was calculated to -1.9 ± 0.05.

Table 2 gives an overview of the measured and derived parameters of the networks for the different scales. The geometrical parameters on different scales vary over several orders of magnitude. The only parameter which appears nearly independent from the scale effect is the connectivity of the networks and hence the percolation parameter, which describes the connectivity in a quantitative way. The percolation parameter leads to similar results for the microscopic and the megascopic scales, whereas it leads to higher values for the stochastic networks (Fig. 5). This discrepancy is due to the differences in the network constructions. The stochastic networks consist of constant fracture lengths and positions guided by a 2-D Poisson distribution. This results in adequate values for the percolation parameter which are consistent with the theoretical value. In our opinion, the low percolation parameter

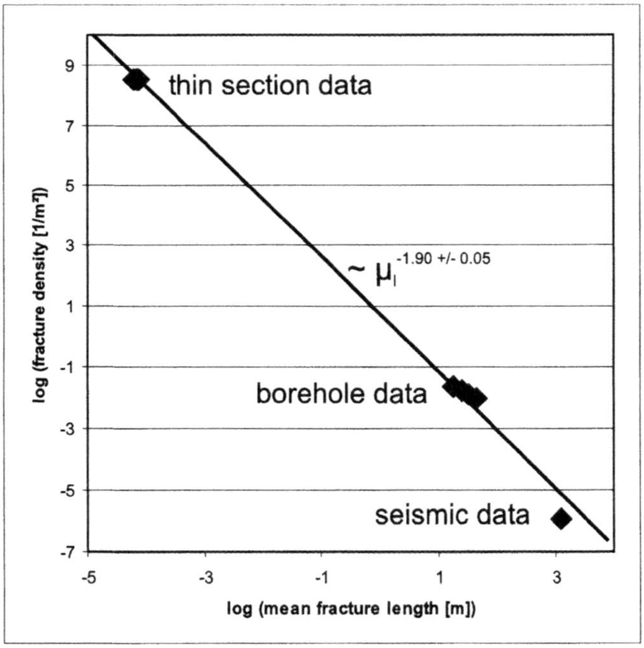

Figure 9
A log-log presentation of fracture density versus mean fracture length for the connected networks (above the percolation threshold) on all investigated scales. The mean fracture length μ_l follows a power-law distribution with an exponent of -1.90 ± 0.05, indicating the fractal nature of fractures in crystalline rock.

Table 2

Comparison of geometrical parameters of the networks on different scales and corresponding calculated percolation parameter p. The percolation parameter leads to similar values for the microscopic and the megascopic scales, whereas this parameter is considerably higher for the macroscopic stochastic network. This discrepancy can be explained by considering that constant fracture lengths were used for the macroscopic network and a 2-D Poisson distribution for the positions of the fractures. Despite these differences, the percolation parameter seems to be a confining parameter between the scales, because this quantity should be scale-invariant from the theoretical point of view, whereas the order network parameters vary extending to several orders in magnitude (a = fracture aperture, k = hydraulic permeability)

	Microscopic Scale	Macroscopic Scale	Megascopic Scale
l[m]	$63.1\ 10^{-6}$	30	1267
$\lambda_L[\mathrm{m}^{-1}]$	$2.21\ 10^{4}$	0.424	$1.43\ 10^{-3}$
$\lambda_A[\mathrm{m}^{-2}]$	$3.35\ 10^{8}$	$1.41\ 10^{-2}$	$1.13\ 10^{-6}$
N	146	824	147
$a[\mu\mathrm{m}]$	0.5	25	100
$k[\mathrm{m}^2]$	$\sim 10^{-19}$	$\sim 10^{-17}$	$\sim 10^{-17}$
p	1.31	12.7	1.82

for the microscopic and megascopic networks results on one hand from a clustering effect, i.e., the positions of the fractures are not randomly distributed. On the other hand, the skew distribution of the fracture lengths leads to an increase of connectivity compared to constant fracture lengths. The main reason is that the influence of a few large fractures is dominating. Despite these differences, the percolation parameter should be scale-invariant from the theoretical point of view, however this is only valid if the origin of the observed structures and discontinuities is similar.

For the macroscopic stochastic fracture systems representing the scale of the borehole, all constructed networks are equivalent concerning the initial data as mentioned. To decide which of these networks represents the hydraulic situation best requires additional information. The calculated permeability of these stochastic networks decreases with decreasing mean fracture length. Additional reduction of the fracture length to 10 m leads to a stochastic network which is not connected in horizontal and vertical directions, i.e., the stochastic network is below the percolation threshold and hence the permeability is zero. The hydraulic condition at the KTB site requires a macroscopic fracture connectivity, which is indicated by the communication experiments between the two boreholes (KESSELS and KÜCK, 1995; HUENGES *et al.*, 1997). These facts restrict the lower boundary of the mean fracture length and lead to a critical value of approximately 15 m at the percolation threshold. Assuming this specific hydraulic circumstance and considering a hydraulic network coherent with the necessary permeability range, the calculated stochastic network with a fracture length of 30 m seems to be the most probable of the three connected networks, because of the lowest percolation parameter. This is in agreement with suggestions of GUÉGUEN *et al.* (1991), who assume that natural media are close to the percolation threshold. ZOBACK and HARJES (1997) reported the existence of critically stressed, permeable fault zones in the crust at the KTB drilling site, even at a depth below the Franconian Lineament which approaches the brittle-ductile transition zone. According to BARTON *et al.* (1995), this is associated with an increase in permeability in these critically stressed faults. They examined the relationship between *in situ* stress and fluid flow in fractured and faulted crystalline rock and revealed that up to 80% of the hydraulically conductive faults and fractures appear to be critically stressed in frictional equilibrium with the *in situ* stress field. Furthermore, ZOBACK and HARJES (1997) demonstrated that micro-earthquakes in these fault zones at the KTB were induced by extremely small pore pressure perturbations (< 1 MPa). This is an indication of a hydraulic state above the percolation threshold with macroscopic hydraulic pathways of more than local extension.

Conclusion

Fundamental understanding of the origin, geometry and extension of fluid pathways in crystalline rock is still incomplete. We analysed a broad spectrum of fracture networks on different scales with varying geometrical parameters to estimate the fracture permeability at the KTB site. The problem is highly underdetermined, therefore, additional assumptions such as the percolation threshold and calibrations using measured permeability from hydraulic tests and core measurements were also included. The hydraulic condition at the KTB site requires a macroscopic fracture connectivity at the edge of the percolation threshold which restricts the lower boundary of the mean macroscopic fracture lengths. The existence of a power-law length distribution for fracture networks could be validated over several orders of magnitude. The percolation parameter as a scale-invariant from the theoretical point of view yields similar values regarding the microscopic and the megascopic scales. For both scales it is lower than expected from theory due to the skew distribution function of the fracture length and a clustering effect, which leads to nonrandom positions of the fractures. For the stochastic networks representing the macroscopic scale, the percolation parameter leads to higher values due to the special construction of the stochastic networks with constant fracture lengths and Poisson distributed positions. Despite these differences, the connectivity of the observed fracture systems seems to be a joint characteristic for all scales of the fractured rock of the KTB.

Acknowledgements

We extend thanks to E. Huenges and A. Gardei (GFZ Potsdam) for essential assistance and constructive discussions concerning microscopy and permeability measurements which were carried out at the GFZ Potsdam, and O. Kolditz (Univ. Hannover) for providing the finite element program *ROCKFLOW*. The stochastic networks were constructed using the program *FracMan* from Golder Associates. C. David and an anonymous reviewer are acknowledged for constructive criticism. This work was financially supported by the Deutsche Forschungsgemeinschaft (DFG) under grant En391/1.

References

ACUNA, J. and YORTSOS, Y. (1995), *Application of Fractal Geometry to the Study of Networks of Fractures and their Pressure Transient*, Water Resour. Res. *31*(3), 527–540.

BALBERG, I., BERKOWITZ, B., and DRACHSLER, G. E. (1991), *Application of a Percolation Model to Flow in Fractured Hard Rocks*, J. Geophys. Res. *96*, B6, 10,015–10,021.

BARTON, C. A., ZOBACK, M. D., and MOOS, D. (1995), *Fluid Flow Along Potentially Active Faults in Crystalline Rock*, Geology *23*(8), 683–686.

BERKOWITZ, B. and BALBERG, I. (1993), *Percolation Theory and Its Application to Groundwater Hydrology*, Water Resour. Res. *29(4)*, 775–794.

BERKOWITZ, B. and EWING, R.P. (1998), *Percolation Theory and Network Modeling Applications in Soil Physics, Surveys in Geophys.*, *19*, 23–72.

BOUR, O. and DAVY, P. (1997), *Connectivity of Random Fault Networks Following a Power-law Fault Length Distribution*, Water Resour. Res. *33*(7), 1567–1583.

BOUR, O. and DAVY, P. (1998), *On the Connectivity of Three-dimensional Fault Networks*, Water Resour. Res. *34*(10), 2611–2622.

BRACE, W. F. (1984), *Permeability of Crystalline Rocks: New in situ Measurements*, J. Geophys. Res. *89*, 4327–4330.

BROADBENT, S. R. and HAMMERSLEY, J. M. (1957), *Percolation processes. I. Crystals and Mazes*, Proc. Cambridge Philosophical Society *53*, 629–641.

BROWN, S. R. and BRUHN, R. L. (1998), *Fluid Permeability of Deformable Fracture Networks*, J. Geophys. Res. *103*, B2, 2489–2500.

BRUDY, M., ZOBACK, M. D., FUCHS, K., RUMMEL, F. and BAUMGÄRTNER, J. (1997), *Estimation of the Complete Stress Tensor to 8 km Depth in the KTB Scientific Drill Holes: Implications for Crustal Strength*, J. Geophys. Res. *102*, B8, 18,453–18,475.

CLAUSER, C. (1992), *Permeability of Crystalline Rocks*, EOS, Trans. AGU *73(21)*, 233–237.

DAVID, Ch. (1993), *Geometry of Flow Paths for Fluid Transport in Rocks*, J. Geophys. Res. *98*, B7, 12,267–12,278.

DAVID, C., GUÉGUEN, Y. and PAMPOUKIS, G. (1990), *Effective Medium Theory and Network Theory Applied to the Transport Properties of Rock*, J. Geophys. Res. *95*, B5, 6993–7005.

DAVY, P. (1993), *On the Frequency-length Distribution of the San Andreas Fault System*, J. Geophys. Res. *98*, B7, 12,141–12,151.

DAVY, P., SORNETTE, A. and SORNETTE, D. (1992), *Experimental Discovery of Scaling Laws Relating Fractal Dimensions and the Length Distribution Exponent of Fault Systems*, Geophys. Res. Lett. *19*, 4, 361–363.

DURHAM, W. B. (1997), *Laboratory Observations of the Hydraulic Behavior of a Permeable Fracture from 3800 m Depth in the KTB Pilot Hole*, J. Geophys. Res. *102*, B8, 18,405–18,416.

EMMERMANN, R. and LAUTERJUNG, J. (1997), *The German Continental Deep Drilling Program KTB: Overview and Major Results*, J. Geophys. Res. *102*, B8, 18,179–18,201.

GAVRILENKO, P. and GUÉGUEN, Y. (1998), *Flow in Fractured Media: A Modified Renormalization Method*, Water Resour. Res. *34*, 177–191.

GUÉGUEN, Y., CHELIDZE, T. and LE RAVALEC, M. (1997), *Microstructures, Percolation Thresholds, and Rock Physical Properties*, Tectonophys., *279*, 23–35.

GUÉGUEN, Y., DAVID, C. and GAVRILENKO, P. (1991), *Percolation Networks and Fluid Transport in the Crust*, Geophys. Res. Lett. *18*, 5, 931–934.

GUÉGUEN, Y., GAVRILENKO, P. and LE RAVALEC, M. (1996), *Scales of Rock Permeability*, Surveys in Geophys., *17*, 245–263.

GUÉGUEN, Y. and PALCIAUSKAS, V., *Introduction to the Physics of Rocks* (Princeton University Press, Princeton, New Jersey, 1994).

HADLEY, K. (1976), *Comparison of Calculated and Observed Crack Densities and Seismic Velocities in Westerly Granite*, J. Geophys. Res. *81*, 3484–3494.

HARJES, H. -P., BRAM, K., DÜRBAUM, H. -J., GEBRANDE, H., HIRSCHMANN, G., JANIK, M., KLÖCKNER, M., LÜSCHEN, E., RABBEL, W., SIMON, M., THOMAS, R., TORMANN, J. and WENZEL, F. (1997), *Origin and Nature of Crustal Reflections: Results from Integrated Seismic Measurements at the KTB Superdeep Drilling Site*, J. Geophys. Res. *102*, B8, 18,267–18,288.

HESTIR, K. and LONG, J. C. S. (1990), *Analytical Expressions for the Permeability of Random Two-Dimensional Poisson Fracture Networks Based on Regular Lattice Percolation and Equivalent Media Theories*, J. Geophys. Res. *93*, B13, 565–581.

HUENGES, E., ERZINGER, J., KÜCK, J., ENGESER, B. and KESSELS, W. (1997), *The Permeable Crust: Geohydraulic Properties down to 910 m Depth*, J. Geophys. Res. *102*, B8, 18,255–18,265.

HUENGES, E. and ZIMMERMANN, G. (1999), *Rock Permeability and Fluid Pressure at the KTB: Implications from Laboratory - and Drill Hole - Measurements*, Oil & Gas Science and Technology - Rev. IFP *54*, 6, 689–694.

KESSELS, W. and KÜCK, J. (1995), *Hydraulic Communication in Crystalline Rocks Between the Two Borehole of the Continental Deep Drilling Programme in Germany*, Int. J. Rock Mech. Min. Sci. Geomech. *32*, 37–47.

KOJIMA, K., TOSAKA, H. and OHNO, H. (1989), *An Approach to Wide-ranging Correlation of Fracture Distributions Using the Concept of Fractal*, 30th U.S. Symposium on Rock Mechanics, Morgantown, West Virginia, 211–218.

KOLDITZ, O., *Kontinuum-Konzept für das geklüftete Medium* (NLfB-Bericht 110327, Hannover 1992).

KOSAKOWSKI, G., *Simulation von Strömung und Wärmetransport im variszischen Grundgebirge: Vom natürlichen Kluftsystem zum numerischen Gitternetzwerk* (NLfB-Bericht 112046, Hannover 1994).

KOSAKOWSKI, G. (1996), *Modellierung von Strömungs- und Transportprozessen in geklüfteten Medien: Vom natürlichen Kluftsystem zum numerischen Gitternetzwerk*, VDI Verlag, *7*, 304, pp. 124.

LEGE, T., KOLDITZ, O. and ZIELKE, W., *Strömungs- und Transportmodellierung, Handbuch zur Erkundung des Untergrundes von Deponien und Altlasten, Bd. 2* (Springer Verlag, Berlin 1996).

NEUMAN, S. P. (1994), *Generalized Scaling of Permeabilities: Validation and Effect of Support Scale*, Geophys. Res. Lett., *21*, 5, 349–352.

PRIEST, S. D., *Discontinuity Analysis for Rock Engineering*, (Chapman & Hall, London 1993).

PYRAK-NOLTE, L. J., MONTEMAGNO, C. D. and NOLTE, D. D. (1997), *Volumetric imaging of Aperture Distributions in Connected Fracture Networks*, Geophys. Res. Lett., *24*, 18, 2343–2346.

ROBINSON, P. C. (1983), *Connectivity of Fracture Systems – A Percolation Theory Approach*, J. Phys. A: Math. Gen. *16*, 605–614

ROBINSON, P. C. (1984), *Numerical Calculations of Critical Densities for Lines and Planes*, J. Phys. A: Math. Gen. *17*, 2823–2830.

SAHIMI, M., *Flow and Transport in Porous Media and Fractured Rock – From Classical Methods to Modern Approaches* (VCH, Weinheim 1995).

SELYAKOV, V. I. and KADET, V. V., *Percolation Models for Transport in Porous Media – with Applications to Reservoir Engineering* (Kluwer Academic Publishers, Dordrecht 1996).

SHAPIRO S. A., AUDIGANE, P. and ROYER, J. -J. (1999), *Large-scale in situ Permeability Tensor of Rocks from Induced Microseismicity*, Geophys. J. Int. *137*, 207–213.

SHAPIRO S. A., HUENGES, E. and BORM, G. (1997), *Estimating the Crust Permeability from Fluid-injection-induced Seismic Emission at the KTB Site*, Geophys. J. Int. *132*, F15–F18.

STAUFFER, D. and AHARONY, A., *Introduction to Percolation Theory* (Taylor & Francis, London 1992).

WAGNER, G. A., COYLE, D. A., DUYSTER, J., HENJES-KUNST, F., PETEREK, A., SCHRÖDER, B., STÖCKERT, B., WEMMER, K., ZULAUF, G., AHRENDT, A., BISCHOFF, R., HEJL, E., JACOBS, J., MENZEL, D., NAND LAL, VAN DEN HAUTE, P., VERCAUTERE, C. and WELZEL, B. (1997), *Post-Variscan Thermal and Tectonic Evolution of the KTB Site and Its Surroundings, J. Geophys. Res. 102*, B8, 18,221–18,232.

WOLLRATH, J., *Ein Strömungs- und Transportmodell für klüftiges Gestein und Untersuchungen zu homogenen Ersatzsystemen* (Dissertation, Universität Hannover 1990).

ZIMMERMANN, G., KÖRNER, A. and BURKHARDT, H. (2000), *Hydraulic Pathways in the Crystalline Rock of the KTB*, Geophys. J. Int. *142*, 4–14.

ZOBACK, M. D. and HARJES, H. -P. (1997), *Injection-induced Earthquakes and Crustal Stress at 9 km Depth at the KTB Deep Drilling Site, Germany*, J. Geophys. Res. *102*, B8, 18,477–18,491.

(Received January 30, 2001, revised July 17, 2001, accepted August 1, 2001)

Pure appl. geophys. 160 (2003) 1087–1102
0033 – 4553/03/061087–16

Pure and Applied Geophysics

Mechanisms of Pore Pressure-stress Coupling which Can Adversely Affect Stress Measurements Conducted in Deep Tunnels

KEITH EVANS,[1,2] TORE DAHLØ,[3,4] and JAN-ATLE ROTI[5]

Abstract—Rock-stress measurements performed in boreholes drilled from a tunnel at a depth of 800 m indicate that the minimum stress has been reduced from pre-excavation levels by drawdown in pore pressure. The disturbance to the ambient stresses extends well beyond two tunnel diameters. A model is developed to explain the inferred strong coupling between the pore pressure and minimum stress in the rock mass around the tunnel. The model includes the effects of poro-elasticity of the intact rock and the stiffness of the fractures. For the field situation under study where the fractures are unusually complaint, both poro-elasticity and fracture compliance could be important contributors to the coupling. Changes in pore pressure are shown to produce reductions in minimum stress that are at least 50% of the drop in pore pressure and probably higher. In more general situations, the poro-elastic component is likely to be the more significant of the two mechanisms and alone accounts for a coupling factor of 50%. In this context, the importance of fractures in crystalline conditions probably lies more in their promotion of penetrative drainage than their contribution to coupling coefficient. The results highlight the importance of establishing the extent and distribution of pore pressure drawdown about galleries from which stress measurements are to be conducted.

Key words: Poro-elasticity, stress perturbation, hydromechanical coupling.

1.1. Introduction

Rock-stress measurements for geotechnical engineering are often made in boreholes drilled from tunnels. It is standard practice to locate test intervals more than 2 tunnel diameters from the tunnel wall to ensure that the measurements are not significantly affected by the stress field arising from the excavation (AMADEI and STEPHANSSON, 1997, p. 424). While this rule is adequate for eliminating the effects of the displacements arising from the free surface of the excavation, it does not consider

[1] ETH Zürich, Geology Institute, CH-8093 Zürich, Switzerland.
E-mail: keith.evans@erdw.ethz.ch

[2] Formerly at: ETH Zürich, Institute of Geophysics, CH-8093 Zürich and Solexperts, Schulstrasse 5, CH-8603 Schwerzenbach, Switzerland.

[3] Sintef, Division of Rock and Mineral Engineering, N-7034 Trondheim, Norway.

[4] Deceased. Tore Dahlø suffered a fatal accident whilst conducting the field measurements. He will be fondly remembered.

[5] Norplan AS, Hydropower Division, Holtaveien 5, N-1401 Ski, Norway.

stresses generated by the drawdown of formation pressure about the tunnel. Practical considerations usually demand that the tests be conducted in boreholes drilled behind the advancing tunnel face, commonly days but often weeks or months after the excavation front has passed. Thus, more often than not, sufficient time has elapsed to permit penetration of significant drawdown beyond two tunnel diameters. Since the stress and pore-pressure fields are coupled, the drop in pore pressure will generate a tensile stress component that will perturb the ambient stress (sometimes referred to as the primary stress) beyond a distance of two tunnel diameters. The problem is particularly acute in deep excavations where predisturbance formation pressures are high, and the drawdown of pore pressure within the local rock volumes sampled by the tests is large. In this and a related paper (DAHLØ *et al.*, 2002) we present an example of the adverse effects of drainage on stress measurements conducted from a tunnel in gneissic rock at a depth of about 800 m. A controlled experiment was performed to compare stress levels measured in boreholes drilled into relatively undrained rock ahead of the face with measurements conducted in an adjacent behind-face borehole. A significant systematic difference in minimum stress levels was found whose sense was consistent with drainage effects. In this paper we examine the mechanisms through which the pore pressure and stress field were coupled at the site with a view towards explaining the observations. The measurements themselves and background to the stress characterisation are described in the related paper.

1.2. Background

The experiment formed part of an extensive stress characterisation program of the Lower Kihansi hydroelectric project which is located on the eastern margin of the Tanzanian Plateau near the town of Mlimba in Tanzania. The project utilises the waters of the Kihansi river by channelling part of the river flow upstream of the Kihansi Falls into an inclined high pressure headrace tunnel leading to an underground powerhouse (SAIDI *et al.*, 2000). The tunnel trajectory was chosen to run under the ridge of a southward-extending promontory of the Plateau, thereby maximising the overburden. The intention was to leave the headrace tunnel unlined, which required that the minimum principal stress along the tunnel trajectory under operational conditions be at least 10 MPa. The stress characterisation was performed to verify that this constraint was met, and also to help orient the powerhouse cavern. Results of early tests conducted along the 5 m diameter Access tunnel suggested that the minimum stress level was slightly lower than required to allow an unlined completion. However, the large spatial variability in the minimum stress estimates with occasional values that were less than the expected predisturbance pore pressure gave reason to doubt that the estimates were true measures of the stresses that prevailed in the rock mass prior to excavation activity. Thus, the test program was extended and the test procedures progressively refined to provide better experimental control. By the end of the campaign, a total of 97 hydrofracture, 25 hydrojacking and 22 triaxial overcoring tests had been performed in 24 different boreholes (DAHLØ *et al.*, 2002). The program culminated in the experiment to assess the impact that drainage was having on the minimum stress levels measured.

The layout of the tunnel complex and test holes in the vicinity of the experiment site is shown in Figure 1. The site is located near the critical point where the headrace pressure tunnel enters the powerhouse complex and is some 2000 m along the Access tunnel that runs underneath the ridge. The local overburden is some 800 m. The promontory is composed of a Precambrian granitic gneiss with variable biotite content. The latter defines a foliation dipping at ~40° to WNW which imparts an elastic anisotropy to the rock. Elastic modulus and anisotropy vary considerably along the tunnel. Tests conducted on core sample from AT-1856 near the experiment site indicated an anisotropy in modulus of 25%. The density of the rock is about 2700 kgm^{-3}. Neglecting topographic effects, the overburden load expected for a depth of 800 m is 21 MPa. The predisturbance formation pressure at the experiment location is estimated as 6.0 MPa, based on hydrojacking tests and other observations discussed in DAHLØ *et al.*, (2001).

Overcoring measurements conducted from the Access tunnel at AT-1856 m indicated that the minimum stress was subhorizontal and oriented N-S, approximately co-axial with the ridge axis and tunnel trajectory. The orientation of induced hydrofractures in tests conducted beyond AT-1500 m support this view. The magnitude of the minimum stress derived from the overcoring measurements was 8.0 MPa, a value confirmed by a hydrojacking test conducted in a neighbouring borehole. The overcoring tests also indicated maximum and intermediate stress magnitudes of 37 MPa and 29 MPa, respectively, with the latter inclined at 47° to vertical. The significance of the deviation from vertical of all principal stresses is uncertain. Nevertheless, taken on face value, these data indicate that the rock mass is supporting very high shear stress levels.

The dominant fracture family throughout the length of the tunnel consists of E-W striking high-angle fractures which for the most part show no evidence of shear, and are commonly filled with soft, seemingly unconsolidated, 'weathered' material. Two members of this set are evident in Figure 1 and are the only persistent set seen beyond AT-1600 where the tunnel intersects a major foliation plane structure. Their spacing varies between 5 and 20 m with an average of 10 m. Since they lie in a plane that is normal to the axis of the Access tunnel, they effectively break the rock mass into somewhat regularly spaced, subvertical slabs which the Access tunnel penetrates almost normally. They are also normal to the minimum stress and thus support a relatively low effective normal stress. This, together with the nature of their filling, suggests that they are mechanically highly compliant to normal loading. Most importantly, they are hydraulically conductive, often significantly so, and have substantial fluid storage, which results in occasional large influxes of water into the tunnel when they are perforated. This fracture family thus imparts a strong

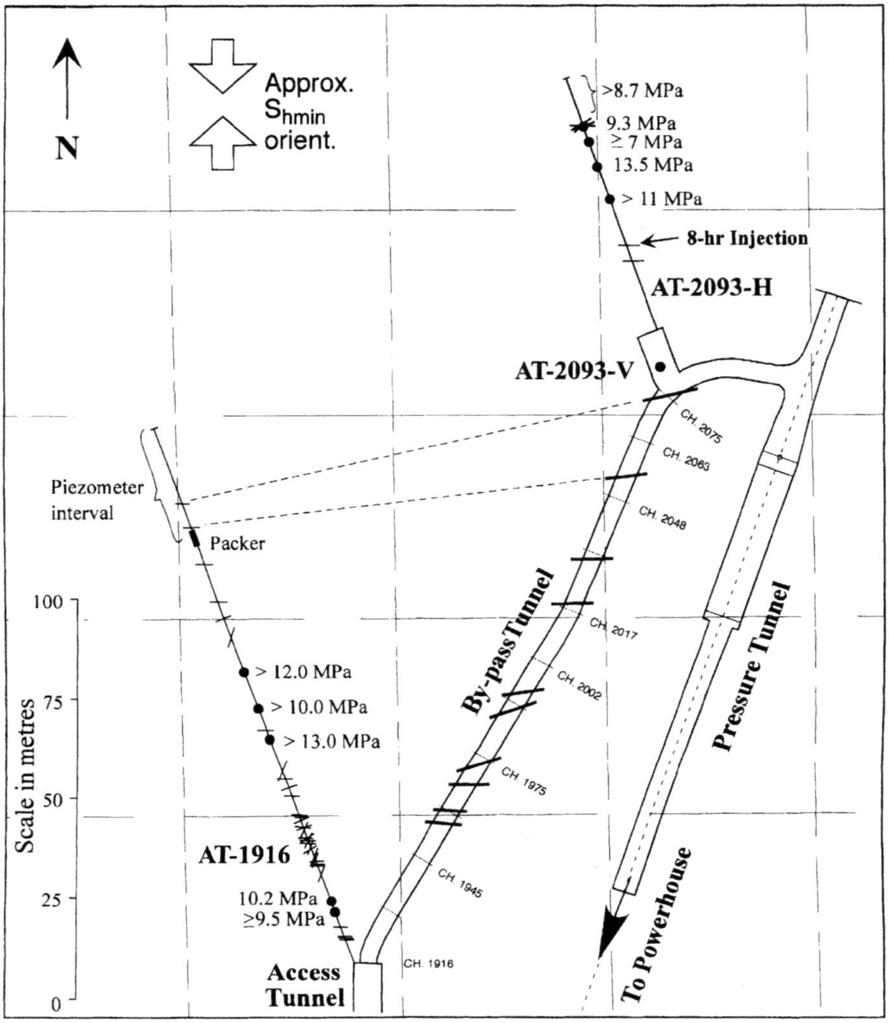

Figure 1
Layout of boreholes and tunnels in the vicinity of the By-pass tunnel. The two horizontal boreholes (AT-1916 and AT-2093H) were drilled ahead of the excavation face into relatively undrained rock whereas the vertical hole (AT-2093V) was drilled downwards from a gallery floor into rock which is likely to have suffered pore pressure depletion. The bold lines along the By-Pass tunnel indicate the intersection of the tunnel with major 'E-W' fractures. Two of these at CH-2055 and -2075 (shown by the dashed lines) show a direct hydraulic connection to the piezometer interval at the end of borehole AT-1916.

permeability anisotropy to the rock mass, leading to rapid, penetrative drainage of pore pressure in the radial direction but little drainage in the axial direction. This is demonstrated by the reaction of the piezometer installed at the end of horizontal borehole AT-1916 to the excavation of the By-pass tunnel (Fig. 1). This hole was drilled ahead of the excavation face into relatively undrained rock before the excavation of the By-pass tunnel, and the piezometer installed after stress tests in the hole had been completed. After sealing, the pressure in the piezometer interval rose to 5.1 MPa, demonstrating that the formation pressure ahead of the rock mass was relatively undrained. However, when the advancing excavation pierced members of the E-W fracture family which passed through the piezometer interval, the interval pressure dropped quickly reaching as low as 2 MPa before grouting of the fractures halted the depressurization (DAHLØ *et al.*, 2002). Thus, formation pressure drawdown from excavations is deeply penetrative in the radial direction and is very severe for ungrouted fractures.

1.3. Observations

To evaluate the impact of pore pressure drainage on the stress measurements, minimum principal stress estimates obtained from tests in the two subhorizontal holes drilled ahead of face were compared with estimates derived from a vertical hole drilled in relatively drained rock behind face. The borehole locations are shown in Figure 1. Borehole AT-1916 was 140 m long, and was drilled, tested and completed as a piezometer prior to excavation of the By-pass tunnel. Some six months later a further subhorizontal borehole, AT-2093H, was drilled 68 m ahead of the face and tested before the vertical hole was drilled 28 m downwards from the tunnel floor AT-2093. Most tests began with a hydrofracturing test on a 1 m interval that was free of natural fractures (as evidenced by intact core samples), and this was immediately followed by a hydrojacking test. Several long intervals which contained natural fractures were also subjected to hydrojacking tests, and a repeat test was performed on one interval after five days drainage. The location of tested intervals along the two horizontal boreholes is indicated on Figure 1 together with the observed jacking pressures in the horizontal holes. Steps were taken to minimise drainage from the horizontal boreholes during testing, although some depressurization of the rock mass about the holes was unavoidable. The test procedures and results are discussed in DAHLØ *et al.*, (2001), and are summarised in Table 1.

The results clearly show that the tests in the vertical hole yielded estimates for the minimum principal stress that were significantly lower than those in the horizontal holes, the difference being some 4 MPa. Whilst the sense of the contrast is consistent with that expected from drainage, it is not clear that it is entirely due to drainage since it could partly reflect natural stress heterogeneity in the rock mass. However, the consistency of the results in the two horizontal holes which measure stress in relatively large rock volumes immediately adjacent to the volume tested in the vertical hole argues against this. More importantly, the minimum principal stress estimates in the vertical hole of $\sim$5.5 MPa are less than the predisturbance pore pressure of 6.0 MPa, implying that the former has been lowered by the excavation. A

Table 1

Summary of minimum stress estimates from tests in boreholes near the bypass tunnel. The estimates reflect the lowest of the reliable jacking pressures observed in the tests in each borehole (see Table 5 of Dahlø *et al. (2002) for more complete listing of results). The formation pressure is a coarse estimate of the pore pressure prevailing in the rock mass near the test intervals at the time of the stress tests*

Borehole ID	Borehole description	Number of successful tests	Minimum stress estimate	Form. pressure during tests
AT-1916	Subhorizontal: ahead of face	5	≥9.5 MPa	Uncertain
AT-2093H	Subhorizontal: ahead of face	5	≥9.5 MPa	3.0–4.0 MPa
AT-2093V	Vertical: behind face	3	5.1–5.7 MPa	1.0–3.0 MPa

rough estimate of the formation pressure prevailing near the boreholes at the time of the stress tests is also listed in Table 1. It must be emphasised that the physical significance of this pressure is somewhat uncertain because the variation in the drawdown distribution about the holes and galleries is unknown. If they are taken as valid measures of the drawdown affecting the minimum principal stress estimates, and the predisturbance pore pressure is 6.0 MPa, then the implied drawdown is 2.0–3.0 MPa about the horizontal hole, and 3.0–5.0 MPa about the vertical, with higher values preferred in the latter case (Dahlø *et al.*, 2002). The contrast in the estimated drawdowns prevailing about the horizontal and vertical holes during the stress tests is thus 0.0–3.0 MPa with higher values preferred.

To quantify the coupling between pore pressure and stress, we define the coupling factor, $F_c = \Delta S_{xx}/\Delta P_p$, as the ratio of the change in minimum principal stress, ΔS_{xx}, induced by a pore pressure drawdown (ΔP_p). If we assume that the stresses in the rock volume under study were essentially homogeneous prior to excavation disturbance, and that the stress contrast is entirely due to drainage, then the results in Table 1 imply coupling factor values in excess of unity. Whether this is real, or an artefact of errors in the estimates of minimum stress and drawdown, or the assumption of predisturbance stress homogeneity is unclear. Nevertheless, the results provide clear evidence of drawdown-induced reductions in minimum stress, and suggest high values for the coupling coefficient, approaching if not exceeding unity. In the following we examine the mechanisms through which pore pressure drawdown leads to reduced minimum stress by developing a model to estimate the pore pressure-minimum stress coupling coefficient in the vicinity of the By-pass tunnel.

2.1. Model to Explain the Coupling of Pore Pressure Drawdown and the Axial Total Stress Along the Tunnel

The model geometry is illustrated in Figure 2. It consists of an infinitely long tunnel penetrating a rock mass that is broken into slices by uniformly-spaced,

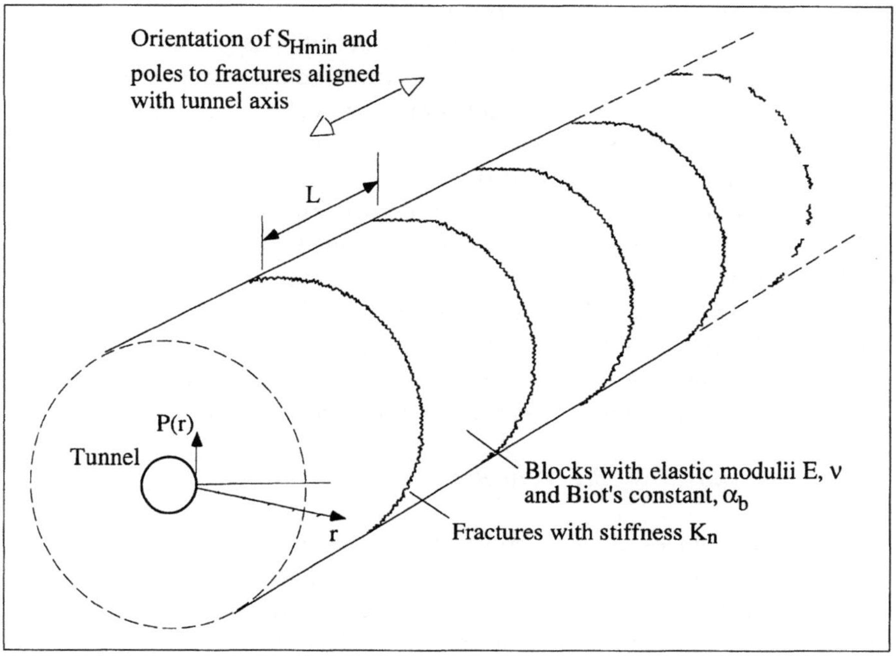

Figure 2
Illustration of the model used to estimate the value of the formation pressure-to-minimum stress coupling coefficient. An isotropic elastic medium is cut by a set of uniformly-spaced fractures representing the E-W, high-angle fracture set seen at Lower Kihansi. The fractures are deformable. An infinitely long tunnel cuts the medium in a direction perpendicular to the fractures. Drainage into the tunnel produces a zone of drawdown in formation pressure that is radially-symmetric about the tunnel. The change in pressure in this zone generates both a poro-elastic stress component along the axis of the tunnel, and also results in an aperture change of the fractures due to the change in effective stress.

elastically-compliant fractures that lie in the plane normal to the tunnel axis. The fractures are considered deformable with stiffness given by K_n, and have a spacing of L. The rock blocks between the fractures are taken to be elastically isotropic and characterised by Young's modulus, E, and Poisson's ratio, ν, both measured under drained conditions. Plane strain elasticity is assumed and compression taken as positive. The tunnel is surrounded by a radially-symmetric zone in which the formation pressure declines from far-field values to zero at the tunnel wall. Changes in the fluid pressure within the fractures and matrix are denoted by ΔP_f and ΔP_p respectively. The variation of formation fluid pressure within the drawdown zone is given by $P(r,t)$, where r is the distance from the centre of the tunnel. The extent out to which significant drawdown occurs is assumed to be large compared to the fracture spacing which field observations suggest is of the order of 10 m.

We seek to determine the effect of drawdown on the minimum principal stress acting about the Access tunnel. On symmetry grounds, the principal axes of the resulting stress perturbation will be oriented radially, circumferentially, and axially with respect to the tunnel. In our situation, the calculation is greatly simplified by the approximate alignment of the tunnel axis, the normal to the E-W fracture set, and the minimum principal stress axis. This implies that the axial principal stress, ΔS_{xx}, of the drawdown-related stress perturbation and the minimum principal stress, S_3, are aligned and thus may be simply added.

There are two contributions to the axial stress perturbation arising from drawdown. The first is a tensile poro-elastic stress developed within the intact rock blocks due to the reduction in pore pressure. The second arises from the reaction of the compliant fractures to changes in the 'effective stress for fracture closure' acting across them. The latter is the difference between the perturbation in total stress acting on the fracture and the internal fluid pressure. In the model formulation, both mechanisms are coupled together so that the axial total stress perturbation is continuous.

2.2. Poro-elastic Contribution

There are two stages to the determination of the poro-elastic stress contribution. The first is to calculate the 'body-force', ΔS_m, that the pore pressure disturbance, ΔP_p, applies to the medium. The second part is to calculate the stress redistribution within the medium that results from this body force. The problem is precisely analogous to the generation of thermo-elastic strains through local heating of a rock volume embedded within an elastic medium. In the following we assume the rock is isotropic. Consideration of the case for anisotropic rock, whilst desirable, is beyond the scope of this study.

The 'body force' generated within a volume of arbitrary geometry by a change in pore pressure, ΔP_p, is given by the change in total stress that would be generated if the rock was prevented from straining. For an isotropic material, (NUR and BYERLEE, 1971) showed that this 'strain-free-stress' is given by,

$$\Delta S_{m|\varepsilon} = -\alpha_b \Delta P_p \ , \tag{1a}$$

where

$$\alpha_b = \{1 - (C_i/C)\} \ . \tag{1b}$$

Here C_i is the intrinsic compressibility of the solid constituent and C is the compressibility of the 'bulk rock' measured under drained (i.e., constant pore pressure) conditions. The constant α_b is known as Biot's constant and physically represents the pore pressure-stress coupling factor that would apply if the rock was prevented from macroscopically deforming.

The next stage of the computation is to determine how much of this strain-free stress becomes relaxed as the material expands against the surrounding medium. Using the solution for the analogous problem in heat given by TIMOSHENKO and GOODIER (1951, eqn. (256), p. 448), it can be shown that the change in axial total stress at a distance r from the tunnel/borehole centre, $\Delta S_{xx}^{(1)}(r)$ generated by a pore pressure perturbation, $\Delta P_p(r)$, under conditions of plane strain is given by,

$$\Delta S_{xx}^{(1)}(r) = \alpha_b \frac{(1-2\nu)}{(1-\nu)} \Delta P_p(r) \ . \tag{2}$$

This shows that the axial stress change at distance r is dependent only on the change in pore pressure at that point. Thus the radial variation of the axial stress perturbation mirrors the axial variation in drawdown. For values of Poisson's ratio of 0.25 and α_b of 0.7, the coupling factor $F_c = \Delta S_{xx}/\Delta P_p$, implied by eqn. (2) is 0.47.

2.3. Fracture Stiffness Contribution

Within the framework of the model, changes in the effective stress acting across the fractures will result in a change in their aperture since they are deformable. The intact blocks are obliged to accommodate this aperture change by straining, and this generates a reactive back stress on the fracture which modifies the effective stress. In this sense, the fracture and blocks are coupled together. The dilation characteristics of the fractures are described by their normal stiffness, K_n. Specifically, the change in aperture (opening positive), Δa, resulting from a change in effective normal stress, σ_{xx}, is given by,

$$\Delta a = -\frac{1}{K_n} \Delta\sigma_{xx} \ . \tag{3}$$

Substituting the effective stress law for fracture closure, $\Delta\sigma_{xx} = \Delta S_{xx} - \Delta P_f$, where ΔP_f is the change in fluid pressure within the fracture, we have

$$\Delta a = -\frac{1}{K_n}\left(\Delta S_{xx} - \Delta P_f\right) \quad \text{opening positive} \tag{4}$$

When the aperture of the fractures changes by Δa, a strain develops in the blocks given by $\Delta a/L$, where L is the spacing between blocks. This strain generates the 'reactive back-stress, $\Delta S_{xx}^{(2)}(r)$, whose magnitude is approximately equal to:

$$\Delta S_{xx}^{(2)}(r) = \frac{E}{L} \Delta a \ , \tag{5}$$

where E is the Young's modulus of the blocks. The equation is approximate because the medium will resist the radial strain in the blocks to some degree. However, the inaccuracy is small (of the order of 10%).

2.4. Coupling of the Two Components

To couple the fracture and block stress components we assume that the radial gradients in both pore pressure and fracture pressure are sufficiently small that we can neglect the shear stresses S_{xr} and S_{rx}. Thus, the change in total axial stress, ΔS_{xx}, that develops within the blocks is given by the sum of the change due to the change in pore pressure within the blocks, $\Delta S_{xx}^{(1)}(r)$ and the reactive backstress from dilation of the fracture, $\Delta S_{xx}^{(2)}(r)$. That is:

$$\Delta S_{xx}(r) = \Delta S_{xx}^{(1)}(r) + \Delta S_{xx}^{(2)}(r) \ . \tag{6}$$

By stress continuity, this is also the total stress that acts on the fracture. Thus eqn. (4) can be written:

$$\Delta a = -\frac{1}{K_n}\left(\Delta S_{xx}^{(1)} + \Delta S_{xx}^{(2)} - \Delta P_f\right) \ . \tag{7}$$

Substituting eqs. (2) and (5) into (7) and rearranging we have

$$\Delta a = -\left\{K_n + \frac{E}{L}\right\}^{-1}\left(\alpha_b \frac{(1-2\nu)}{(1-\nu)}\Delta P_p(r) - \Delta P_f\right) \ . \tag{8}$$

Finally, the required expression for the change in total stress is obtained by substituting eqn. (8) into (5), and then substituting eqs. (5) and (3) into (6) to give,

$$\Delta S_{xx}(r) = \alpha_b \frac{(1-2\nu)}{(1-\nu)}\Delta P_p(r) - \left\{\frac{K_n L + E}{E}\right\}^{-1}\left(\alpha_b \frac{(1-2\nu)}{(1-\nu)}\Delta P_p(r) - \Delta P_f\right) \ . \tag{9}$$

This can be rearranged to show the contributions to the net stress change that separately arise from the changes in pore pressure within the rock blocks, ΔP_p, and fracture fluid pressure, ΔP_f:

$$\Delta S_{xx}(r) = \alpha_b \frac{(1-2\nu)}{(1-\nu)}\left\{1 - \frac{E}{(K_n L + E)}\right\}\Delta P_p(r) + \frac{E}{(K_n L + E)}\Delta P_f \ . \tag{10}$$

Equation (10) is valid provided the radial gradients in both pore and fracture fluid pressures are small. There are two cases of interest where this condition is reasonably satisfied which we shall refer to as the short-term and long-term solutions. The short-term can be thought of as representing the conditions prevailing shortly after the perforation of a fracture during borehole drilling or tunnel excavation. The major fractures are so permeable that significant drawdown will rapidly penetrate them, whereas the pore pressure within the blocks will remain essentially undisturbed. The expected change in axial total stress for this case is given by eqn. (10) with ΔP_p set to zero. The long-term solution represents the conditions that prevail once the drawdown has pervaded the blocks. The predicted axial stress change in this case is obtained by setting $\Delta P_p = \Delta P_f$. Later we estimate characteristic times for these two end-member cases.

3.1. Parameter Values Used in the Calculations

Biot's constant, α_b, generally takes values between 0 and 1. Perhaps surprisingly, low porosity crystalline rocks appear to take values in excess of 0.5, at least at confining pressures of the order of 10 MPa. The data given by BRACE (1965) suggest a value of α_b for Westerly granite (porosity = 0.01) of 0.62, and for charcoal granite (porosity = 0.003) of 0.73, both at 10 MPa confining pressure. Unfortunately no data are available for the granitic/biotite gneisses of Lower Kihansi. In calculations we shall use a value 0.7, and note that the very high deviatoric stress suggested by the triaxial overcoring results would be expected to promote microcracking and hence higher rather than lower values of Biot's constant.

The normal stiffness, K_n, of fractures generally increases with effective normal stress. For present purposes we shall assume that the E-W striking high-angle fractures in question are supporting an effective normal stress of 4 MPa under ambient conditions (i.e., $S_{\min}$ of 10 MPa and P_p of 6 MPa). In the case of natural fractures subjected to cyclic loading in the laboratory, plots of normal stiffness versus effective normal stress are often approximately linear and pass through the origin inasmuch as zero stiffness is associated with zero effective stress (BANDIS *et al.*, 1983; EVANS *et al.*, 1992; WALSH and GROSENBAUGH, 1979). This allows us to scale published values of *in situ* stiffness to values appropriate for a 5.0 MPa effective confining stress. Unfortunately there are rather few published determinations of normal stiffness from *in situ* tests. WALSH and GROSENBAUGH (1979) measured the normal stiffness of a 9 m^2 fresh vertical joint in Sherman (Wyoming) granite loaded with flat jacks. Extrapolated to 4 MPa, the stiffness is $\approx$40 GPa/m. EVANS and WYATT (1984) inferred a value of less than 20 GPa/m for a 100 m deep horizontal weathered fracture in granite. Scaled to 4 MPa, this value becomes $<$53 GPa/m. JUNG (1989) measured the stiffness of an extensive fracture in granite under 1.5 MPa confining stress and found 2.5 GPa/m. Scaled to 4.0 MPa this becomes 6.5 GPa/m. We consider it highly likely that the E-W, high-angle fractures at Kihansi are more compliant than the fractures in these examples because of the thick, unconsolidated, nature of the filling that is typically present. Thus, in later calculations we adopt extreme values for normal stiffness of 1 and 30 GPa/m with the preferred range being 3–10 GPa/m.

The Young's modulus that applies to the blocks between the fractures is estimated from laboratory test on samples of biotite gneiss from a borehole at AT-980 which yielded a value of 3e10 Pa. This is almost certainly higher than the value appropriate for deformation on the 10 m scale of the blocks: For although the blocks are relatively intact, they undoubtedly contain populations of fractures and cracks at the intermediate scale which are not represented in the core samples. In view of this we take the laboratory value as the upper bound, and ascribe a lower bound which is a factor of three smaller. Poisson's ratio is taken as 0.25.

3.2. Coupling Factors Computed from Parameter Values

The coupling factor, F_c, between the change in minimum stress and the change in formation pressure is given by $F_c = \Delta S_{xx}/P_f$. In Table 2 we list F_c values computed

Table 2

Values of the minimum stress-formation pressure coupling factor for short- and long-term solutions and various values of E and K_n. Other parameter values used in the calculations are $\alpha_b = 0.7$, $\nu = 0.25$ and a fracture spacing, $L = 10$ m

	Short-term ($\Delta P_p = 0$)		Long-term ($\Delta P_p = \Delta P_f$)	
	E = 1e10 Pa	E = 3e10 Pa	E = 1e10 Pa	E = 3e10 Pa
K_n = 1 GPa/m	0.5	0.75	0.73	0.87
K_n = 3 GPa/m	0.25	0.50	0.60	0.73
K_n = 10 GPa/m	0.09	0.23	0.52	0.59
K_n = 30 GPa/m	0.03	0.09	0.48	0.52

for the short-term and long-term drainage cases using the aforementioned bounds for the individual parameters.

The short-term solution represents the effects of depressuring the fractures alone. The predicted values of the coupling factor depend strongly on the values ascribed to both fracture stiffness and Young's modulus, and range between 0.03 and 0.75 for extreme values. For the preferred range of stiffness of 3–10 GPa/m, the predicted F_c values range between 0.09 and 0.5 depending upon the Young's modulus used. Thus, for the situation under consideration, it is unclear whether fracture compliance contributed significantly to the high coupling suggested by the field observations, although that is certainly possible. Once sufficient time has elapsed for the pore pressure in the blocks themselves to become drained, the coupling factor for the preferred range of fracture stiffness values rises to 0.52–0.73 depending on Young's modulus. It is of note that the long-term solution predicts a coupling factor in excess of ~0.5 regardless of the stiffness of the fracture.

The effect of varying the fracture spacing and Biot's constant is shown in Table 3 using median values for the other parameters. Not surprisingly, the coupling factor for the short-term solution is strongly dependent upon the assumed fracture spacing. However, the long-term solution is not, and converges to the value of 0.47 when fracture spacing is set to infinity. This corresponds to the coupling factor arising from poro-elasticity alone. Since fracture spacing is likely to vary from point to point along the gallery, the effect of closer fracture spacing on the local coupling factor will be to more greatly augment the local poro-elastic component arising from long-term drainage.

3.3. Time for Drawdown Penetration of the Blocks

It is of interest to develop some sense for the time required for significant drainage of the 'intact' blocks to occur. The major E-W striking, high-angle fractures are sufficiently permeable to justify the assumption that penetrative drainage occurs immediately after they are punctured by the tunnel or borehole. Once this occurs, the pressure-depletion front will move into the intact block from the bounding fracture

Table 3

Dependence of coupling factor on the assumed value of fracture spacing. Medium values are taken for all other parameters (i.e., K_n = 6 GPa/m; E = 2e10 Pa; ν = 0.25)

Frac. Spacing	Short-term ($\Delta P_p = 0$)	Long-term ($\Delta P_p = \Delta P_f$)	
	E = 2e10 Pa	$\alpha_b = 0.4$	$\alpha_b = 0.9$
L = 5 m	0.40	0.56	0.76
L = 10 m	0.25	0.45	0.70
L = 20 m	0.14	0.37	0.66

faces, as well as from the tunnel wall. Since we are interested in estimating the time required for significant pore-pressure drainage to penetrate to a depth that is at least as large as the fracture spacing, the scale length governing the diffusion time is essentially the fracture spacing. Moreover, the depletion of pore pressure within the 'intact' blocks occurs primarily through linear diffusion from the fracture faces, augmented by radial diffusion from the tunnel wall. We will consider only linear diffusion from the fracture faces, and estimate the time required for the diffusing pressure waves from each fracture face to produce a change in pore pressure in the centre of the block that is 90% of the pressure change in the fractures. The primary quantity governing this time is the diffusion constant, D, given by,

$$D = K/S_s \ , \tag{11}$$

where K is the hydraulic conductivity of the intact blocks and S_s is the coefficient of specific storage. The available hydro-test data from Kihansi do not place useful bounds upon the permeability or specific storage of the 'intact' blocks. Thus, for the scoping calculations we will consider permeability values ranging from 1e-16 m^2 to 1e-18 m^2, which are reasonable values for intact, core-size specimens of low-porosity crystalline rock. The corresponding range of hydraulic conductivities is 1e-9 to 1e-11 m/s. A reasonable value for the specific storage coefficient of intact, low-porosity crystalline rock is given by 1e-11 Pa^{-1}, mostly arising from microcrack dilation (EVANS *et al.*, 1992). We take this as the lower bound. The effect of macroscopic fractures within the blocks will be to add compliant porosity to the system thereby increasing the storage coefficient. Thus we set the upper bound an order of magnitude higher at 1e-10 Pa^{-1} (or 1e-6 m^{-1}), but note that such a high value is unlikely to coincide with a low hydraulic conductivity of 1e-11 m/s. Excluding this case (i.e., K = 1e-11 m/s and S_s = 1e-6 m^{-1}), the range of values of hydraulic diffusivity obtained from eqn. (11) is 1e-2 $m^2/s \leq D \leq$ 1e-4 m^2/s.

CARSLAW and JAEGER (1973, p. 91, eqn. (3.3.9)) give an expression for the pressure distribution as a function of time in a slab of thickness L subject to a step pressure change, $\Delta P(L/2,0)$, at both boundaries at zero time. The solution is in the form of an infinite series. Using this solution, the pressure change at the centre of the slab at time t, $\Delta P(0,t)$ is given by:

$$\Delta P(0,t) = \Delta P(L/2,0) \cdot 2 \cdot \sum_{n=0}^{\infty}(-1)^n\left\{\mathrm{erfc}\left((2n+1)\cdot\frac{L/2}{2\sqrt{Dt}}\right)\right\} , \tag{12}$$

where erfc( ) is the complementary error function. For most cases of practical interest, only the first two terms of this series are significant. If we consider the time required for the pressure drawdown at the depth $L/2$ (i.e., the 'centre' of the block) to reach 90% of the fracture drawdown, eqn. (12) becomes,

$$\frac{\Delta P(0,t)}{\Delta P(L/2,0)} = 0.9 = 2 \cdot \sum_{n=0}^{\infty}(-1)^n\left\{\mathrm{erfc}\left((2n+1)\cdot\frac{L/2}{2\sqrt{Dt}}\right)\right\} . \tag{13}$$

Retaining the first two terms of eqn. (13), we have for 90% drawdown in the slab centre,

$$0.45 = \left\{\mathrm{erfc}\left(\frac{L/2}{2\sqrt{Dt}}\right) - \mathrm{erfc}\left(\frac{3L/2}{2\sqrt{Dt}}\right)\right\} . \tag{14}$$

From published tables of the erfc function (e.g., CARSLAW and JAEGER, 1973) it can be seen that eqn. (14) is satisfied when,

$$t = t_{90\%} = \frac{1.03(L/2)^2}{D} , \tag{15}$$

which gives the time required for pore pressure drawdown at the centre of the slabs of intact rock between the fractures to attain 90% of the drawdown in the fractures. In Table 4 we list the 90%-drawdown times for the range of fracture spacing and diffusivity values considered appropriate. The times for 50% and 70% drawdown at the slab centre can be obtained by simply dividing the listed times by 2.7 and 1.75, respectively.

For a fracture spacing of 10 m the results indicate that near-complete drawdown of pore pressure in the intact blocks is likely to occur within hours to

Table 4

Time required for the pressure drawdown in the centre of the slab of rock between two parallel fractures a distance L apart to attain 90% of the drawdown in the fractures assuming 1-D linear diffusion from both faces of the fracture and a simultaneous step change in fracture pressure. The time is shown for different diffusivity constants and fracture spacings. The times to attain 50% and 70% drawdown are obtained by dividing the listed times by 2.7 and 1.75, respectively

Fracture spacing	D = 1e-2 m^2/s	D = 1e-3 m^2/s	D = 1e-4 m^2/s
L = 5 m	0.2 hrs	1.8 hrs	18 hrs
L = 10 m	0.7 hrs	7.1 hrs	71 hrs
L = 20 m	2.8 hrs	28 hrs	286 hrs
L = 40 m	11.5 hrs	115 hrs	1146 hrs

days. Only in the case where fracture spacing is greater than 20 m (in which case the tunnel wall is the dominant diffusion boundary) and the rock has either a very low permeability and/or a high storage coefficient will the penetrative drainage of the pore pressure within the blocks take place over periods of days to weeks. Thus, in most situations encountered along the Access tunnel, significant poro-elastic stresses are likely to have developed by the time stress measurements were conducted in 'behind-face' holes. In these cases the coupling factors obtained from the long-term solution to eqn. (10) are considered more appropriate for correcting the minimum stress estimates for the effects of drainage.

4.1. Discussion and Conclusions

The modelling results indicate that the minimum principal stress around the tunnels at Kihansi are likely to be reduced by at least 50% of the reduction in pore pressure, and that the disturbance extends well beyond 2 tunnel diameters. This supports the inference from tests near the By-pass tunnel that the coupling is very strong. The coupling coefficients computed from the model are very sensitive to the assumed value of fracture normal stiffness which we are unable to constrain precisely. For the preferred range of values of 3–10 GPa/m, and median values for the other parameters, the range of coupling coefficients is 0.17–0.4 for the short-term and 0.56–0.68 for the long-term solutions. The latter is the more appropriate for correcting the measured minimum stress estimates for the effect of drawdown. In the case of the Kihansi site, both poro-elasticity and fracture compliance could be important contributors to the coupling, although this is a consequence of the unusually high compliance of the E-W high-angle fracture set. More generally, we expect the poro-elastic component to be the more significant.

As to the general relevance of stress-pore pressure coupling for stress measurements conducted in other deep situations, the results demonstrate the importance of determining the depth to which significant depletion of pore pressure occurs about the excavation. If it is penetrative, then the zone of significant stress disturbance is likely to be similarly penetrative, even in crystalline rocks with low fracture density (provided drainage is penetrative). The depth of significant pore pressure depletion must be evaluated for each field situation, either by measurement, or through consideration of the local hydrogeology. Factors which promote deep penetration are high intrinsic permeability of the rock, and/or the presence of permeable fractures in the rock mass. Indeed, the importance of fractures in crystalline states probably lies more in their promotion of penetrative drainage than their contribution to the coupling coefficient through their normal compliance. It should also be borne in mind that drainage of the rock around the test holes themselves is difficult to avoid during drilling and testing, and this will serve to reduce the pore pressure within and about the rock volumes in which the tests are conducted.

Acknowledgements

The authors express their appreciation to the diligent efforts of all Norplan, Sintef and Solexperts field staff concerned with collecting the data that motivated this work. We are grateful to Tanesco for permission to publish this paper. ETH Institute of Geophysics contribution number 1191.

References

AMADEI, B. and STEPHANSSON, O. *Rock Stress and its Measurement* (Chapman and Hall 1997).

BANDIS, S. C., LUMSDEN, A. C. and BARTON, N. R. (1983), *Fundamentals of Rock Joint Deformation*, Int. J. Rock Mech. Min. Sci. and Geomech. Abstr. *6*, 249–268.

BRACE, W. F. (1965), *Some New Measurements of Linear Compressibility in Rocks*, J. Geophys. Res. *70*, 391–398.

CARSLAW, H. S. and JAEGER, J. C. *Conduction of Heat in Solids* (Oxford University Press, Oxford, UK 1973).

DAHLØ, T., EVANS, K. F., MYRVANG, A., and HALVORSEN, A. (2003), *Adverse Effects of Pore-pressure Drainage on Stress Measurements Performed in Deep Tunnels: An Example from the Lower Kihansi Hydroelectric Power Project*, Tanzania, Int. J. Rock Mech. Min. Sci. and Geomech. Abs. *40*, 65–93.

EVANS, K. F., KOHL, T., HOPKIRK, R. J., and RYBACH, L. (1992), *Modelling of Energy Production from Hot Dry Rock systems*, Final Report to Swiss National Energy Research Fund, Swiss Federal Institute of Technology/Polydynamics, Nationaler Energie-Forchungs-Fonds Projekt 359, 316 p. April 1992.

EVANS, K. F. and WYATT, F. (1984), *Water Table Effects on the Measurement of Earth Strain*, Tectonophysics *108*, 323–337.

JUNG, R. (1989), *Hydraulic in situ Investigations of an Artificial Fracture in the Falkenberg Granite*, Int. J. Rock Mech. Min. Sci. and Geomech. Abstr. *26*, 301–308.

NUR, A. and BYERLEE, J. D. (1971), *An Exact Effective Stress Law for Elastic Deformation of Rock with Fluids*, J. Geophys. Res. *76*, 6414–6419.

SAIDI, F. X., LINDEMARK, J., and WILHELM, V. C. (2000), *Kihansi Falls Pave the Way for Power Development*, Int. Water Power and Dam Construction, March, 18–22.

TIMOSHENKO, S. and GOODIER, J. N. *The Theory of Elasticity* (McGraw-Hill, New York 1951).

WALSH, J. B. and GROSENBAUGH, M. A. (1979), *A New Model for Analysing the Effects of Fractures on Compressibility*, J. Geophys. Res. *84*, 3532–3536.

(Received February 6, 2001, revised/accepted October 17, 2001)

Pure appl. geophys. 160 (2003) 1103–1124
0033–4553/03/061103–22

Pure and Applied Geophysics

Role of Stress-controlled Flow Pathways in HDR Geothermal Reservoirs

TAKATOSHI ITO[1] and KAZUO HAYASHI[1]

Abstract—We addressed effects of *in situ* stress on the formation of flow pathways in fractured rocks in geothermal reservoirs, especially for HDR projects. Here we focused on fractures which are critically-stressed, causing shear slip in a current stress field. The sliding is likely to break sealing in the fractures and, as a result, to increase their permeability. Such a mechanism is possibly significant under high-temperature conditions at geothermal fields because of temperature enhancement on chemical reactions for the sealing. We present a procedure to estimate the orientation of the critically-stressed fractures relative to axes of *in situ* principal stress with the aid of the Mohr diagram. The procedure allows us to evaluate intuitively how the orientation changes with factors such as magnitude of *in situ* principal stresses and pore pressure. We applied the procedure to estimate possible orientations of the critically-stressed fractures in major HDR test sites. Results show that overall alignments of microseismicity during hydraulic stimulation are within predicted ranges for possible orientations of the critically-stressed fractures. Furthermore, it was found that if the state of *in situ* stress is not favorable to cause sliding of natural fractures, it tends to lead a high wellhead pressure at hydraulic stimulation and a high recovery rate at circulation. On the other hand, if the state of *in situ* stress is favorable for sliding, it tends vice versa to lead a low wellhead pressure at hydraulic stimulation and a low recovery rate at circulation.

Key words: *In situ* stress, fluid flow, Coulomb criterion, Mohr diagram, HDR, geothermal energy.

1. Introduction

The HDR (Hot Dry Rock) heat mining technology, which utilizes the natural heat contained in the earth's crust, can provide a widely available source of nonpolluting energy. The system is composed of injection well(s), underground heat exchanger, i.e., reservoir, and production well(s). Water is pumped down the injection well, flows into the reservoir to be heated, and is back up through the production well. In the original HDR concept, disk-shaped fractures were supposedly created artificially by hydraulic stimulation to be used as the heat exchanger. However, according to field observations, it has become apparent that the hydraulic stimulation activates pre-existing natural fractures, and that injected water flows into

---

[1] Institute of Fluid Science, Tohoku University, Sendai 980-8577, Japan.
E-mails: ito@ifs.tohoku.ac.jp, hayashik@ifs.tohoku.ac.jp

them. If there is no natural fracture across a pressurized interval of an injection well, pressurized water creates and extends new fractures, however they soon meet with natural fractures within a short distance, i.e., possibly few tens of meters at most, from the injection well. The new fractures then stop growing, and afterwards injected water flows into the natural fractures. Thus it is now recognized that the newly created fractures contribute to connect injection/production wells with natural fractures rather than a role as a heat exchanger. As a result, for construction of engineered geothermal reservoirs, it becomes important to be able to predict flow pathways of pressurized water in dispersed natural fractures with various orientations. We present here a concept which would give bases for predicting the flow pathways. The concept was verified through comparing its prediction with results of hydraulic stimulation carried out in major HDR test sites, the Fenton Hill, Rosemanowes and Soultz sites.

2. *Critically-stressed Fracture*

It is often assumed for fluid flow in fracture systems that the direction of maximum bulk permeability is closely aligned with the dominant trend of fractures. However, it is not always appropriate, since all fractures are not equally permeable, i.e., some fractures are considerably more permeable than others. Regarding this problem, BARTON *et al.* (1995) reported from field observations that the permeability of critically-stressed fractures is substantially higher than that of fractures which are not optimally oriented in a current stress field. This finding is supported by other case studies (HICKMAN *et al.*, 1997, ITO *et al.*, 2000). The critically-stressed fractures are fractures which are in a critical condition to slide against friction between the fracture planes. They are represented as fractures which lie close to the Coulomb failure line on 3-D Mohr diagrams. The Coulomb criterion is defined as

$$\tau = \mu(S_n - P_p) \ , \tag{1}$$

where τ and S_n are shear and normal stresses on a fracture respectively, μ is the coefficient of friction along the fracture plane, and P_p is pore pressure.

Mechanics behind the Barton's finding could be drawn as follows. Most fractures may be permeable immediately following their formation. However, over time, healing and sealing of the fractures occur due to water/rock chemical reactions which reduce fracture-induced permeability. Even in such a case, if shear sliding occurs on fractures by a change in the state of *in situ* stress and/or an increase in pore pressure associated with certain kinds of geological event, it regains back high permeability of the fractures. Namely, brecciation and increases in porosity are expected to occur along the shear plane due to the shear sliding and there will also be an increase in the permeability due to increases of the surface roughness of the fracture plane (YEO *et al.*, 1998) as well as the breakdown of seals in fractures (OLSEN *et al.*, 1998). As a result, the fractures under critically-stressed conditions for frictional sliding may tend to have high permeability in comparison to other fractures. The tendency should be enhanced especially in geothermal fields, since high temperature will activate the water/rock chemical reaction and cause healing and sealing of fractures. Thus it seems logical to consider that flow pathways in the HDR reservoir are formed along the critically-stressed fractures.

3. *Orientation of the Flow Pathways*

Now we examine a relationship in orientation between *in situ* stresses and fractures which are critically-stressed for sliding. It is assumed here that the stress field and pore pressure are uniform throughout the rock volume in question, that is, uniform at the scale of the HDR reservoir. We discuss this assumption later in this section. It is also assumed that poles of fracture planes are inclined to principal axes of the *in situ* stresses. The *in situ* principal stresses are denoted as $S_i(i = 1, 2, 3, |S_1| \geq |S_2| \geq |S_3|)$, and direction cosines between normal to a fracture plane and axes of S_1, S_2 and S_3 are denoted as L, M and N, respectively. Stress vector (resultant stress) acting on the fracture plane is denoted as $\boldsymbol{p}$ which has components $p_i(i = 1, 2, 3)$ for the principal stress axes. Then, taking Cauchy's formula into account, the magnitude p of $\boldsymbol{p}$ is given by

$$p^2 = p_1^2 + p_2^2 + p_3^2 = L^2S_1^2 + M^2S_2^2 + N^2S_3^2 \ . \tag{2}$$

$\boldsymbol{p}$ is also resolved into two components that are normal and parallel with the fracture, i.e., the normal stress S_n and the shear stress τ, respectively, so that

$$p^2 = S_n^2 + \tau^2 \ . \tag{3}$$

The normal stress S_n is

$$S_n = Lp_1 + Mp_2 + Np_3 = L^2S_1 + M^2S_2 + N^2S_3 \ . \tag{4}$$

Substituting Eqs. (2) and (4) into Eq. (3) gives

$$|\tau| = \left\{L^2M^2(S_1 - S_2)^2 + M^2N^2(S_2 - S_3)^2 + N^2L^2(S_3 - S_1)^2\right\}^{\frac{1}{2}} \ . \tag{5}$$

The stress components S_n and τ on a fracture can be plotted as a point in the shaded area on a 3-D Mohr diagram as shown in Figure 1a, where the abscissa is taken as the effective normal stress, $S_n - P_p$. On the diagram, the Coulomb criterion, Eq. (1), is represented as a straight line assuming an appropriate value of the friction coefficient μ, so that the hatched area represents stress states which satisfy the criterion. If S_n and τ on a fracture is estimated as a point in the hatched area by Eqs. (4) and (5), the fracture is to be under the critical condition for sliding.

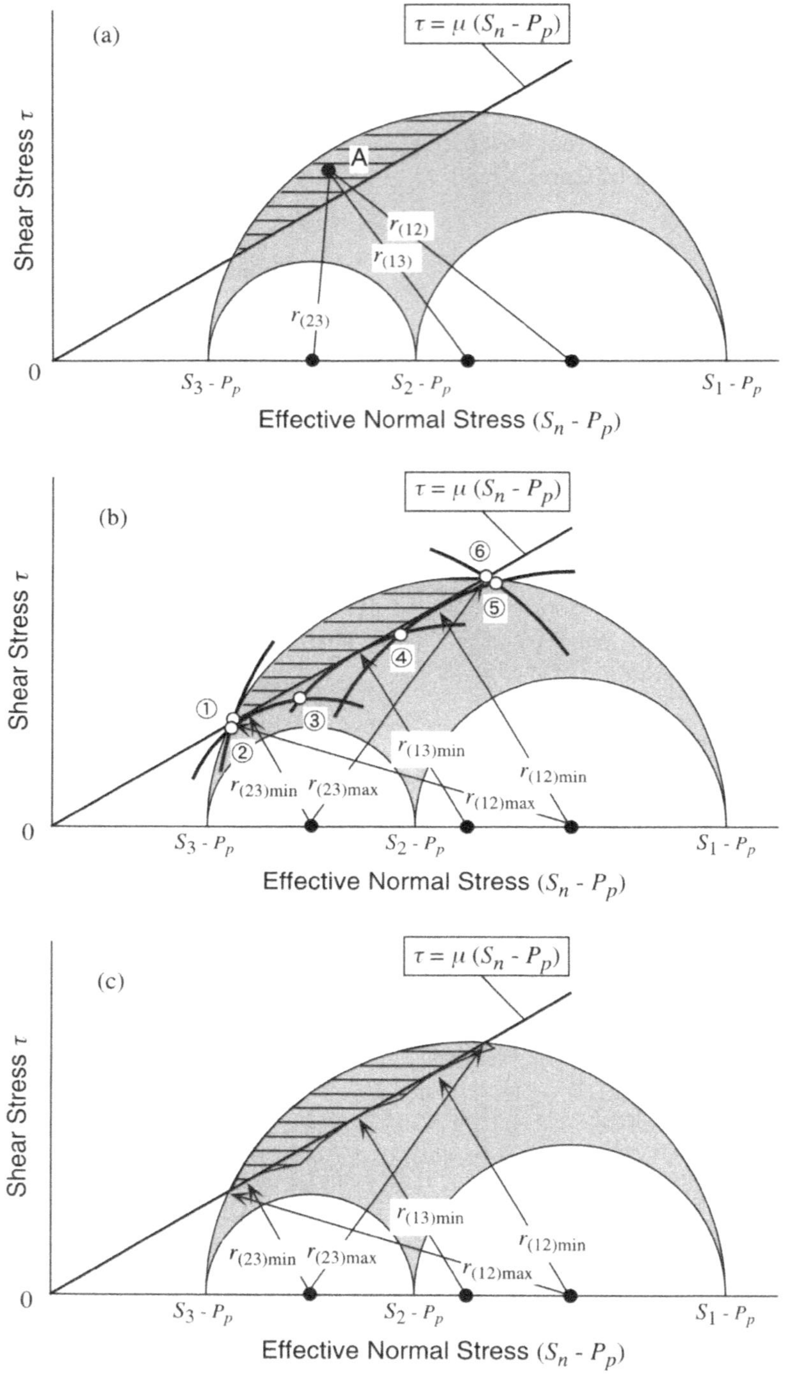
(a)
τ = μ (Sn - Pp)
A
r(12)
r(13)
r(23)
Shear Stress τ
0
S3 - Pp
S2 - Pp
S1 - Pp
Effective Normal Stress (Sn - Pp)
(b)
τ = μ (Sn - Pp)
①
②
③
④
⑤
⑥
r(13)min
r(12)min
r(23)min
r(23)max
r(12)max
Shear Stress τ
0
S3 - Pp
S2 - Pp
S1 - Pp
Effective Normal Stress (Sn - Pp)
(c)
τ = μ (Sn - Pp)
r(13)min
r(12)min
r(23)min
r(23)max
r(12)max
Shear Stress τ
0
S3 - Pp
S2 - Pp
S1 - Pp
Effective Normal Stress (Sn - Pp)

It should be noted here that the sliding will change the stress state in the vicinity of the fracture as relaxing the critical condition for sliding. However, such a phenomenon is not taken into account for deriving Eqs. (4) and (5). This fact raises the potential problems as follows. First, if the sliding has occurred for a fracture and it has completely relaxed the stress state causing the fracture sliding, the fracture will no longer slide. In addition, if a fairly long time has elapsed after the sliding, the elapsed time might allow the sealing and healing to be reformed in the fracture. Such a situation might result in low permeability of the fracture at present, while Eqs. (1), (4) and (5) indicate that the fracture should be under the critical condition for sliding. However, McGARR (1999) reported from the analyses of both mining-induced and natural earthquakes that the fracture/fault sliding drops the average shear stress loading the fracture/fault plane to cause slip by just a few percent, i.e., around 6%. Thus McGarr's finding suggests that even after a fracture has slid, the fracture still may be almost under the critical condition for sliding currently, as long as the state of *in situ* stresses, i.e., S_1, S_2 and S_3, causing the fracture slip remains unchanged. On the other hand, all critically-stressed fractures will not slide simultaneously. Once one fracture has slid, the sliding will change the stress state in proximity to the fracture, and it might inactivate the adjacent fractures for sliding. However, taking into account the theoretical stresses around a fully-shearing fracture (POLLARD and SEGALL, 1987), the McGarr's finding suggests that the stress change caused by sliding is limited in a region within a distance from the fracture plane which is considerably smaller than the fracture size, i.e., the distance possibly within 10–20% of the fracture size according to the amount of sliding-induced drop in the shear loading on the fracture. Hence, to the degree of the fracture in question is separated from the previously-sliding fracture by a distance more than 10–20% of the size of the latter fracture, the sliding of the former fracture may not be interrupted by the sliding of the latter fracture. Thus, Eqs. (4) and (5) possibly do not cause large errors in evaluating the trend of the critically-stressed fractures among dispersed natural fractures, even though those equations tell us just the stresses on the fractures before sliding.

Next, we consider the orientation of the critically-stressed fracture. For a point A on the Mohr diagram of Figure 1a, the point location is related to the fracture orientation as follows (e.g., ATSUMI, 1978);

$$r_{(12)} = \sqrt{\left(\frac{S_1 - S_2}{2}\right)^2 + N^2(S_1 - S_3)(S_2 - S_3)} \tag{6}$$

◀

Figure 1
Representation of stress states acting on fracture planes using Mohr circles. The shaded areas show the possible stress state for fractures arbitrarily oriented to the principal stresses S_1, S_2 and S_3. The hatched areas show the stress state on fractures under the critical condition for sliding in accordance with the Coulomb criterion. See text in more detail.

$$r_{(13)} = \sqrt{\left(\frac{S_1 - S_3}{2}\right)^2 - M^2(S_2 - S_3)(S_1 - S_2)} \tag{7}$$

$$r_{(23)} = \sqrt{\left(\frac{S_2 - S_3}{2}\right)^2 + L^2(S_1 - S_2)(S_1 - S_3)} \ , \tag{8}$$

where $r_{(12)}, r_{(13)}$ and $r_{(23)}$ are the distances between the point A and centers of the Mohr circles with diameters of $(S_1 - S_2)$, $(S_1 - S_3)$ and $(S_2 - S_3)$, respectively. Note that the distance $r_{(ij)}(i, j = 1, 2, 3)$ has the same dimension with S_i and P_p, and it must be defined on the Mohr diagram drawn by using scales for τ and $(S_n - P_p)$ with the same magnification. As can be seen from Eqs. (6)–(8), a unique orientation of a fracture, i.e., a unique set of (L, M, N), corresponds to a unique location of the point A, i.e., a unique pair of S_n and τ on the fracture, one by one for a certain set of S_1, S_2 and S_3. This fact means that from the location of a point on the Mohr diagram, we can determine the orientation of the fracture which is under the stress state denoted by the point. From Eqs. (6)–(8) we find that L, M and N are given as functions of the distances $r_{(12)}, r_{(13)}$ and $r_{(23)}$ by

$$|L| = \sqrt{\frac{r_{(23)}^2 - \left(\frac{S_2 - S_3}{2}\right)^2}{(S_1 - S_2)(S_1 - S_3)}} \ , \tag{9}$$

$$|M| = \sqrt{\frac{\left(\frac{S_1 - S_3}{2}\right)^2 - r_{(13)}^2}{(S_2 - S_3)(S_1 - S_2)}} \ , \tag{10}$$

$$|N| = \sqrt{\frac{r_{(12)}^2 - \left(\frac{S_1 - S_2}{2}\right)^2}{(S_1 - S_3)(S_2 - S_3)}} \ . \tag{11}$$

These equations show that, for example, the maximum and minimum values of L for the critically-stressed fractures are limited by the maximum and minimum distances between a center of the Mohr circle with diameter of $(S_2 - S_3)$ and a point contained in the hatched area shown in Figure 1b, where the maximum and minimum distances are denoted as $r_{(23)\max}$ and $r_{(23)\min}$, respectively. Namely, a range of L for the critically-stressed fractures is given by

$$\sqrt{\frac{r_{(23)\max}^2 - \left(\frac{S_2 - S_3}{2}\right)^2}{(S_1 - S_2)(S_1 - S_3)}} \geq |L| \geq \sqrt{\frac{r_{(23)\min}^2 - \left(\frac{S_2 - S_3}{2}\right)^2}{(S_1 - S_2)(S_1 - S_3)}} \ . \tag{12}$$

In the same way, ranges of M and N for the critically-stressed fractures are given by

$$\sqrt{\frac{\left(\frac{S_1-S_3}{2}\right)^2-r_{(13)\min}^2}{(S_2-S_3)(S_1-S_2)}} \geq |M| \geq 0 \ , \tag{13}$$

$$\sqrt{\frac{r_{(12)\max}^2-\left(\frac{S_1-S_2}{2}\right)^2}{(S_1-S_3)(S_2-S_3)}} \geq |N| \geq \sqrt{\frac{r_{(12)\min}^2-\left(\frac{S_1-S_2}{2}\right)^2}{(S_1-S_3)(S_2-S_3)}} \ , \tag{14}$$

where $r_{(13)\min}$ is the minimum distance between a center of the Mohr circle with diameter of $(S_1 - S_3)$ and the hatched area, and $r_{(12)\max}$ and $r_{(12)\min}$ are the maximum and minimum distances between centers of the Mohr circle with diameter of $(S_1 - S_2)$ and the hatched area respectively (see Fig. 1b). Note that since the maximum distance between the center of the Mohr circle with diameter of $(S_1 - S_3)$ and the hatched area is $(S_1 - S_3)/2$, the minimum absolute value of M is zero as can be seen from Eq. (10). Equations. (12)–(14) allow us to define the orientations of the critically-stressed fractures from the boundary values of the three radii, $r_{(12)}, r_{(23)}$, and $r_{(13)}$, on the Mohr diagram. However, from the mathematical standpoint, Eqs. (12)–(14) correspond to the necessary conditions but not the sufficient conditions for a fracture with an orientation of (L, M, N) to be critically-stressed for sliding. Therefore, the fracture orientations defined by Eqs. (12)–(14) are slightly overestimated as the orientations of the critically-stressed fractures. Namely, in Figure 1c, the hatched region manifests the stress states on the fractures whose orientations are defined by Eqs. (12)–(14), however, the hatched region is slightly larger than the region in which the Coulomb criterion is exactly satisfied, i.e., the hatched region in Figure 1b. In other words, Eqs. (12)–(14) define the orientations of the critically-stressed fractures for the case that the Coulomb failure line on the Mohr diagram is approximated by a series of circles with radii of $r_{(12)\min}, r_{(12)\max}, r_{(13)\min}, r_{(23)\min}$ and $r_{(23)\max}$ (see Fig. 1b).

The ranges of L, M and N given by Eqs. (12)–(14) are schematically illustrated in lower-hemisphere equal-area stereonets as Figures 2a, b and c, respectively, where the rock volume is assumed to be in a normal faulting stress regime. Namely, it is assumed that $S_1 = S_v, S_2 = S_H$ and $S_3 = S_h$, where S_v is the vertical stress, S_H and S_h are the maximum and minimum horizontal stresses, respectively. Possible ranges in orientation of poles to the critically-stressed fractures are determined approximately as overlaps between the areas shown in Figures 2a–c. In Figure 2d, the overlaps between the areas shown in Figures 2a–c are illustrated as the areas bounded by bold lines. The poles at the corners of the areas, e.g., the poles which are numbered 1, 2,..., 6 in Figure 2d, are defined as combinations of the maximum and minimum values of L, M and N (see Eqs. (12)–(14)), taking into account the relation $L^2 + M^2 + N^2 = 1$. For example, the pole 3 in Figure 2d is defined as $(L, M, N) = \left(L_{\min}, M_{\max}, \sqrt{1 - L_{\min}^2 - M_{\max}^2}\right)$, where $L_{\min}$ and $M_{\max}$ are the minimum and maximum values of L and M, respectively. The stress states on the fractures with

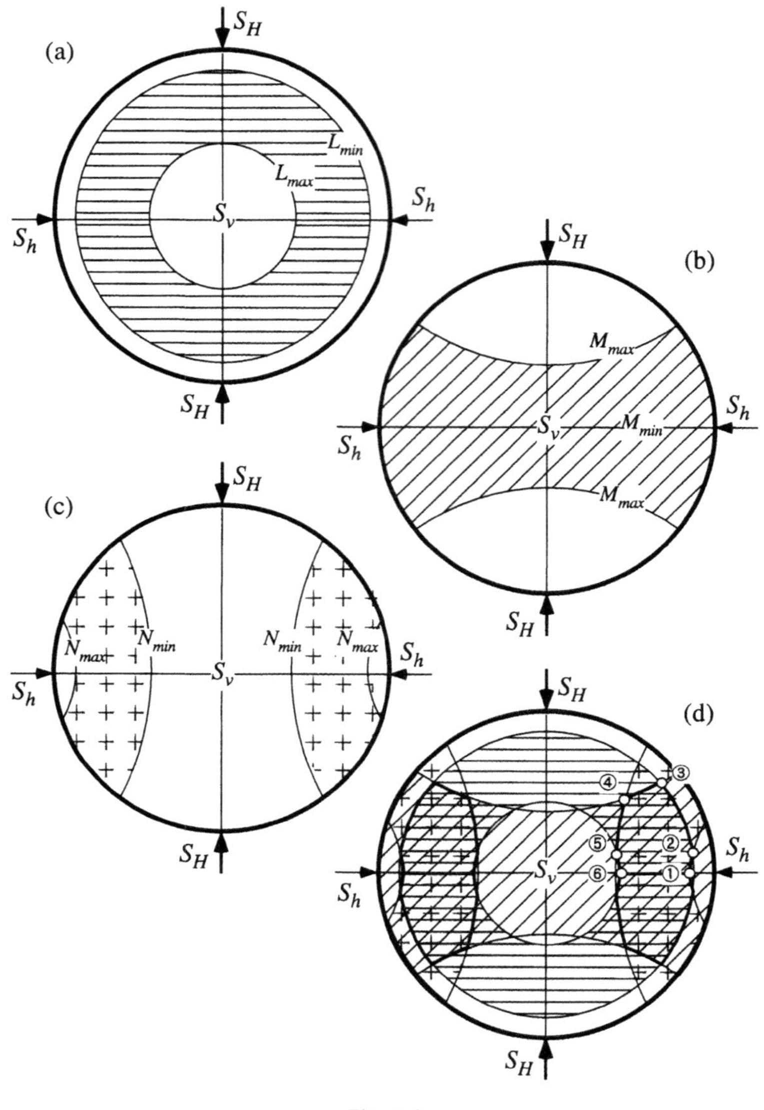

Figure 2
Schematically illustrated possible range in orientation of poles to the critically-stressed fractures in the case of a normal faulting stress regime. Lower-hemisphere equal-area stereonets are used. Each figure shows the possible range which satisfies each condition for (a) L, (b) M and (c) N. In (d), overlaps of those ranges are illustrated as areas bounded by bold lines. The stress states on the fractures with the numbered poles in (d) are represented by the points with the same numbers in Figure 1b.

the poles 1, 2,. . ., 6 in Figure 2d, are represented as the points with the same numbers respectively on the Mohr diagram of Figure 1b. As shown in Figure 1b, the numbered points are close to but not exactly on the Coulomb failure line except for the points 1 and 6. This is because, as discussed above, Eqs. (12)–(14) overestimate the range in orientation of the critically-stressed fractures. For example, the point 3 in Figure 1b will move to the Coulomb failure line if both $r_{(13)\min}$ and $r_{(23)\min}$ become larger by a few percent. As both $r_{(13)\min}$ and $r_{(23)\min}$ become larger, the values of $L_{\min}$ and $M_{\max}$ become larger and smaller, respectively in accordance with Eqs. (12) and (13). This means that if the point 3 in Figure 1b moves toward the Coulomb failure line, the pole 3 moves inward of the area bounded by the bold line in Figure 2d, where the pole's movement can be estimated from the point's movement, Eqs. (12) and (13). Therefore, the actual boundary on stereonets in orientation of the critically-stressed fracture should be represented as contours which are approximately equal to the bold lines in Figure 2d but are slightly rounded at the corners of those lines.

On the other hand, Figure 3 indicates the great circles of fractures whose poles are located at the corners of the bold lines in Figure 2d. Those great circles indicate fractures exposed to stress states on the boundary in which the Coulomb criterion is satisfied, while they are slightly overestimated as discussed above. In this sense, those great circles are considered to show representative orientations of the critically-stressed fractures. On the hypothesis that the fractures under the critical condition for sliding in a current stress field are permeable, if a fracture exists whose orientation

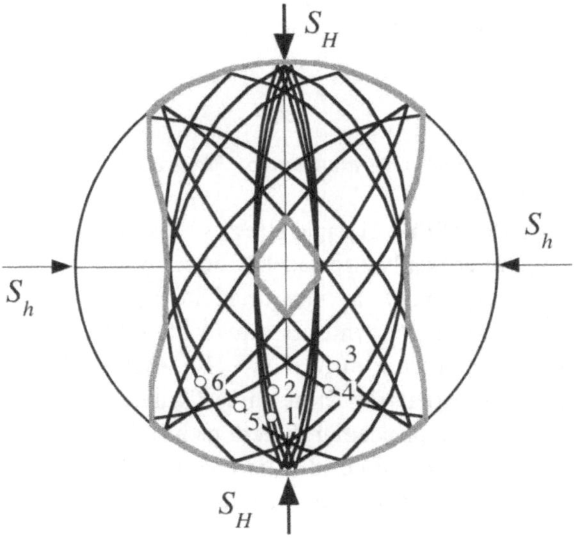

Figure 3
Great circles of fracture planes whose poles are defined as combinations of the maximum and minimum values of L, M and N given by Eqs. (12)–(14), respectively, where a normal faulting stress regime is assumed. The numbered great circles correspond to the poles with the same numbers in Figure 2d. The stress states on the fractures shown as the numbered great circles are represented by the points with the same numbers in Figure 1b.

is close to one of such great circles, it would be so permeable as to form flow pathways.

Here we discuss one of the assumptions for the present analysis, i.e., the state of stress and pore pressure are assumed uniform at the scale of the HDR reservoir. Spatial distribution of the microseismic events observed during hydraulic stimulation, suggests that the HDR reservoir spreads to about 1 km^3 around the injection point of the hydraulic stimulation. The HDR reservoir is so large that the state of stress and pore pressure possibly changes locally within the reservoir. The main factor for the variation is usually the difference in depth of each location. It is expected that the magnitude of stresses and pore pressure at the top of the reservoir is different to those at the bottom of the reservoir by a few tens of Megapascals due to the difference in depth, i.e., 1 km. However, even in such a case, the orientations of the critically-stressed fractures do not change throughout the reservoir, as long as the magnitude of stress and pore pressure is directly proportional to depth, i.e., $S_H = az, S_h = bz, S_v = cz, P_p = dz$, where a, b, c, d are constant and z is depth, and furthermore the stress orientation, i.e., the azimuth of S_H, is constant independently of depth. This is because, in this case, the location of the Coulomb failure line relative to the Mohr circles representing the stress state at a depth on the Mohr diagram, is the same as that for the stress state at any depth, as can be seen easily. As a result, the orientations of the critically-stressed fractures defined by Eqs. (12)–(14) become uniform independently of depth. On the other hand, if the state of stress and pore pressure changes locally in other manners, the orientations of the critically-stressed fractures should be estimated locally from the state of stress and pore pressure at each location.

4. *Factors that Control Flow Pathways*

In Figure 2, the order of the principal stresses is assumed to be $S_v > S_H > S_h$. The pole orientation of the critically-stressed fractures moves of course in accordance with the order of S_H, S_h and S_v as illustrated in Figure 4. Figures 4a–c are the stereographic projections of the contours drawn on a sphere which were estimated using Eqs. (12)–(14) for the case of $S_1 - P_p = 100$ (MPa), $S_2 - P_p = 40$ (MPa) and $S_3 - P_p = 25$ (MPa). Note that for those three figures, the original contours on a sphere are the same but the projecting directions are different.

Even for a certain stress regime, drastic changes occur in the pole orientation depending on the magnitude of the intermediate principal stress S_2 relative to S_1 and S_3. This is in particular true when S_2 is very close to S_1 or S_3. Such an effect of S_2 is demonstrated in Figure 5 for a normal faulting stress regime. When the Mohr circle with the diameter of $(S_1 - S_2)$ crosses the Coulomb failure line as shown in Figure 5c, the radius $r_{(12)\min}$ becomes equal to $(S_1 - S_2)/2$, and therefore, the minimum absolute value of N for the critically-stressed fractures, $N_{\min}$, becomes equal to zero (see Eq. (14)). For this reason, the critical regions are connected in Figure 5c, although they

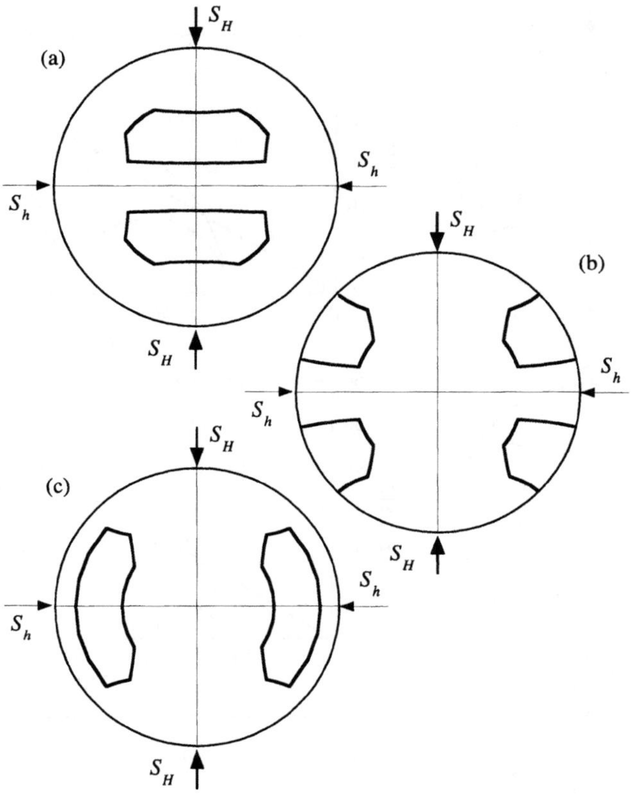

Figure 4
Orientation of poles to the critically-stressed fractures in the cases of (a) a reverse faulting stress regime, i.e., $S_H > S_h > S_v$, (b) a strike-slip faulting stress regime, i.e., $S_H > S_v > S_h$, and (c) a normal faulting stress regime, i.e., $S_v > S_H > S_h$.

are separated in Figures 5a and b. Oppositely, when the Mohr circle with diameter of $(S_2 - S_3)$ crosses the Coulomb failure line as shown in Figure 5d, the radius $r_{(23)\min}$ becomes equal to $(S_2 - S_3)/2$, and therefore, the minimum absolute value of L for the critically-stressed fractures, $L_{\min}$, becomes zero (see Eq. (12)). As a result, the critical regions touch the boundary of the stereonet, and some vertical fractures become critically-stressed.

The range of possible orientations of the poles to the critically-stressed fractures varies also with the pore pressure P_p as shown in Figure 6. With increasing P_p, the Mohr circles shift leftward on the diagram while keeping their shape. As a result, the larger P_p, the possible range in orientation of the critically-stressed fractures becomes wider. The variation of P_p occurs due to geological events or artificial hydraulic stimulation. Let us consider the latter case in more detail. We assume a single fracture with small permeability and no pore pressure under natural

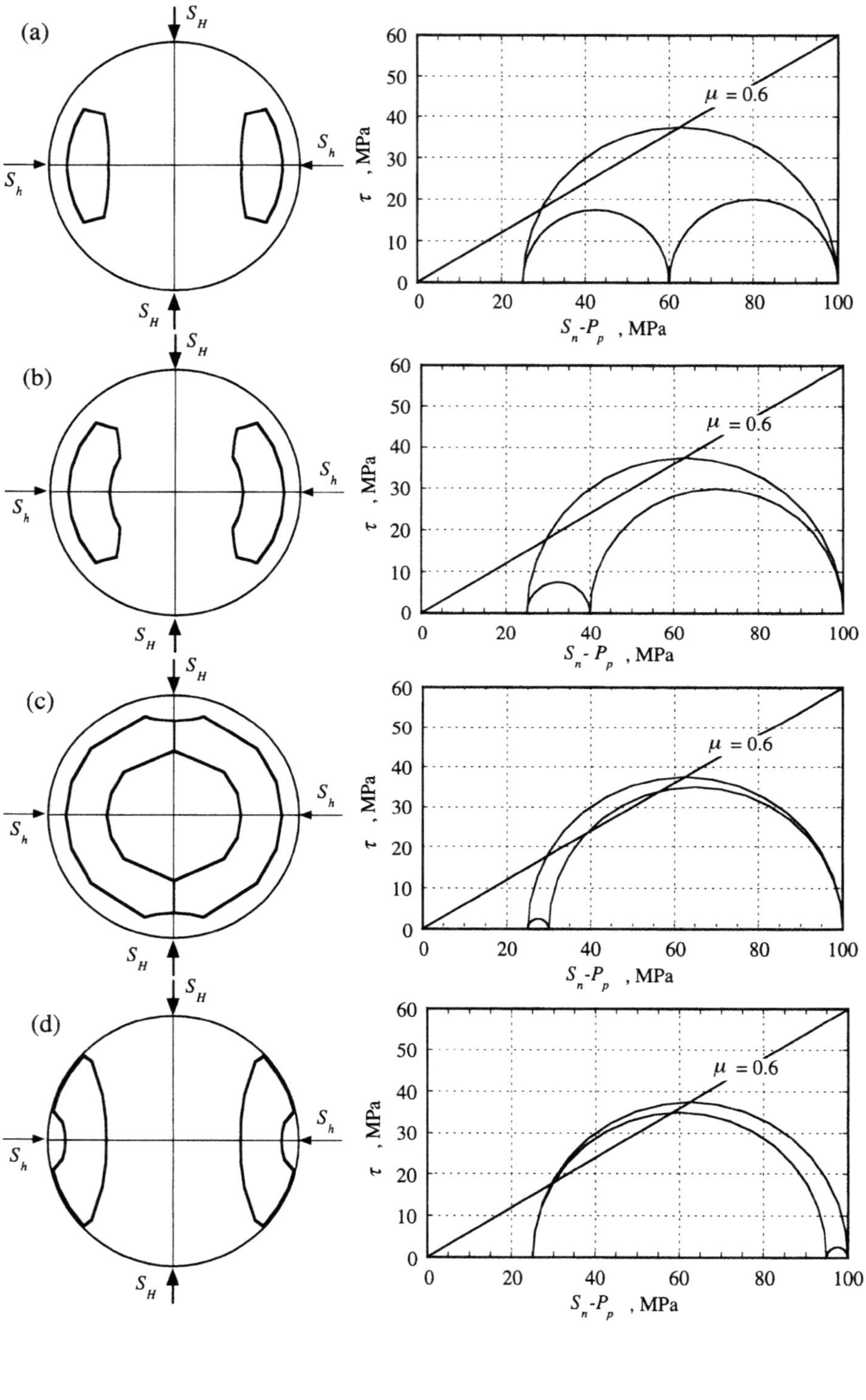
(a)
(b)
(c)
(d)
S_H
S_h
$\mu = 0.6$
τ , MPa
S_n-P_p , MPa
0
10
20
30
40
50
60
80
100

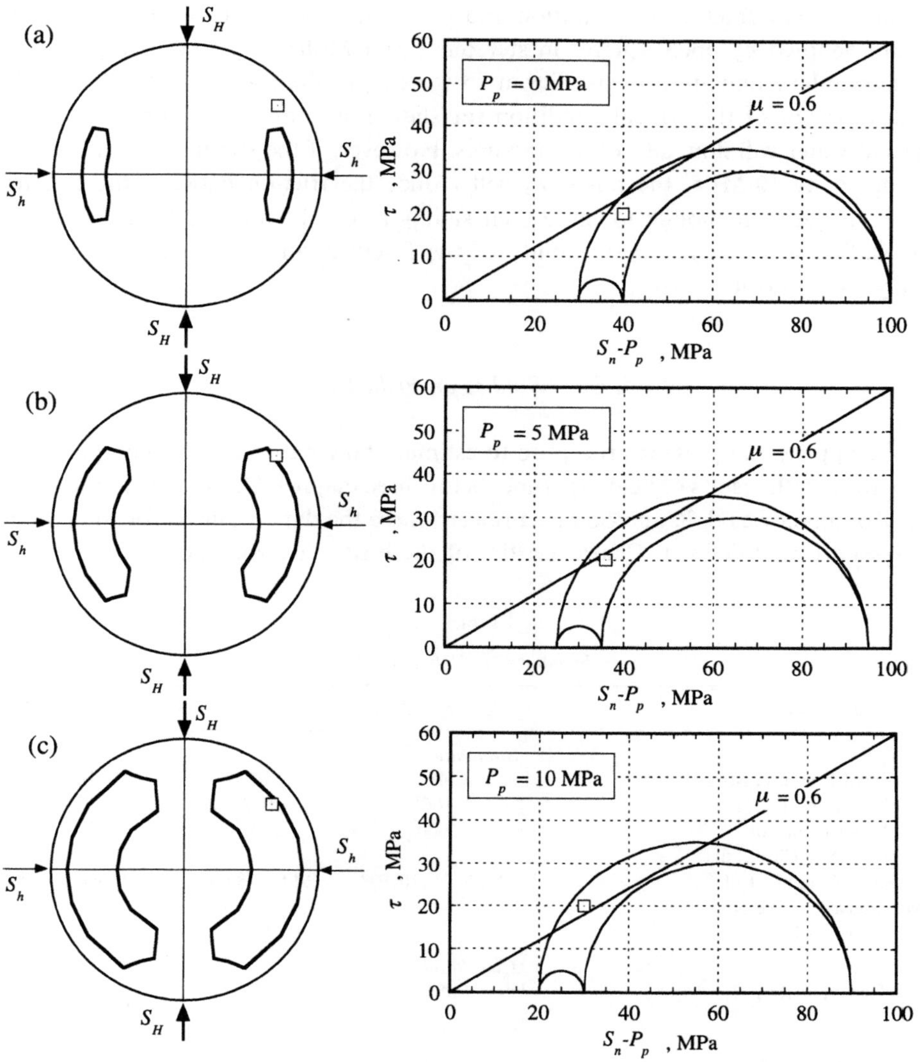

Figure 6
Effect of pore pressure P_p on possible orientations of the critically-stressed fractures. A normal faulting stress regime is assumed. The larger P_p, the wider the possible range in orientation of poles to the critically-stressed fractures. See text regarding the symbols of the open square.

◀

Figure 5
Effect on possible orientations of the critically-stressed fractures of the magnitude of the intermediate principal stress S_2 relative to the magnitude of the maximum and minimum principal stresses S_1 and S_3. A normal faulting stress regime is assumed.

conditions. The fracture's orientation and stress state estimated from Eqs. (4) and (5) are denoted by open squares in stereonets and Mohr diagrams of Figure 6. If the hydraulic stimulation is insufficient to raise P_p in the fracture, i.e., $P_p = 0$, the fracture is out of the critical condition for sliding as shown in Figure 6a, so that injected water will flow into other fractures. However, if the stimulation succeeds in raising P_p by 10 MPa, the raised P_p will reduce the friction between the fracture planes in question enough to cause their sliding as can be inferred from Figure 6c. The sliding will enhance permeability of the fracture, and the injected water will finally rush into it.

5. *Case Studies on Field Data*

We applied the present procedure to estimate flow pathways and compared the estimation with results of field experiments for three major HDR sites as Fenton Hill, US, Rosemanowes, UK and Soultz, France, where the data required for the present analysis are well defined. Characteristics of these sites are summarized in Table 1,

Table 1

Summary of HDR test sites

Name of Site		Fenton Hill, US	Rosemanowes, UK	Soultz, France
		Stimulation		
Well-head Pressure, MPa		48[a]	14[e]	10[h]
Flow Rate, liters/s		106[a]	100[e]	37[h]
Mean Depth of Microseismic Cloud, m		3600[b]	2500[f]	2500[h]
Overall Alignment of Microseismic Envelope, strike (dip)		N170°E (70°E)[b]	N140°E (80°W)[f]	N160°E (90°)[h]
		Circulation		
Recovery Rate		85%[c]	75%[c]	N/A[i]
		Stress State		
Stress Regime		NF	SS	NF
Magnitude of Stresses at the Mean Depth of Microseismic Cloud, MPa	Maximum	94 (S_v)[d]	85 (S_H)[g]	65 (S_v)[j]
	Intermediate	65 (S_H)[d]	65 (S_v)[g]	59 (S_H)[j]
	Minimum	55 (S_h)[d]	36 (S_h)[g]	31 (S_h)[j]
Azimuth of Maximum Horizontal Stress		N14°E[d]	N130°E[g]	N170°E[j]

[a] Estimated from "Fig. 9.26" in ARMSTEAD and TESTER (1987), [b] Estimated from "Fig. 3" in FEHLER (1989), [c] Estimated from "Tables 1, 2" in WILLIS-RICHARD *et al.* (1995), [d] Estimated from the stress state at a depth of 3000 m reported in FEHLER (1989) assuming that *in situ* stresses and pore pressure are proportional to depth, the increasing rate of S_v to depth is 26 MPa/km and P_p is hydrostatic, [e] Reported in PARKER (1999), [f] Estimated from "Figs. 3–5" in PINE and BATCHELOR (1984), [g] Estimated from "Eqs. (1)–(3)" in PINE and BATCHELOR (1984), [h] Estimated from "Figures 6, 8" in BARIA *et al.* (1995), [i] See comments in text, [j] Estimated from equations in BARIA *et al.* (1995).

where the recovery rate is defined as the ratio of the amount of water pumped down injection wells to that of heated water reemerging up production wells. It has been reported that the recovery rate in Rosemanowes decreased in the case that the injection pressure at wellhead exceeded about 10 MPa (PARKER, 1999). The recovery rate in Soultz is unknown, since a downhole pump was used to extract heated water from the subsurface. Impedance for fluid flow in rock mass must be very low in order for the extraction to be carried out by using the downhole pump. With respect to microseismic events observed during hydraulic stimulation, Table 1 indicates only the mean depth and the overall alignment in space. See details of the microseismic events in e.g., FEHLER (1989) for the Fenton Hill site, PINE and BATCHELOR (1984) for the Rosemnowes site and BARIA *et al.* (1995) for the Soultz site. In all the

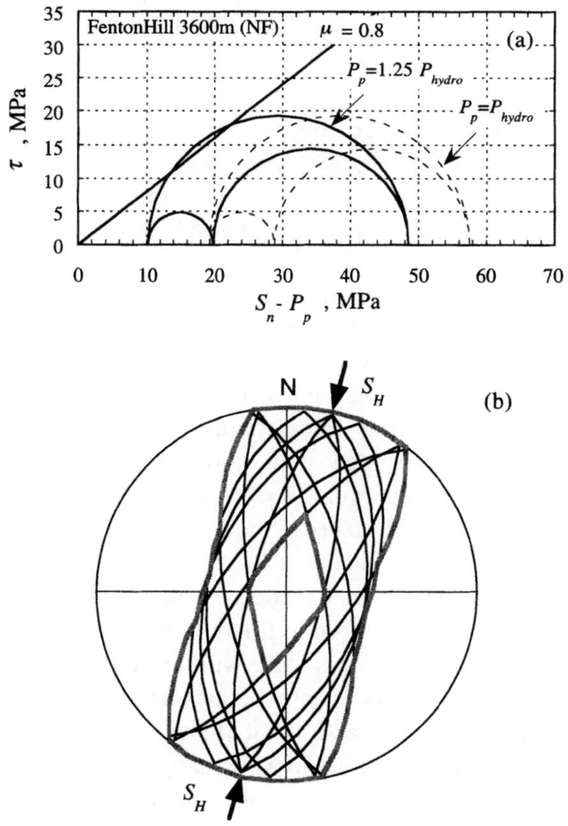

Figure 7
(a) Mohr diagram showing the state of *in situ* stress at Fenton Hill, where the circles drawn by solid and dashed lines correspond to different pore pressures. (b) Great circles showing representative orientations of the critically-stressed fractures estimated from the *in situ* stress state.

following analyses, the friction coefficient μ is assumed to be 0.8, since μ is thought to be in the range of $\mu = 0.6 - 1$ (BYERLEE, 1978).

First, Figure 7a shows Mohr circles for stress state at the mean depth of the microseismic envelope in Fenton Hill. In this case, the Coulomb failure line does not cross nor touch the Mohr circles, if P_p is assumed to be hydrostatic (see the circles drawn by dashed lines in Fig. 7a). This result suggests that there are less flow pathways under natural conditions, and injected water which lost flow pathway should contribute to raise P_p in a rock volume surrounding the injection well. Thus we assumed that P_p was elevated in some degree due to hydraulic stimulation. We examined increasing P_p from the hydrostatic head by 5%, and found that the largest Mohr circle begins to cross the Coulomb failure line at the P_p elevated up to 125% of the hydrostatic head (see the circles drawn by solid lines in Fig. 7a). This assumption is harmonic with the fact that wellhead pressure at the stimulation was considerably higher in Fenton Hill than other sites. The elevated P_p will cause frictional sliding of optimally-oriented fractures and create flow pathways along the fractures. However, distant from the injection well, P_p will remain hydrostatic independently of the stimulation, and fractures with any orientation will not slide. For this reason, the injected fluid cannot recede so much from the injection well, and this possibly resulted in the high recovery rate of circulation in Fenton Hill. Figure 7b shows the predicted great circles of the representative fractures which are under the critically-stressed condition for sliding. This result indicates that any critically-stressed fracture is inclined to vertical direction striking in azimuth between NNW and NE. The orientation of the microseismic envelope (Strike N170°E and Dip 70°E, see Table 1) is within the predicted range in orientation of the critically-stressed fractures. It should be noted here that microseismicity may or may not occur at hydraulically conductive zones (PHILLIPS *et al.*, 1997). However, microseismicity is certainly induced by an increase in pore pressure associated with hydraulic stimulation. This requires seismically active areas to be hydraulically connected to injection wells. For this reason, major fluid pathways are expected to be along overall alignments of the microseismicity.

On the other hand, in the case of Soultz, fractures with a wide variety of orientation are under the critical condition for sliding as can be seen from Figure 8, even if P_p is hydrostatic. Such a stress state will sustain high permeability of the fractures with various orientations. Those permeable fractures will result in high bulk-permeability of the rock mass, and it will allow us to extract heated water from the subsurface by using a downhole pump as actually applied to this site. Figure 8b shows the predicted great circles of the critically-stressed fractures. Even though the Soultz site is in a normal faulting stress regime, $S_v > S_H > S_h$, like the Fenton Hill site, there potentially exists the critically-stressed fractures in a vertical direction, in other words, a direction parallel with the maximum principal stress, in contrast to the Fenton Hill site. This is because the intermediate principal stress, $S_2(S_H)$, is close to

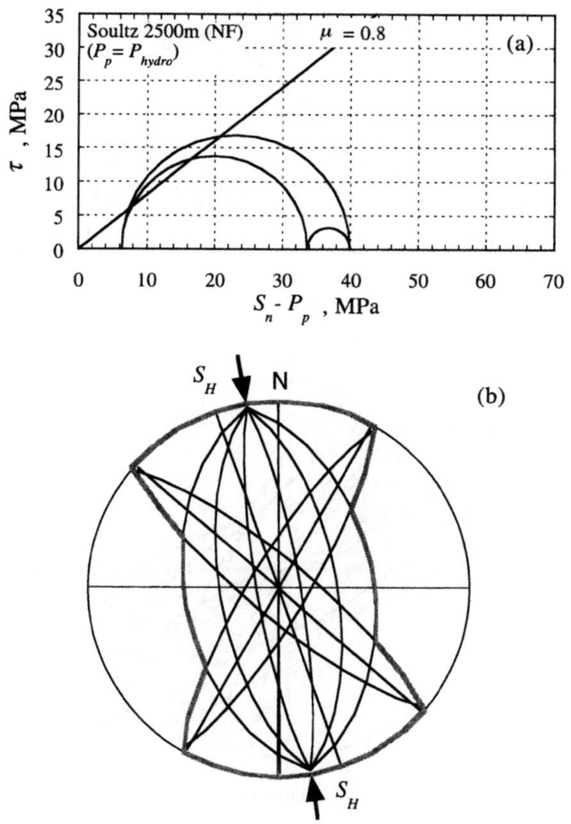

Figure 8
(a) Mohr diagram showing the state of *in situ* stress at Soultz. (b) Great circles showing representative orientations of the critically-stressed fractures estimated from the *in situ* stress state.

the maximum principal stress, $S_1\,(S_v)$, in this site (see Fig. 5d). Actually the observed microseismic envelope is almost vertical, and the azimuth is within the predicted range in orientation of the critically-stressed fractures.

In the case of Rosemanowes, the Mohr diagram shows that when P_p is hydrostatic, the critically-stressed fractures exist although their orientation is limited (see Fig. 9). However, the possible range in orientation of the critically-stressed fractures will expand easily with a slight increase in P_p. These mechanisms possibly explain the results of the high and low recovery rates at low and high injection pressures, respectively. Furthermore, the orientation of the microseismic envelope is again within the predicted range in orientation of the critically-stressed fractures shown in Figure 9b.

Those considerations suggest that site characteristics such as wellhead pressure required for hydraulic stimulation and the recovery rate can be classified according to

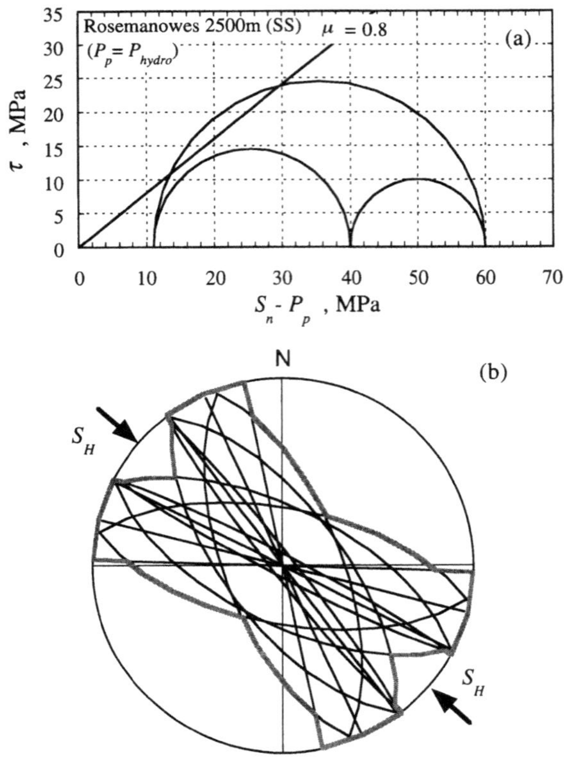

Figure 9
(a) Mohr diagram showing the state of *in situ* stress at Rosemanowes. (b) Great circles showing representative orientations of the critically-stressed fractures estimated from the *in situ* stress state

the location of the Coulomb failure line relative to Mohr circles on the Mohr diagram, i.e., the line being apart from, touching or crossing the circles, as summarized in Table 2. If the state of *in situ* stress is not favorable to induce sliding of natural fractures as the Coulomb failure line is apart from the Mohr circles, it may result in a high wellhead pressure at hydraulic stimulation and a high recovery rate at circulation. Conversely, if the state of *in situ* stress is favorable for sliding as the Coulomb failure line fully crosses the Mohr circles, it may result *vice versa* in a low wellhead pressure at hydraulic stimulation and a low recovery rate at circulation. Based on these considerations, we introduce a nondimensional parameter R which allows us to estimate quantitatively the relative location of the Coulomb failure line. The R is defined as

$$R = \frac{P_{\text{hydro}} - P_c}{S_1 - S_3} , \qquad (15)$$

Table 2

Effects of in situ stress state on the performance of HDR reservoirs

Stress State	Fracturing Pressure	Flow System	Recovery Rate at Circulation	Example	R
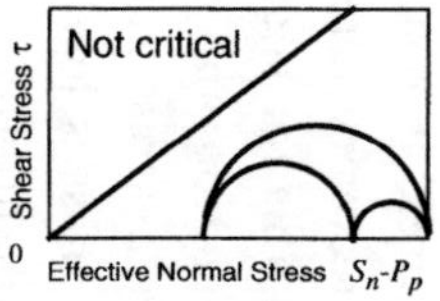	High	Closed	High	Fenton-Hill	−0.18
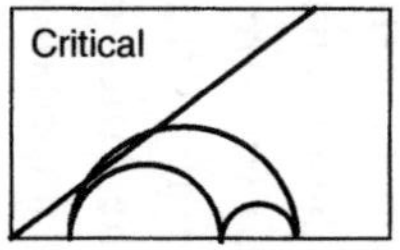	Low	Closed @ Low pres. Open @ High pres.	High @ Low pres. Low @ High pres.	Rosemanowes	+0.08
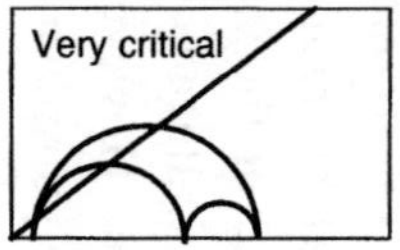	Low	Open	Low	Soultz	+0.21

where P_{hydro} is hydrostatic head. The Mohr circles shift leftward and rightward on the Mohr diagram with increasing and decreasing P_p, respectively, and P_c in Eq. (15) is a value of P_p at which the largest Mohr circle touches the Coulomb failure line at a point. In the case of Fenton Hill, the Coulomb failure line is apart from the Mohr circles at $P_p = P_{\text{hydro}}$ as described above (see the dashed circles in Fig. 7a). This fact results in a negative value of $R = -0.18$, since P_c is larger than P_{hydro} in this case. In the case of Soultz, R takes a positive value of $R = +0.21$ *vice versa*, since the Coulomb failure line fully crosses the Mohr circles at $P_p = P_{\text{hydro}}$. As compared with this, in the case of Rosemanowes, the Coulomb failure line is almost in touch with the Mohr circles at $P_p = P_{\text{hydro}}$ (see Fig. 9a) so that P_c is nearly equal to P_{hydro}, and R takes a value close to zero, $R = +0.08$. We can estimate the value of R in advance of hydraulic stimulation, to the extent we know *in situ* stresses and hydrostatic head at the depth of concern. If the estimated R takes negative, zero and positive values, the site characteristics at the depth would be expected similar to those at the Fenton Hill, Rosemanowes, and Soultz sites, respectively.

6. Conclusions

Since natural fractures in geothermal field are under a desirable condition to be sealed or healed due to chemical reactions enhanced by a high temperature, they should be basically less permeable. There must be specific reasons for some fractures to keep their high permeability in geothermal field. *In situ* stress may be one of the candidates. A certain state of *in situ* stress causes frictional sliding of the fractures, and the sliding breaks the seals in fractures and increases their permeability. The sliding is obeyed by the Coulomb criterion. Thus natural fractures which satisfy the criterion in a current stress field are likely to hold high permeability. Orientation of the fractures that are critically-stressed for sliding can be estimated geometrically from the location of the Coulomb failure line relative to the Mohr circles representing the *in situ* stress state on the Mohr diagram. The geometrical method allows us to understand intuitively how the orientation changes in accordance with types of stress regimes, pore pressure and the magnitude of intermediate principal stress relative to the maximum and minimum principal stresses. The critically-stressed fractures are likely to form flow pathways in geothermal reservoirs created by HDR projects. If the state of *in situ* stress is not favorable to produce sliding of natural fractures, it results in a high wellhead pressure at hydraulic stimulation and a high recovery rate at circulation. On the other hand, if the state of *in situ* stress is favorable for sliding, it results *vice versa* in a low wellhead pressure at hydraulic stimulation and a low recovery rate at circulation. However, in the latter case, such a stress state allows us to use a downhole pump, since flow impedance in rock mass is low. Finally, we introduced a nondimensional parameter R in order to quantitatively evaluate the effect of *in situ* stress on the characteristics of reservoir. The R takes negative or positive values depending on whether the state of *in situ* stress is favorable or not for fracture sliding.

Acknowledgements

We thank M.D. Zoback, Stanford University for his helpful discussions and comments. We also thank M.J. BoutÉca and an anonymous referee for their reviews of this manuscript and H.J. Kümpel for considerable editorial assistance. This work was supported by the MURPHY/MTC project of the New Energy and Industrial Development Organization of Japan and partly by JSPS – Research for Future Program.

References

Armstead, H. C. H. and Tester, J. W., *Heat Mining: A New Source of Energy* (E. & F. N. Spon, London 1987) Chap. 9, pp. 161–240.

Atsumi, A., *Introduction to Mechanics of Solids* (in Japanese) (Corona Publ. Co., Tokyo 1978) Chap. 2.12, pp. 26–30.

Baria, R., Garnish, J., Baumgärtner, J., Gérald, A., and Reinhard, J., *Recent Development in the European HDR Research Programme at Soultz-Sous-Forets (France), Proc. World Geoth. Cong.* (eds. E. Barbier, G. Frye, E. Iglesias and G. Pálmason) (Int. Geoth. Assoc., Inc., Auckland 1995) Vol. 4, pp. 2631–2635.

Barton, C. A., Zoback, M. D., and Moos, D. (1995), *Fluid Flow along Potentially Active Faults in Crystalline Rock*, Geology *23*(8), 683–686.

Byerlee, J. (1978), *Friction of Rocks*, Pure and Appl. Geophys. *116*, 615–626.

Fehler, M. C. (1989), *Stress Control of Seismicity Patterns Observed during Hydraulic Fracturing Experiments at the Fenton Hill Hot Dry Rock Geothermal Energy Site, New Mexico*, Int. J. Rock Mech. Min. Sci. and Geomech. Abstr. *26*(3/4), 211–219.

Hickman, S., Barton, C. B., Zoback, M. D., Morin, R., Sass, J., and Benoit, R. (1997), *In situ Stress and Fracture Permeability in a Fault-Hosted Geothermal Reservoir at Dixie Valley, Nevada,* Geoth. Res. Coun. Trans. *21*, 181–189.

Ito, T. and Zoback, M. D. (2000), *Fracture Permeability and In Situ Stress to 7 km Depth in the KTB Scientific Drillhole*, Geophys. Res. Lett. *27*(7), 1045–1048.

McGarr, A. (1999), *On Relating Apparent Stress to the Stress Causing Earthquake Fault Slip*, J. Geophys. Res. *104*(B2), 3003–3011.

Olsen, M. P., Scholz, C. H., and Léger, A. (1998), *Healing and Sealing of a Simulated Fault Gouge under Hydrothermal Conditions: Implications for Fault Healing*, J. Geophys. Res. *103*(B4), 7421–7430.

Parker, R. (1999), *The Rosemanowes HDR Project 1983–1991*, Geothermics *28*(4/5), 603–615.

Phillips, W. S., House, L. S., and Fehler, M. C. (1997), *Detailed Joint Structure in a Geothermal Reservoir from Studies of Induced Microearthquake Clusters*, J. Geophys. Res. *102*(B6), 11,745–11,763.

Pine, R. J. and Batchelor, A. S. (1984), *Downward Migration of Shearing in Jointed Rock during Hydraulic Injections*, Int. J. Rock Mech. Min. Sci. and Geomech. Abstr. *21*(5), 249–263.

Pollard, D. D. and Segall, P., *Theoretical displacements and stresses near fractures in rock: With applications to faults, joints, veins, dikes, and solution surfaces,* In *Fracture Mechanics of Rocks* (ed. B. K. Atkinson) (Academic Press Inc., London 1987), pp. 277–349.

WILLIS-RICHARDS, J., GREEN, A. S. P., and JUPE, A., *A Comparison of HDR Geothermal Sites, Proc. World Geoth. Cong.* (eds. E. Barbier, G. Frye, E. Iglesias and G. Pálmason) (Int. Geoth. Assoc., Inc., Auckland 1995) Vol. 4, pp. 2601–2605.

YEO, I. W., DE FREITAS, M. H., and ZIMMERMAN, R. W. (1998), *Effect of Shear Displacement on the Aperture and Permebility of a Rock Fracture*, Int. J. Rock. Mech. Min Sci. *35*(8), 1051–1070.

(Received January 31, 2001, revised July 3, 2001, accepted August 1, 2001)

Pure appl. geophys. 160 (2003) 1125–1136
0033–4553/03/061125–12

Pure and Applied Geophysics

Porosity and Thermal Conductivity of the Soultz-sous-Forêts Granite

FABRICE SURMA[1] and YVES GERAUD[1]

Abstract—The success of the Soultz-sous-Forêts Hot-Dry-Rock project depends on the ability to maintain fluid circulation in a fractured granite. Fractures represent the main fluid pathways. To understand the behavior of this granite in respect to thermal fluid-rock interaction the important aspects are (1) the porous network around these fractures and (2) the thermal conductivity of the rock. This granite is altered and composed of different weathered facies. Variations of porosity and thermal conductivity take place in regard to the alteration and fracturing of the granite. Two types of porosity measurements were performed, mercury injection and water porosity on two samples sizes. The two methods give similar porosity values between 0.3% and 10%. Thermal conductivity measurements were performed in two perpendicular directions to look at anisotropy with two methods at different scale and value ranges from 2.3 to 3.9 $W.m^{-1}.K^{-1}$. Optical scanning provides us with a good knowledge of local increase of thermal conductivity due to sealed fracture or quartz-cemented matrix.

The relationship between porosity and thermal conductivity is not obvious and has to be studied in details, and results show three cases:

(1) a relationship between conductivity and porosity (increase of conductivity with a decrease of porosity),
(2) a relationship between conductivity and sealed fractures (increase of conductivity related to an increase of fracture density),
(3) and a combination of the two previous ones.

The results are carefully compared for different types of granite: alterated, fractured or both. These first results indicate that parameters such as thermal conductivity are linked to the porous medium, the structure and the mineralogy of the rock.

Key words: Porosity, thermal conductivity, granite, Soultz-sous-Forêts, optical scanning, weathering.

1. Introduction

Pores, fractures and cracks form the damaged zone around main fractures and are potential sites for fluid circulation (CAINE *et al.*, 1996). Physical properties of rocks are strongly related to precipitation and dissolution of minerals. This mineral precipitation is due to fluid circulation and is strongly dependent on chemical components of the fluid and thermodynamic conditions. Therefore, physical

[1] Centre de Geochimie de la Surface (CNRS-ULP), 1 rue blessig 67084 Strasbourg CEDEX.
E-mail: surma@illite.u-strasbg.fr, ygeraud@illite.u-strasbg.fr

properties of rocks associated with chemical and mineralogical components are important to understand thermal and chemical fluid-rock interactions.

Porosity and thermal conductivity play an important role in transfer properties of fluid-rock interactions. The understanding of the relationship between these two parameters could be a good way to understand the fluid-rock interactions between the porous and the fracture network and to predict the porosity of a rock reservoir.

Soultz-sous-Forêts, located in the Rhine Graben, is one of the deepest geothermal 'Hot Dry Rock' test sites (GENTER, 1990). The aim of this project is at once to make a constant fluid flow circulation in the fractured hot rock to recover energy with hot fluid. Temperature exchange between fluid and rock is located in the rock at a depth between 3.5 and 5 km (GERARD *et al.*, 1998). Conductivity in the fracture network depends on chemical fluid-rock interactions that can occur in the damaged zone.

We worked on one of the three boreholes, EPS1, which has been fully cored. We took core samples to make porosity and thermal conductivity measurements. This granite is strongly altered and fractured in different facies (GENTER and TRAINEAU, 1992). The physical, mineralogical and chemical parameters of these facies vary considerably (LEDESERT, 1996; SAUSSE *et al.*, 1999).

The aim of this study is to understand the relationship between heat transfer properties and porous media. The first step is to characterize thermal conductivity and porosity. We used two methods for porosity: water porosity and mercury injection porosity, and two methods to characterize thermal conductivity: line source (LS) and optical scanning (OS). The second step is to see the effect of the porous networks induced by fracturing and weathering on measured thermal conductivity.

2. Methods

2.1. Porosity

Water porosity and mercury injection porosity measurements were performed on samples of different alteration facies originating from the borehole. Water total porosity (Nt) determination is based on the measurements of dry sample weight, wet sample weight and the weight of the sample under hydrostatic conditions (SIZUN, 1995; MERTZ, 1991). Measurements are performed on dm^3 samples.

The second technique used is mercury injection porosity. A well known volume of mercury is injected in the sample for a well known applied pressure. The investigated volume is between 2 and 5 cm^3. The connected porosity (Nhg) provides an idea of the structure of the porous media (PELLERIN, 1981).

Porosity estimations were also made on the same samples using LS thermal conductivity measurements for dry and wet samples. Assuming the geometric mean model for an air-saturated sample (PRIBNOW and SASS, 1995) the porosity is calculated as follows (PRIBNOW *et al.*, 1996):

$$\phi = \frac{\ln(\lambda_{\text{sat}}) - \ln(\lambda_{\text{dry}})}{\ln(\lambda_{\text{wat}}) - \ln(\lambda_{\text{air}})}$$

where ϕ is the porosity, λ_{sat}, λ_{wat} and λ_{dry} are the thermal conductivity of saturated samples, water and dry samples, respectively.

2.2. *Thermal Conductivity*

We use two methods to study thermal conductivity: line source (LS) and optical scanning (OS). The first one is a classical method based on the theory of a line source in an infinite medium (CARSLAW and JAEGER, 1947). It is a needle probe embedded in and flush with the surface of a material of very low thermal conductivity (Plexiglas) (described by HUENGES *et al.*, 1990). This thermal conductivity measurement is local, and the random error is about 5%. The length of the thermal source is 5 cm. Conductivity determinations are made at a number of azimuths on sawed and polished surfaces both parallel and perpendicular to the axis of the core. Conductivity was measured in two directions to determine the possible anisotropy (Fig. 1).

The second method, optical scanning (OS), is based on scanning a sample surface with a focused, mobile, and continuously operated constant heat source in combination with a temperature sensor. The heat source and the temperature sensor move with the same speed and to a constant distance to each other (Fig. 1, POPOV *et al.*, 1999). The size of the heating spot is 1 mm in diameter. Thermal conductivity normal to the heated surface is determined as an arithmetic mean of local conductivities along the entire scanning line. The basic measurements relative error is not more than 2%.

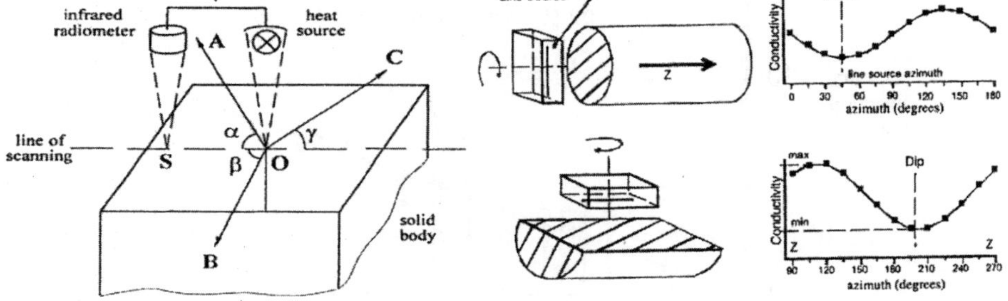

Figure 1
Left. Principle of optical scanning method. V velocity; O area of heat spot; S detection area radiometer; A, B, C main axes of thermal conductivity with angle α, β and γ to line of scanning (POPOV *et al.*, 1999). **Right**. Determination of thermal conductivity components with the half-space line source (POPOV *et al.*, 1999).

A set of profiles or scanning lines is made along two orthogonal directions and one oblique direction on one sample of the granite half core. Thermal conductivity gaps measured at the intersection point of two scanning profiles and for each profile is within the error bar of the technique.

3. Results

3.1. Porosity

Porosity varies considerably between 0.3 and 10%. Results obtained by the two techniques are well correlated (Fig. 2) on two orders of magnitude of porosity. For the selected samples it seems that there is no scale effect (from cm^3 to dm^3) on the investigated volume.

Petrographical analysis provides classification of the samples in three different facies: porphyric fresh granite, fractured granite and altered granite, corresponding to a porosity increase. Fractured samples are characterized by a crack network cemented by quartz.

3.2. Porosity and Thermal Conductivity

Figure 3 represents thermal conductivity (measured by LS) in relation to porosity for all the samples. Thermal conductivity ranges from 2.3 $W.m^{-1}.K^{-1}$ to 3.9 $W.m^{-1}.K^{-1}$. Measurements performed on two directions are similar, showing the material isotropy.

There is no good correlation, even if we observe a global trend: decrease of thermal conductivity with an increase of porosity between 0.9 and 10%. GUEGUEN and PALCIAUSKAS (1992) show the same relationship.

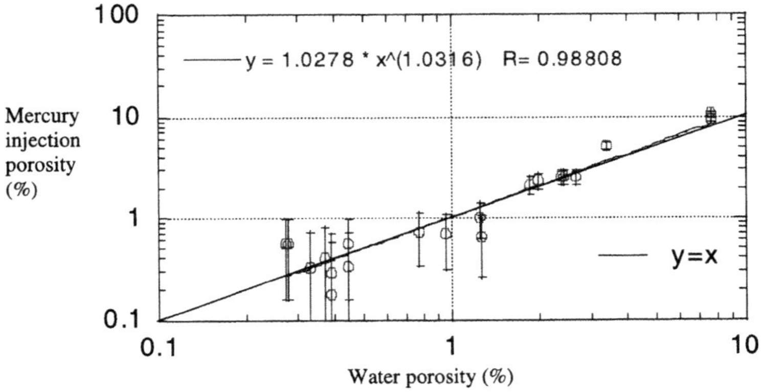

Figure 2
Water and mercury porosity relationship.

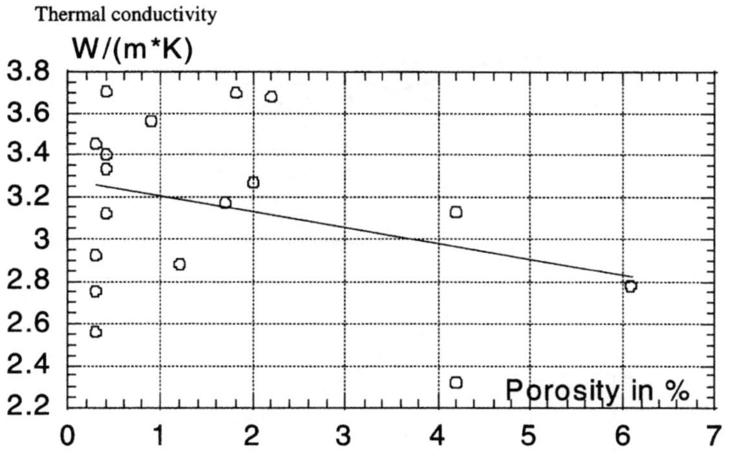

Figure 3
Thermal conductivity and porosity for all samples.

The petrographical classification helps to interpret the relationship between porosity and thermal conductivity. The sample sets are specifically located on the diagram (Fig. 4). The fresh granite samples are located for low porosity and thermal conductivity values. Sealed fractured samples are localized for low porosity and high thermal conductivity, and finally altered samples are characterized by a mean thermal conductivity and a large porosity variation (Fig. 4). Between groups, transition samples could be identified (Fig. 4). High thermal conductivity values in very low porosity samples are probably due to cementation of the crack network. Indeed, thermal conductivity of quartz, reaching 6.5 $W.m^{-1}.K^{-1}$, is higher than the average value of the whole rock, 3.2 $W.m^{-1}.K^{-1}$. Samples with a relatively high porosity and high thermal conductivity are samples containing quartz-sealed fractures and quartz-cemented matrix and open cracks.

The comparison of calculated porosity from LS measurements and porosity measurements seems to be good, but the number of samples is insufficient to be sure (Fig. 5) and the porosity shape must be taken into account for a better fit.

3.3. Local Thermal Conductivity: Results and Discussion

Optical scanning is used to define more local thermal conductivity values. Two samples were tested: K108 (1630,5 m of depth) and K076 (1432,8 m of depth).

The first sample shows the quartz cementation effect in and around a fracture. In Figure 6, the black lines represent the thermal conductivity profiles. We can clearly see the effect of a sealed fracture on lines 2, 3 and 4 with the high thermal conductivity value. On these lines, thermal conductivities vary between 2.4 $W.m^{-1}.K^{-1}$, in the feldspar and 3.4 $W.m^{-1}.K^{-1}$ in the vein quartz. Thermal conductivity values of these mineral species are respectively: 2.3 $W.m^{-1}.K^{-1}$ and 6.5

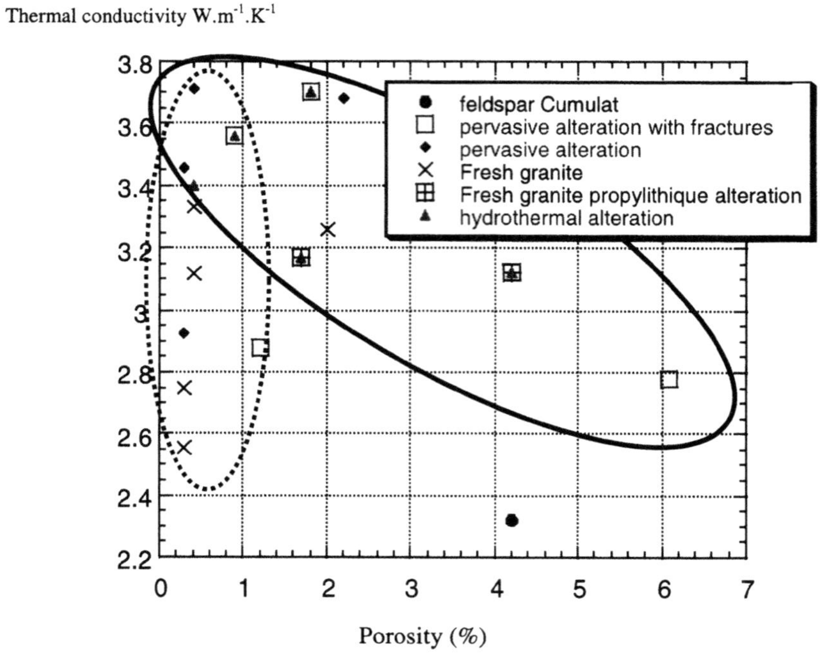

Figure 4
Thermal conductivity versus porosity relationship to the granite facies.

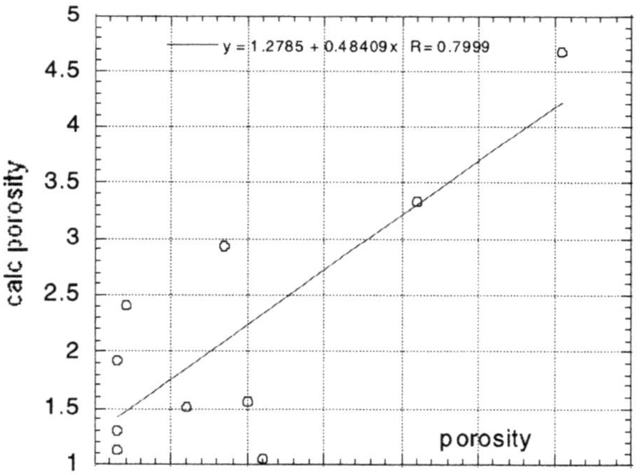

Figure 5
Estimation of porosity in % calculated from thermal conductivity versus measured porosity in %.

$W.m^{-1}.K^{-1}$ (CLAUSER and HUENGES, 1995). We can also see that quartz-sealed fractures (right of Fig. 6) have a high thermal conductivity in contrast to feldspars (left of Fig. 6). The effect of secondary quartz also could be measured in the matrix

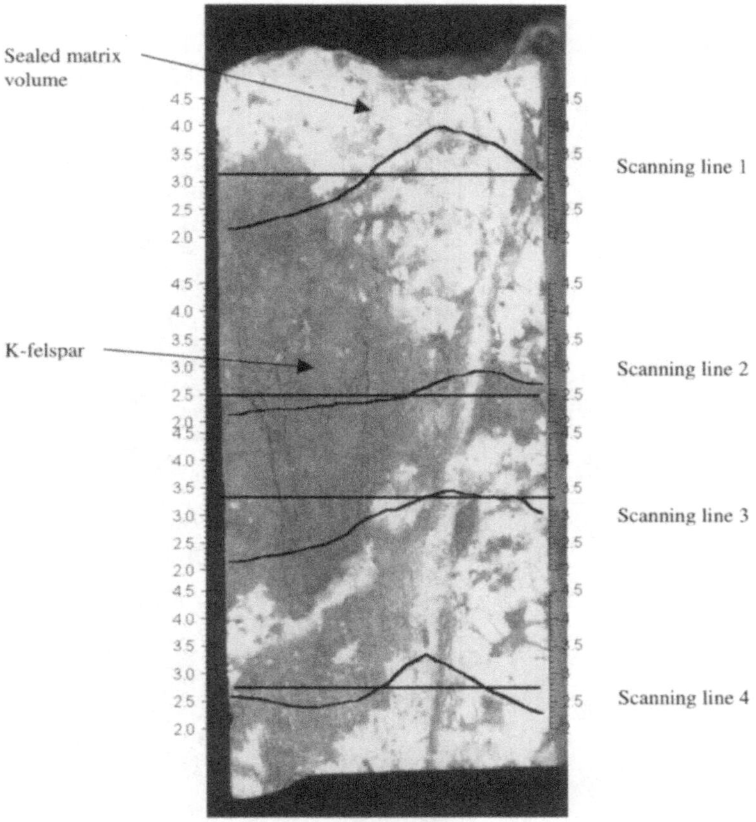

Figure 6
OS scanning profile (in black) of sample K108 (see text).

around the fracture on scanning line 1; the maximum thermal conductivity value on line 1 is located into the altered matrix and not in the fracture.

The second sample allows us to characterize the effect of alteration in feldspars on thermal conductivity. On Figure 7a, we can distinguish two zones. Zone (3) with low thermal conductivity is an altered feldspar zone opposite to zones (1) and (2) with higher thermal values, high strain materials and fairly altered feldspars. Zone (1) is characterized by a high thermal conductivity response of a first sealed fracture (F1 on Fig. 7a). Zone (2) has a high thermal conductivity induced by a second fracture (F2 on Fig. 7a).

A set of six profiles is described to characterize each compartment of the material in regard to their main petrographical and structural parameters. The six profiles are in white on Figs. 7a and 7b. We have different parameters on these profiles:

λ_{min} and λ_{max} minimum and maximum thermal conductivity values on the profile, λ_{par} and λ_{per} average thermal conductivity value on the parallel and perpendicular

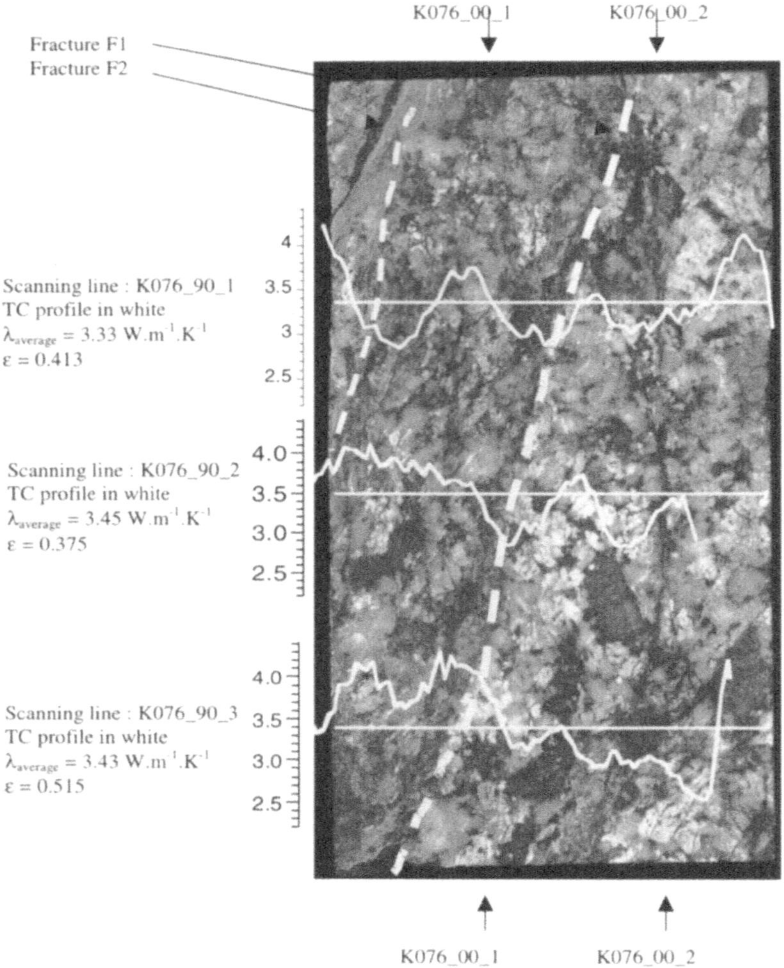

Figure 7a
Sample K076: TC profile from OS in white, Average TC value ($\lambda_{average}$) of the profile represented by a white line. ε is the heterogeneity factor (see text for interpretation).

profile to fractures, $\lambda_{average}$ is the mean value of the whole sample, $K = \lambda_{par}/\lambda_{per}$ is the anisotropy and $\varepsilon = (\lambda_{max} - \lambda_{min})/\lambda_{average}$ is the inhomogeneity factor.

The different parameters of the profiles are summarized in Table 1. The average value ($\lambda_{average}$) of the scanning lines is 3.39 $W.m^{-1}.K^{-1}$ and is comparable to the value 3.23 $W.m^{-1}.K^{-1}$ obtained with the line source method within the error bars.

The K value is around 1, that is to say, there is no anisotropy in agreement with the line source measurements. The inhomogeneity factor determines the thermal conductivity heterogeneity which can be related to the mineralogical or structural heterogeneity of the sample.

Figure 7b
Sample K076: TC profile from OS in white, Average TC value ($\lambda_{average}$) of the profile represented by a white line. ε is the heterogeneity factor (see text for interpretation)

We can clearly see on Figure 7a the heterogeneity of the material. The horizontal scanning lines 2 and 3 are divided in two parts. At the right of the figure, thermal conductivity values are high above 3.5 $W.m^{-1}.K^{-1}$, and at the left of the figure values are low under 3.5 $W.m^{-1}.K^{-1}$. The average value of these profiles is: 3.45 and 3.43 $W.m^{-1}.K^{-1}$. The inhomogeneity factor for these lines are 0.375 and 0.515. Inhomogeneity is low in both cases, but the third line is more heterogeneous with a

Table 1

Parameters of the scanning profiles of sample K076. K076_00_1 is the vertical profile number 1. K076_45_1 is the oblique one and K076_90_1 is one of the horizontal one. λ_{min} and λ_{max} minimum and maximum thermal conductivity values on the profile, λ_{par} and λ_{per} average thermal conductivity values on the parallel and perpendicular profile to the fracturation, $\varepsilon = (\lambda_{max} - \lambda_{min})/\lambda_{average}$ is the inhomogeneity factor and $\lambda_{average}$ is the mean value of the whole sample

Scanning lines	**K076_00_1**	**K076_00_2**	**K076_45_1**	**K076_90_1**	**K076_90_2**	**K076_90_3**
λ_{min}	2.61	2.38	2.22	2.84	2.81	2.51
λ_{max}	4.39	4.02	4.04	4.22	4.11	4.28
$\lambda_{average}$	3.54	3.01	3.18	3.33	3.45	3.43
		$\lambda_{par} = 3.27$			$\lambda_{per} = 3.40$	
			K = 0.96			
ε	0.488	0.542	0.640	0.413	0.375	0.515

value of 0.6. The profile crosses big quartz and feldspar grains which increases and decreases thermal conductivity, respectively.

The sample is divided in two parts. At the right of the fracture in the middle of the sample, the sample is highly fractured and deformed. The oblique scanning line on Figure 7b clearly depicts the two parts, one with high thermal conductivity at the left of fracture F2 and the other with low thermal conductivities at the right. The inhomogeneity factor (0.640) of this line is high and shows the heterogeneity of the sample. Mean thermal conductivity is high due to an augmentation of the amount of quartz. At the left of this fracture, the material is altered and not deformed as the right side is. There are more feldspars than in the right side, and thermal conductivity is lower. The variation of thermal conductivity is due to the presence of quartz grain which increases the conductivity.

In conclusion, OS allows us to characterize our sample in terms of alteration and fracture because these two phenomena are directly connected to porosity and mineralogical changes. These changes influence the variation of thermal conductivity of the sample. When water invades this system, porosity and thermal conductivity of the rock will play an important role in the temperature and mass exchanges.

4. Conclusion

The use of two methods to measure porosity allows us to deal with two volumes of analysis dm^3 and cm^3. In the case of water porosity measurements, the analyzed volume takes into account the whole rock with all the structures. In the case of mercury injection porosity measurements, attention must be paid to the porosity values and the location of the samples because of the small size of the sample.

Usually porosity is related to internal rock structure and geometry. We saw in this study that it can also be related to thermal conductivity that is directly correlated to mineralogy.

Porosity varies from 0.3% to 10%. For low porosity samples, there is no relationship between porosity and thermal conductivity. For a given value of porosity, thermal conductivity varies considerably. The use of petrographical and structural characteristics helps to define the relationship between porosity and thermal conductivity. Three sets of samples are defined: the fresh sample set, the quartz-cemented samples with fractures set and the altered sample set.

The first set is characterized by low porosity and low thermal conductivity values.

Samples of the second set have a high thermal conductivity and low porosity. This increase of thermal conductivity is induced by quartz in small cracks. Some samples have a relatively high porosity (4%) induced by alteration and cracks. Fractured and cemented samples have low porosity and high thermal conductivity values. However, these samples are not only altered granites but also fractured. Samples with high porosity values contain fractures that can be sealed by quartz; they have high thermal conductivity values that increase the bulk thermal conductivity of the rock.

The last set of altered samples shows a large variation of porosity and thermal conductivity and a global decrease of the conductivity with the increase of porosity.

The optical scanning analysis allows definition of relationships between porosity and thermal conductivity at the millimeter scale, to point out the effect of a fracture zone as a thermal drain at different scales, and the effects of weathering on thermal conductivity

The relationship between porosity and thermal conductivity reflects the petrography and the structural characteristics of the rock. We have to improve this relation for low porosity samples.

The next step of this study is to examine the samples by SEM, to have a good location of porosity and minerals, and to see the spatial organization of them. The multiplication of such a study on more samples could also provide us new informations.

Acknowledgements

The authors would like to thank Daniel Pribnow for the constitutive reading, criticism and valuable suggestions. Thanks to the GGA in Hannover which permits us to perform all the thermal conductivity measurements. This is EOST contribution no. 2002-401

References

Caine, J. S., Evans, J. P., and Forster, C. B. (1996), *Fault Zone Architecture and Permeability Structure*, Geology *24*, 11, 1025–1028.

CARSLAW, H. and JAEGER, J. C., *Conduction of Heat in Solids* (Clarendon, Oxford 1947).

CLAUSER, C. and HUENGES, E. (1995), *Thermal conductivity of rocks and mineral.* In *AGU Handbook of Physical Constants* (ed. Ahrens T. J.), pp. 105–126.

GENTER, A. (1990), *Géothermie roches chaudes sèches: le granite de Soultz-Sous-Forêts (Bas-Rhin, France).* Thesis of the University of Orléans, 201 pp.

GENTER, A. and TRAINEAU, H. (1992), *Borehole EPS1, Alsace, France: Preliminary Results from Granite Core Analysis for Hot Dry Rock Research.*, Scientific Drilling *3*, 205–214.

GERARD, A., JUNG, R., BAUMGÄRTNER, J., BARIA, R., GANDY, T., and TRAN-VIET, T. (1998), *Essais de circulation moyenne durée conduits à Soultz-sous-Forêts en 1997*, Socomine Internal Report, 20 pp.

GUEGUEN, Y. and PALCIAUSKAS, V., *Introduction à la physique des roches* (Hermann, Paris 1992).

HUENGES, E., BURKHARDT, H., ERBAS, K. (1990), *Thermal Conductivity Profile of the KTB Pilote Corehole*, Scientific Drilling *1*, 224–230.

LEDESERT, B., DUBOIS, J., GENTER, A., and MEUNIER, A. (1993), *Fractal analysis of Fractures Applied to Soultz-sous-Forêts Hot Dry Rock Geothermal Program*, J. Volcanology and Geothermal Res. *57*, 1–17.

MERTZ, J. D. (1991), *Structures de porosité et propriétés de transport dans les grès*, Sciences Géologiques Strasbourg, Mem., 149 pp.

PELLERIN, F. M. (1981), *La porosité mercure appliquée à l'étude géotechnique des sols et des roches*, Bull. Liaison Lab. Ponts et Chaussées *106*, 106–116.

POPOV, Y. A., PRIBNOW, D., SASS, J. H., WILLIAMS, C., and BURKHARDT, H. (1999), *Characterisation of Rock Thermal Conductivity by High-resolution Optical Scanning*, Geothermics *28*, 253–276.

PRIBNOW, D. and SASS, J. H. (1995), *Determination of Thermal Conductivity for Deep Boreholes*, J. Geophys. Res. *100*, B6, 9981–9994.

PRINOW, D., WILLIAMS, C., SASS, J. H., and KEATING, R. (1996), *Thermal Conductivity of Water-saturated Rocks from KTB Pilot Hole at Temperatures of 25 to 300°C*, Geophys. Res. Lett., *23*, 4, 391–394.

SAUSSE, J., GENTER, A., LEROY, J., and LESPINASSE, M. (1998), *Description et quantification des altérations filoniennes. paléoécoulements fluides dans les granites de Soultz-sous-Forêts (Bas-Rhin, France).* Bull. Soc. Géol. France. *169*, 5, 655–664.

SIZUN, J. P. (1995), *Modification des structures de porosité dans les grès lors de transformations pétrographiques dans la diagenèse et l'hydrothermalisme: application au Trias de la marge ardéchoise et au fossé rhénan*, Thesis of the University of Strasbourg I, 256 pp.

(Received February 6, 2001, revised July 4, 2001 accepted August 1, 2001)

Pure appl. geophys. 160 (2003) 1137–1161
0033–4553/03/061137–25

Pure and Applied Geophysics

Interrelations Between Thermal Conductivity and Other Physical Properties of Rocks: Experimental Data

YURI POPOV,[1] VLADIMIR TERTYCHNYI,[1] RAISA ROMUSHKEVICH,[1] DMITRI KOROBKOV,[1] and JEAN POHL[2]

Abstract—A new non-contact and non-destructive optical scanning instrument provided a large number of high-precision measurements of thermal conductivity tensor components in samples of sedimentary and impact rocks, as well as new insights into interrelations between thermal conductivity and other physical properties. More than 800 core samples (dry and fluid-saturated) of sedimentary rocks from different Russian oil-gas deposits and impact rocks from the well "Nördlingen 1973" drilled in the Ries impact structure (Germany) were studied using optical scanning technology. It was established that the thermal conductivity parallel to the stratification is more informative for petrophysical investigations than the thermal conductivity perpendicular to the layering. Different approaches were developed to estimate porosity, permeability, pore space geometry, and matrix thermal conductivity with a combination of thermal conductivity measurements in dry and fluid-saturated samples and mathematical modelling. These approaches allow prediction of the rock porosity and permeability and their spatial distribution along a well using thermal conductivity measurements performed with the optical scanning instrument directly applied to cores. Conditions and constraints for using Lichtenecker-Asaad's theoretical model for the estimation of porosity and thermal conductivity of sedimentary rocks were determined. A correlation between thermal conductivity and acoustic velocity, porosity, density, and electric resistivity of impact rocks was found for different rock types. New relationships between permeability, electrical and thermal conductivity found for sedimentary rocks are described.

Key words: Thermal conductivity, porosity, permeability, anisotropy, physical properties.

1. Introduction

As shown by numerous theoretical and experimental studies, the thermal conductivity of rocks depends on various factors including porosity, fracturing, mineral composition and structure. These studies have demonstrated the significance of porosity and fracturing on rock thermal conductivity, and have shown that between these properties there exists a complicated relationship, which is particularly dependent on the pore space structure. Detailed data on this relationship were

[1] Moscow State Geological Prospecting Academy, Miklukho-Maklai str. 23, 117873 Moscow, Russia. E-mail: yupopov@dol.ru

[2] Institut für Allgemeine und Angewandte Geophysik, Ludwig-Maximilians-Universität, Theresienstr. 41, D-80333 München, Germany.

presented by SOMERTON (1958, 1992), ROBERTSON (1979), ZIMMERMAN (1989), BRIGAUD (1989), HORAI (1991), WILLIAMS *et al.* (1995), MCKENNA *et al.* (1996), and SCHOEN (1996).

However, many serious methodical problems (rock anisotropy influence, insufficient precision of the measurements, significant discrepancy between theoretical models of rock thermal conductivity and real rock conditions, etc.) inhibit our understanding of the relationship between thermal conductivity and porosity and limit our knowledge of the effect of the abovementioned factors on thermal conductivity. Even the best traditional methods and instruments (divided bar, line source instruments, etc.) cannot provide the necessary quality of measurement on dry and fluid-saturated rock samples (POPOV *et al.*, 1999). For example, in most cases anisotropy and inhomogeneity of sedimentary rocks was not taken into account. In addition available theoretical models of effective thermal conductivity often do not include possible variations in the internal geometry of the porous medium. Finally, the statistical representation of rock collections under study was insufficient. For example, the measured thermal conductivity was often separately reported only for dry or only for fluid-saturated samples (BRIGAUD, 1989; MCKENNA *et al.*, 1996), and in some cases, the rock collections were not geologically homogeneous, e.g., in Brigaud's study. In general, there exist few experimental studies of the relationship between thermal conductivity and rock permeability. This study reports an extensive investigation of this relationship.

2. *Rock Collections*

The following six collections of rocks were studied experimentally.

Collection I. 143 samples of polymictic, arkose, and mica-quartz sandstones and siltstones from the Upper Jurassic and Lower Cretaceous were taken from a depth interval of 1968–3005 m in the Middle Ob' Petroliferous Province and Vartovsk region, (West Siberian Platform). The sandstones are mainly fine-grained and massive while the siltstones were well bedded.

Collection II. 30 samples of fine-grained and slightly bedded quartz sandstones of Cambrian age were collected in the Baltic Petroliferous Province (the Baltic Syneclise) from a depth interval of 1637–2572 m. The rocks have quartz and occasionally quartz and calcite pisolitic cement.

Collection III. 71 cores were collected in the Volga-Urals Petroliferous Province from a depth of 1859–1948 m. The rocks are Paleozoic in age and consist of organogenic limestones. The rock texture is organogenic detrital, relic organogenic detrital, and cloddy. The structure is bedded, banded, or massive. Clay-rich intercalations and stylolitites filled with a black clayey material are clearly seen.

Collection IV. 121 rock samples were collected in Middle-Ob' district, Vartovsk region (West Siberia Petroliferous Province) from a depth interval of 1364–1423 m.

The siltstones are of Cretaceous age and have uniform mineral composition and structure. These low-permeable samples are characterized by an intercalation of shale and siltstone. Texture is layered, cryptolayered, and rarely massive. The rock samples studied were prepared as cylinders, 2.5 cm in diameter and 3.0 cm in length. Two cylinders were prepared from each core: one was oriented parallel to the core axis (61 sample, the so-called Vertical Orientation Collection – VOC) and the second one perpendicular to the core axis (60 samples, the so-called Horizontal Orientation Collection – HOC).

Collection V. The collection comprised 54 limestone samples from the Timan-Pechora Petroliferous Province (depth interval of 2797–3444 m). The algal limestones of Frasnian age are pelitomorphic. Epigenetic processes have significantly modified the pore space of the rocks. In the highly porous varieties, the pores occur between the algal pellets and are rounded or elongated in shape. The coarse-pellet varieties are highly and coarsely porous (voids exceed 100 μm); samples with fine pellets have a high or low porosity and respectively fine pores (less than 100 μm), often with indications of leaching on their walls. Each sample was provided as a cube with 3 cm long edges.

Collection VI. The thermal conductivity was measured on more than 400 cores (8–41 cm in length) from the scientific well "Nördlingen 1973" (1206 m in deep) drilled in the Ries impact structure (Germany) formed as the result of a meteorite impact on Lower Permian, Mesozoic, and Tertiary horizontally layered sediments that overlie crystalline basement of pre-Permian age. The well penetrated the post-impact lake sediments, suevite impact breccia, suevite intercalating with crystalline rocks, amphibolites, gneisses, granites, ultramafic rocks, and breccia dikes (POHL *et al.*, 1977).

3. Methods

A non-contact optical scanning instrument (two versions: a laser laboratory instrument and a portable field instrument) (POPOV, 1983, 1997) provided in principle a new approach to the study of the relationship between the thermal conductivity and other physical properties of rocks. The optical scanning method is based on scanning a sample's surface with a focussed, movable heat source in combination with a temperature sensor. Scanning along three (for the case of 3-D anisotropy) or two (2-D anisotropy) directions provides information about the thermal conductivity tensor components. As described in POPOV *et al.* (1999) the advantages of this instrument are (1) high precision (1.5%) and accuracy (1.5%) of measurements within the thermal conductivity range of 0.2–50 W/m·K for dry and fluid-saturated samples; (2) determination of thermal conductivity tensor components for each rock sample under study; (3) nondestructive numerous measurements on cylindrical or flat core surfaces with no additional mechanical treatment of cores; (4) measurements can be made on rock samples of different sizes, from 1 cm up to 70 cm in a length; (5) the thermal conductivity distribution along a scanning line for each sample can be recorded to detect the rock inhomogeneity; (6) short time of measurement (10–30 s) for a sample.

The thermal conductivity tensor components $\lambda_{\|}$ and $\lambda_{\perp}$ (parallel and perpendicular to rock bedding, respectively) were measured for all samples of Collections I–VI on dry samples (thermal conductivity of air is 0.024 W/m·K) and on samples after saturation with water (brine) (thermal conductivity of water (brine) is 0.60 W/m·K). The measurement procedure was described in detail previously (POPOV and MANDEL, 1998). Collections IV and V were also studied after ethylene glycol-saturation (thermal conductivity 0.27 W/m·K). Rock saturation with water (brine) and ethylene-glycol was carried out under vacuum except for the samples of Collection VI which were saturated under atmospheric pressure.

The porosity of rock samples from Collections I–V was determined using the water saturation method. The permeability was measured with a helium method. Physical properties of rock samples of Collection VI were determined from the logging data.

Each core sample from Collections I–III, 3.0–11.5 cm long, was cut along its axis into two parts. One part of the core was used to prepare a cylindrical sample (2.5 cm in diameter and 3.0 cm in length) for measurements of porosity and permeability. The second part was used for measurements of thermal conductivity on the flat core surface. For the rock samples of Collections IV–V, the thermal conductivity, porosity, and permeability were measured on the same samples.

We compared the experimental data obtained by means of the optical scanning instrument with theoretical models for the effective thermal conductivity of rocks. A new theoretical model of effective thermal conductivity based on petrophysical data and mineralogical content was used for the calculations (TERTYCHNYI *et al.*, *this issue*). Solutions for heat flow density $\vec{q}(\vec{r})$ and temperature gradient $\nabla T(\vec{r})$ could be obtained using information on geometry, orientation, and thermal properties of inclusions in granular composites. An effective thermal conductivity is derived from the equation:

$$<q_i> = \lambda_{ij}^{\text{eff}} <\nabla_j T>, \tag{1}$$

where brackets mean ensemble averaging. In the frame of effective medium theory, the effective thermal conductivity tensor of the granular medium with ellipsoidal inclusions is obtained by:

$$\lambda^{\text{eff}} = \left\langle \lambda^{(a)} \left[\lambda^{(0)} \left(I - F^{(a)} \right) + \lambda^{(a)} F^{(a)} \right]^{-1} \right\rangle \left\langle \left[\lambda^{(0)} \left(I - F^{(a)} \right) + \lambda^{(a)} F^{(a)} \right]^{-1} \right\rangle^{-1}, \tag{2}$$

where

$$\lambda^{(0)} = \left\langle \left[\lambda^{(0)} \left(I - F^{(a)} \right) + \lambda^{(a)} F^{(a)} \right]^{-1} \right\rangle^{-1}, \tag{3}$$

I is a unit tensor, $\lambda_{ij}^{(a)}$ is the thermal conductivity and $F_{ij}^{(a)}$ is a depolarization tensor of component a. The model includes an arbitrary number of rock components, varying mineral grain shapes, thermal anisotropy caused by mineral anisotropy, grain or crack ordering, and spatial fracture distribution.

4. Results

Statistical data on thermal conductivity tensor components, thermal anisotropy coefficient $K = \lambda_{||}/\lambda_{\perp}$, porosity Ψ and permeability P values measured for Collections I–VI are presented in Table 1.

Most rock samples of Collections I–IV were found to be anisotropic (for dry samples $K_d = \lambda_{||d}/\lambda_{\perp d} = 1.03$–$2.1$). For Collection V the anisotropy was very small. To exclude a significant influence of rare but thick cracks on thermal conductivity results for Collection V, the thermal conductivity tensor components were averaged (as well as the permeability components measured for three orthogonal directions) before estimation of the correlation between thermal conductivity and reservoir properties. No significant thermal anisotropy was found for impact formations of Collection VI. This is consistent with the results for impact formations of the Puchezh-Katunk impact structure (POPOV *et al.*, 1998).

The correlation between the tensor component $\lambda_{||}$ and other physical properties for all the collections turned out to be much closer than for the tensor component $\lambda_{\perp}$. This finding is opposite to Brigaud's results (Brigaud is the only researcher who carefully took into account the thermal anisotropy of sedimentary rocks in petrophysical research). This difference may be due to the significant influence of rare cracks parallel to the bedding plane on the determination of $\lambda_{\perp}$.

(a) Thermal Inhomogeneity Factor β Variations as a Result of Small-scale Variations of Porosity within Cores

The thermal inhomogeneity factor β ($\beta = \lambda_{max} - \lambda_{min})/\lambda_{aver}$, where λ_{max}, λ_{min} and λ_{aver} are maximum, minimum, and average thermal conductivities recorded along each scanning line) was determined for each scanning line and for each core sample to characterize the thermal inhomogeneity of rocks. The results show that the variations in β for different rock samples provide information that is similar to the standard deviation (SD) of λ along the scanning line on the sample. The interrelations of inhomogeneity factors β_d and β_s for dry and saturated sedimentary rock samples for Collection III are illustrated in Figure 1. The results are ordered here with increasing values of β_{pd} values. As can be seen from Figure 1, if $\beta_{d,s} < 0.22$,

Table 1

Thermal conductivity, porosity and permeability measured for rock samples studied

Collection	Rock type	Thermal conductivity, W/m*K				Anistropy coeffecient		Porosity, %	Permeability, mD	N
		Dry samples		Saturated samples		K_d	K_s			
		$\lambda_{\parallel d}$	$\lambda_{\perp d}$	$\lambda_{\parallel s}$	$\lambda_{\perp s}$					
I	Sandstone	1.90 (0.33)* 1.42–2.74**	1.76 (0.31) 1.21–2.62	3.01 (0.27) 2.35–3.80	2.85 (0.20) 2.17–3.41	1.08 (0.07) 1.00–1.26	1.06 (0.07) 1.00–1.31	17.0 (4.70) 3.10–25.0	4.47 (10.0) 0.01–199.5	99
	Siltstone	2.07 (0.20) 1.65–2.76	1.78 (0.25) 1.35–2.76	2.82 (0.27) 2.37–3.34	2.54 (0.33) 1.79–3.34	1.18 (0.10) 1.00–1.43	1.12 (0.09) 1.00–1.35	14.8 (2.66) 8.00–20.2	0.26 (4.47) 0.01–7.41	44
II	Sandstone	3.45 (0.72) 2.18–5.60	3.57 (0.82) 2.32–6.09	6.21 (0.48) 5.49–7.38	6.08 (0.34) 5.49–6.62	1.00 (0.04) 0.89–1.00	1.00 (0.04) 0.86–1.00	11.3 (3.45) 1.40–16.6	234.4 (3.80) 11.0–1949.8	30
III	Limestone	2.42 (0.20) 1.76–2.98	2.34 (0.21) 1.77–2.94	2.74 (0.12) 2.38–3.14	2.66 (0.14) 2.18–2.99	1.04 (0.04) 1.00–1.15	1.04 (0.03) 1.00–1.18	6.69 (4.40) 0.30–14.0	0.53 (5.24) 0.01–70.8	71
IV	Intercalating Siltstone and Shale	1.78 (0.17) 1.52–2.07	1.19 (0.16) 0.88–1.35	2.90 (0.22) 2.46–3.18	2.20 (0.46) 1.34–2.81	1.54 (0.34) 1.18–2.36	1.37 (0.29) 1.10–2.19	22.1 (2.40) 18.0–26.1	0.62 (1.01) 0.04–3.40	12
	Siltstone	1.52 (0.23) 1.30–2.47	1.33 (0.19) 1.07–2.24	3.10 (0.08) 2.92–3.27	2.87 (0.16) 2.28–3.21	1.14 (0.10) 0.99–1.61	1.08 (0.06) 0.96–1.29	26.3 (4.09) 9.60–30.4	11.9 (11.1) 0.03–62.2	49
V	Limestone	2.29 (0.08) 1.23–2.83	2.24 (0.23) 1.27–2.83	2.70 (0.13) 2.40–3.08	2.68 (0.15) 2.23–3.08	1.02 (0.08) 0.77–1.23	1, 02 (0.03) 0.95–1.12	8.00 (2.53) 0.40–24.8	105.0 (111.2) 0.001–1552.0	54

VI	High-temperature Suevite	1.20 (0.18) / 0.88–1.88	1.73 (0.24) / 1.41–2.73	–	–	–	–	55
	Low-temperature Suevite	1.05 (0.23) / 0.71–1.54	1.96 (0.14) / 1.77–2.33	–	–	–	–	15
	Grey Breccia dike	1.11 (0.17) / 0.91–1.53	1.94 (0.14) / 1.74–2.41	–	–	–	–	15
	Brown Breccia dike	1.77 (0.14) / 1.39–2.17	2.29 (0.22) / 1.99–2.76	–	–	–	–	21
	Granite	1.88 (0.40) / 0.96–2.80	2.52 (0.31) / 1.85–3.27	–	–	–	–	38
	Ultrabasite	2.27 (0.27) / 1.69–2.70	2.44 (0.24) / 1.91–2.85	–	–	–	–	22
	Amphibolite	1.88 (0.26) / 1.41–2.50	2.24 (0.20) / 1.98–2.78	–	–	–	–	56
	Gneiss	2.11 (0.29) / 1.28–2.74	2.44 (0.26) / 1.93–3.19	–	–	–	–	190

Commentary: *Average value (RMS) / ** Min value – Max value, N – number of samples

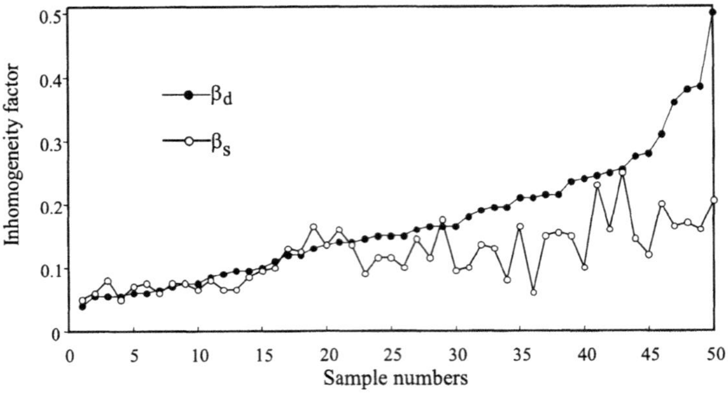

Figure 1
Interdependence of inhomogeneity factors β_d and β_s for Collection III. Rock samples are ordered according to increasing β_d. The error of inhomogeneity factor determination is $\pm 5\%$.

there is no significant difference between β_d and β_s. However, for $\beta_d > 0.22$ the curves for β_d and β_s diverge. This effect is explained by a significant variation of porosity within each sample with $\beta_d > 0.22$, except for a few samples. After water-saturation, the effect of porosity on thermal conductivity becomes insignificant, which is in agreement with a decrease in β after saturation. The thermal inhomogeneity when β_d and β_s are <0.22 can be interpreted as a mineralogical inhomogeneity remaining unchanged under water-saturation. A similar inhomogeneity factor decrease under water-saturation conditions as a result of significant variations in porosity and fracturing within the sample was also observed for impact rocks (Collection VI). The change in β after saturation indicates that significant variations of the porosity Ψ and hence of the permeability, P, can effect the relationship between thermal conductivity and reservoir properties of rocks when the measurements are not confined to the same samples. Therefore, we can state that our investigations on Collections IV and V (where all measurements were carried out on the same samples) are the most reliable.

(b) Correlation between Thermal Conductivity and Permeability

A close correlation (correlation coefficient r ranges from 0.7 to 0.9) was observed between the relative change of rock thermal conductivity after water-saturation of samples and the logarithm of permeability P ($\log P$) of sedimentary rocks. In order to characterize the change of thermal conductivity after water-saturation, we introduce the parameter $\delta\lambda_{||} = (\lambda_{||s} - \lambda_{||d})/\lambda_{||d}$ where $\lambda_{||d}$ and $\lambda_{||s}$ correspond to dry and water-saturated rock samples. The relationship between $\delta\lambda_{||}$ and $\log P$ for Collection IV is shown in Figure 2. The vertical zoning of parameters $\delta\lambda_{||}$ and $\log P$ along the wells generally shows a good correlation for all collections of sedimentary rocks (Figs. 3 and 4). As follows from Figure 4, where the samples of Collection IV were ordered

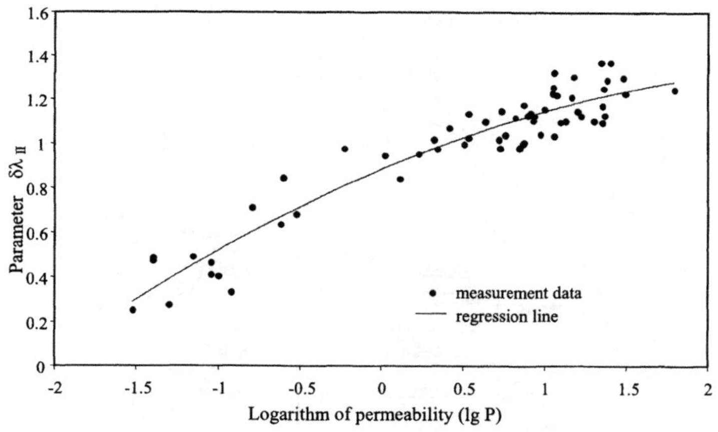

Figure 2
Correlation between parameter $\delta\lambda_{||}$ and permeability for Collection IV. The error of measurements is $\pm 4\%$ for $\delta\lambda_{||}$ and $\pm 5\%$ for permeability.

according to their increasing anisotropy coefficient, the permeability decreases with increasing thermal anisotropy coefficient. It is important to note that the permeability and $\lambda_{||}$ were measured for the same direction parallel to the bedding. Taking into account the rapid nature of the thermal conductivity measurements using the optical scanning instrument, these results yield detailed rapid information on permeability from thermal conductivity measurement. This provides an opportunity to estimate the permeability and formation zoning from the empirical relationship between $\delta\lambda_{||}$ and log P. It should be noted that in all cases the correlation between $\delta\lambda_{||}$ and log P is better than between log P and porosity. This can be regarded as evidence for the direct relationship between parameter $\delta\lambda_{||}$ and permeability.

(c) Correlation between Thermal Conductivity and Porosity

A very close correlation between $\lambda_{||d}$ of dry samples and porosity was found for all collections where both parameters were measured (Table 2). Experimental results for Collections I, IV and V are shown in Figure 5. They demonstrate that the effect of mineralogical composition of dry rocks is small and the porosity is a predominant factor. There is no evident difference in the correlation for low- and high-anisotropic rocks. The thermal conductivity versus porosity correlation for Collections IV and V, where both parameters were measured on the same samples, is better than for Collections I–III where $\lambda_{||}$ and Ψ were measured on two halves of cores (Fig. 5). This confirms a significant variation of $\lambda_{||}$ and Ψ within particular cores. According to correlation between $\lambda_{||}$ and Ψ for Collection V (Fig. 5), the limestones can be subdivided into three groups: (1) porous limestone with a very close correlation between $\lambda_{||}$ and Ψ, (2) vesicular (cavernous) and fractured limestone with a low

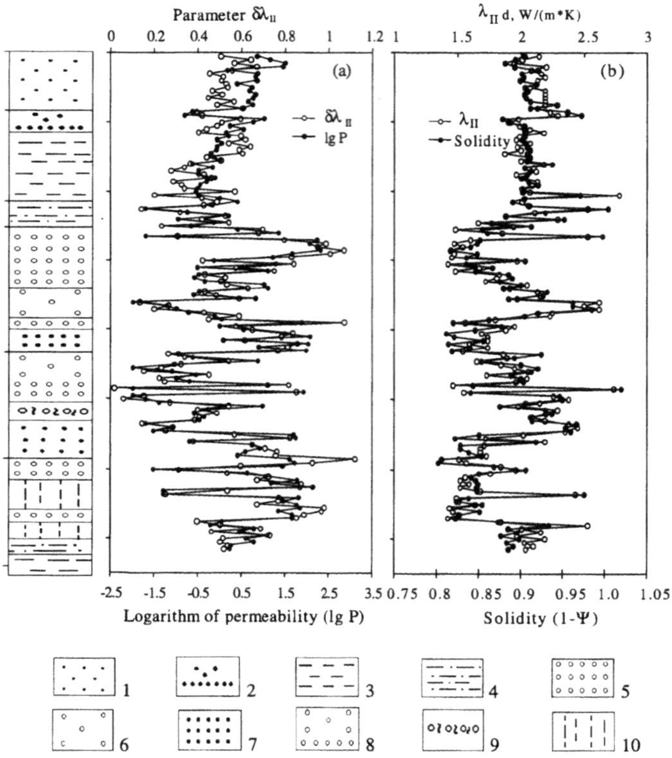

Figure 3

Logs of parameter $\delta\lambda_{\|}$ and logarithm of permeability log P (a) and thermal conductivity $\lambda_{\|d}$ and solidity (b) for Collection I. Rock samples are ordered according to their well depths. Accuracy and precision of $\lambda_{\|d}$ measurements are 1.5% each (confidence probability of 0.95). Legends: 1 — fine-grained massive polymictic sandstone with clayey cement; 2 — massive and bedded polymictic sandstone with clayey and clayey-carbonate cement; 3 — bedded polymictic and arkosic siltstone with clayey cement; 4 — intercalated sandstone and siltstone; 5 — fine-grained bedded mica-quartz sandstone with clayey cement; 6 — fine-grained massive mica-quartz sandstone with clayey cement; 7 — medium- and coarse-grained bedded polymictic sandstone; 8 — fine- and coarse-grained massive and bedded mica-quartz sandstone with clayey cement; 9 — fine- and coarse-grained bedded mica-quartz sandstone with carbonate cement; 10 — bedded mica-quartz siltstone with clayey cement.

correlation between $\lambda_{\|}$ and Ψ, and (3) fractured limestone with a lack of correlation between $\lambda_{\|}$ and Ψ.

A concomitant vertical zoning of $\lambda_{\|d}$ with rock solidity, introduced as $(1 - \Psi)$, is observed for all collections and shown for Collections I and IV in (Figs. 3b and 6).

The correlations between $\lambda_{\|d}$ and Ψ can be used to determine the thermal conductivity λ_M of the rock matrix for $\Psi \to 0$ (Table 2) as discussed hereafter.

It is evident that the relative change of thermal conductivity under water-saturation conditions should depend upon porosity. Figure 7 illustrates a close

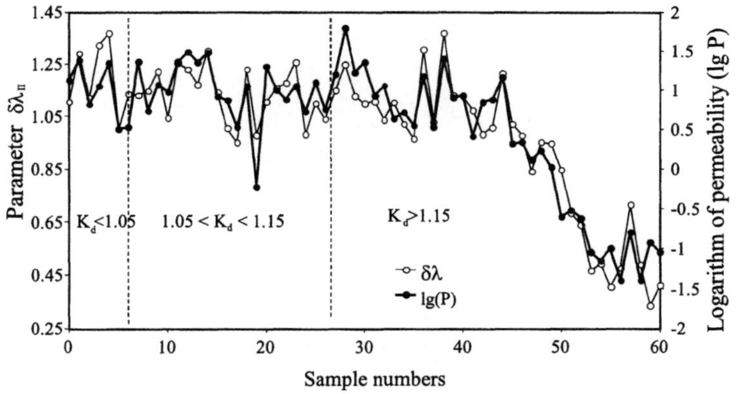

Figure 4
The zoning of parameters $\delta\lambda_{||}$ and permeability for Collection IV. The rock samples are ordered according to their anisotropy coefficient increase.

correlation between $\delta\lambda_{||}$ and Ψ for Collections I and V. The regression equations are given in Table 2. It is important to note that the regression equations are quite similar for different rock types indicating a possible universal character of the correlation between $\delta\lambda$ and Ψ for sedimentary rocks.

There is virtually no correlation between λ and Ψ for water (brine)-saturated samples (Figs. 5a,b) (except for limestones from Collection V, see Fig. 5c), which can be explained by a variable thermal conductivity of the mineral matrix and also, in particular by the influence of the clay content. Similar complicated situations, with an absence of correlation between λ_s and Ψ for sandstones, were also observed by ROBERTSON (1979) and MCKENNA (1996). In water-saturated rocks, the contrast in thermal conductivity of the rock matrix and the filling fluid is thought to be much smaller than for dry rocks. Other factors including the quartz content (quartz has a high thermal conductivity of 7.7 W/(m·K)), the difference in porous medium geometry, and clay content may suppress the effect of porosity.

(d) Estimation of Porosity and Thermal Conductivity

Lichtenecker's theoretical model of effective thermal conductivity is the most popular in geothermics and petrophysics and its use for different rock types has been recommended by many researchers (PRIBNOW and SASS, 1995; WILLIAMS *et al.*, 1995; SCHOEN, 1996, etc):

$$\lambda_{\text{eff}} = \lambda_M^{(1-\Psi)} \lambda_{\text{fluid}}^{\Psi}. \tag{4}$$

ASAAD (1955) specified a formula for the pore space geometry:

$$\lambda_{\text{eff}} = \lambda_M^{(1-f\Psi)} \lambda_{\text{fluid}}^{f\Psi}. \tag{5}$$

Table 2

Regression parameters for thermal conductivity and porosity of the sedimentary rock samples

Collection	λ_{matr}	Regression equation for $\lambda_{\parallel}$ and Ψ	Remarks	r	Regression equation for $\delta\lambda_{\parallel}$ and Ψ	Remarks	r
I	3.23	$\lambda_{\parallel} = 3.2316 \cdot e^{-0.0322 \cdot \Psi}$		0.90	$\delta\lambda_{\parallel} = 0.0411 \cdot \Psi - 0.0655$	$K_d < 1.10$	0.89
					$\delta\lambda_{\parallel} = 0.0433 \cdot \Psi - 0.2041$	$K_d > 1.10$	0.83
II	6.23	$\lambda_{\parallel} = 6.2268 \cdot e^{-0.0538 \cdot \Psi}$		0.94	$\delta\lambda_{\parallel} = 0.1759 \cdot e^{0.1283 \cdot \Psi}$		0.93
III	2.73	$\lambda_{\parallel} = 2.7285 \cdot e^{-0.0174 \cdot \Psi}$	$K_d < 1.05$	0.82	$\delta\lambda_{\parallel} = 0.0481 \cdot e^{0.1227 \cdot \Psi}$	$K_d < 1.05$	0.90
	2.36	$\lambda_{\parallel} = 2.3617 \cdot e^{-0.004 \cdot \Psi}$	$K_d > 1.05$	0.42	$\delta\lambda_{\parallel} = 0.0661 \cdot e^{0.0589 \cdot \Psi}$	$K_d > 1.05$	0.59
IV	3.57	$\lambda_{\parallel} = 3.5663 \cdot e^{-0.0327 \cdot \Psi}$	VOC	0.97	$\delta\lambda_{\parallel} = 0.0953 \cdot e^{0.0895 \cdot \Psi}$	VOC	0.93
	3.43	$\lambda_{\parallel} = 3.4299 \cdot e^{-0.0312 \cdot \Psi}$	HOC	0.96	$\delta\lambda_{\parallel} = 0.1339 \cdot e^{0.0763 \cdot \Psi}$	HOC	0.95
V	2.75	$\lambda_{\parallel} = 2.7529 \cdot e^{-0.0437 \cdot \Psi}$	Fractured	0.78	$\delta\lambda_{\parallel} = 0.0607 \cdot e^{0.1869 \cdot x}$	Fractured	0.44
	2.75	$\lambda_{\parallel} = 2.7450 \cdot e^{-0.0009 \cdot \Psi}$	Porous-fractured	0.02	$\delta\lambda_{\parallel} = 0.0859 \cdot e^{0.1034 \cdot x}$	Porous–fractured	0.25
	2.71	$\lambda_{\parallel} = 2.7146 \cdot e^{-0.0238 \cdot \Psi}$	Porous	0.97	$\delta\lambda_{\parallel} = 0.0712 \cdot e^{0.0983 \cdot \Psi}$	Porous	0.94

$\lambda_{\parallel}$ — thermal conductivity component measured parallel to bedding plane (dry samples), λ_{matr} (W/m·K) — thermal conductivity of rock mineral skeleton, Ψ — porosity, K_d — thermal conductivity anisotropy coefficient of dry rock samples, r — correlation coefficient

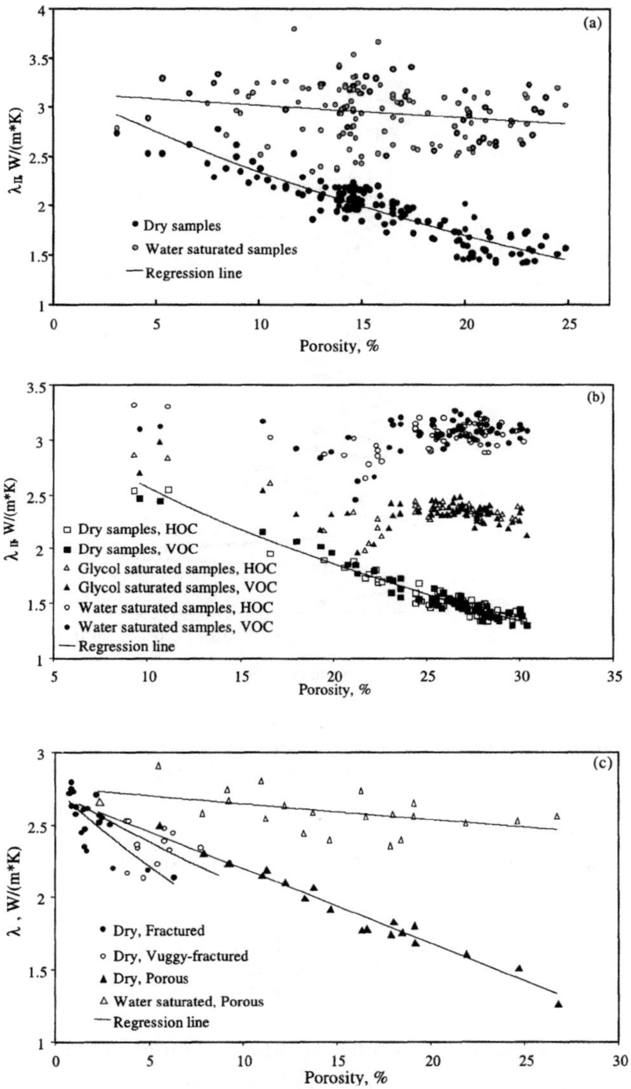

Figure 5

Correlations between thermal conductivity components $\lambda_{\|}$ and porosity. The total absolute error of the porosity measurements is $\pm 0.5\%$. a — Collection I, b — Collection IV, c — Collection V.

where f is the correlation factor that describes a degree of sedimentary rock lithification. The factor f is considered to be a structural characteristic of the rock similar to the Archie cementation factor used in well-log interpretation.

We used model (5) with a variable coefficient f (called Lichtenecker-Asaad's model) for the following purposes: (1) to calculate λ_{eff} for dry and fluid-saturated

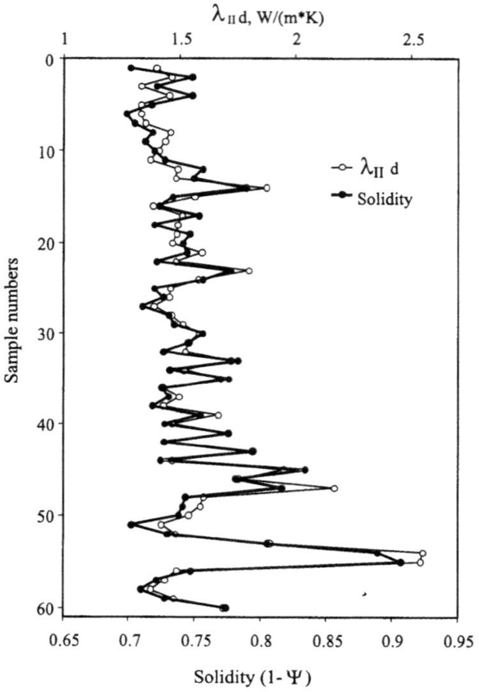

Figure 6
Vertical zoning of thermal conductivity and rock solidity for Collection IV. Rock samples are ordered according to their depth in the well.

samples when λ_M and porosity are known from measurements, (2) to calculate the porosity:

$$\Psi = \frac{1}{f} \bullet \frac{\ln \frac{\lambda_{\text{eff}}}{\lambda_M}}{\ln \frac{\lambda_{\text{air}}}{\lambda_M}} \tag{6}$$

derived from formula (5) when λ_{eff} and λ_M are measured values. In both cases, the results are compared with our experimental data. It was established that for each rock type a stable value f could be determined only for the thermal conductivity component $\lambda_{||d}$. Values of the coefficient f determined for different collections are shown in Table 3.

The consistence in measured and calculated effective thermal conductivity of dry samples (formula (5)) and porosity (formula (6)) is good (Figs. 8a,b, Collection IV as an example). No substantial influence of anisotropy was found. This implies that the parameter $\lambda_{||}$ is affected by fracturing to a much lesser extent than $\lambda_{\perp}$. No satisfactory f values were found which provide even a rough coincidence with thermal conductivity measurements of fluid-saturated samples. These results

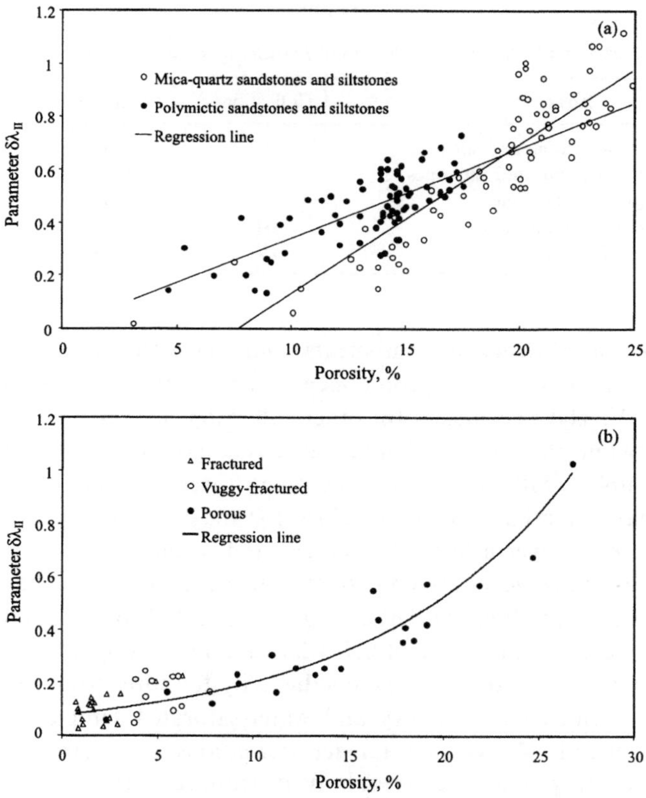

Figure 7
Correlation between parameter $\delta\lambda_{||}$ and porosity for Collections I (a) and V (b).

show that the computational determination of thermal conductivity for fluid-saturated rocks, i.e., for *in situ* conditions, is a substantially more complicated problem than the study of dry samples because of clay content and other physical reasons.

We tested the possibility of determining the porosity from thermal conductivity values for dry and fluid-saturated samples using the formula derived from equation (5) with a correction coefficient c proposed by SOMERTON (1958):

$$\Psi = \log(\lambda_s/\lambda_d)/(c \cdot \log(\lambda_{\text{water}}/\lambda_{\text{air}})). \tag{7}$$

As was shown for the rock collections under study, we were able to determine the c coefficient that provides the best agreement of the porosity measurement results and the porosity calculations from formula (7). Once again we found that the $\lambda_{\perp s}$ and $\lambda_{\perp d}$ components yield considerably more unstable, random results; thus, the $\lambda_{||s}$ and $\lambda_{||d}$ components are to be preferred for the application.

Table 3

Correction factors for calculations of porosity and thermal conductivity

Rock	Collection	*f* coefficient	*c* coefficient
Mica-quartz sandstone and siltstone	I	0.50	0.87
Polymictic sandstones and siltstones	I	0.60	0.50
Quartz sandstone	II	0.82	1.7
Organogenic limestone	III	0.35	0.65
Siltstone	IV	0.64	0.82

The results of porosity measurements and calculations (formula (7)) with experimental data for $\lambda_{\|s}$ and $\lambda_{\|d}$ components with $c = 0.82$ are shown in Figure 8c for Collection IV. The c coefficient for other collections is presented in Table 3. The agreement between the measured and calculated porosity is the best for low-anisotropic samples (with $K_d < 1.15$). In many cases, the high-anisotropic rocks ($K_d > 1.15$) have markedly different values for the "c" coefficient than the low-anisotropic rocks. Nevertheless, the values and zoning of the measured and calculated porosity are generally consistent with each other in spite of significant variations in anisotropy (Fig. 8c). The c range of 0.5–1.7 (Table 3) obtained from our research greatly exceeds the range of 0.9–1.2 recommended by SOMERTON (1958).

The measurements of rock thermal conductivity for cores from the "Nördlingen 1973" well were carried out on dry and water-saturated samples (Table 1). The petrophysical analysis shows that impact formations are characterized by high fracturing. Rock fragments are irregularly distributed throughout the matrix of mineral grains. The pores were considered to be a set of randomly distributed, non-aligned ellipsoidal cracks. The estimation of porosity, matrix thermal conductivity, and crack aspect ratio is based on the equations (2, 3) for air- and water-filled rock samples. The rock matrix was approximated by an agglomerate of spherical grains. The inclusion shape is characterized by the non-dimensional parameter aspect ratio, i.e., ratio of vertical (axis of axial symmetry) to horizontal ellipsoid semi-axis.

The theoretical model developed for the interpretation of experimental data on thermal conductivity has a very significant limitation: the number of unknown parameters is larger than the number of equations. The problem of inconsistency of equation number and number of parameters could be partially settled by the introduction of additional relations between the parameters. We have shown the clear correlation between the relative changes in thermal conductivity after water-saturation ($\delta\lambda = (\lambda_s - \lambda_d)/\lambda_d$) and the thermal conductivity λ_d of dry samples for different rock types of Collection VI (Fig. 9). Based on this correlation, we derived the empirical regression equations from the data shown in Figure 9 in addition to the equations from the theoretical model for dry and water-saturated samples. The mean correlation coefficient for different petrophysical groups exceeds 0.8. The set of equations has the form:

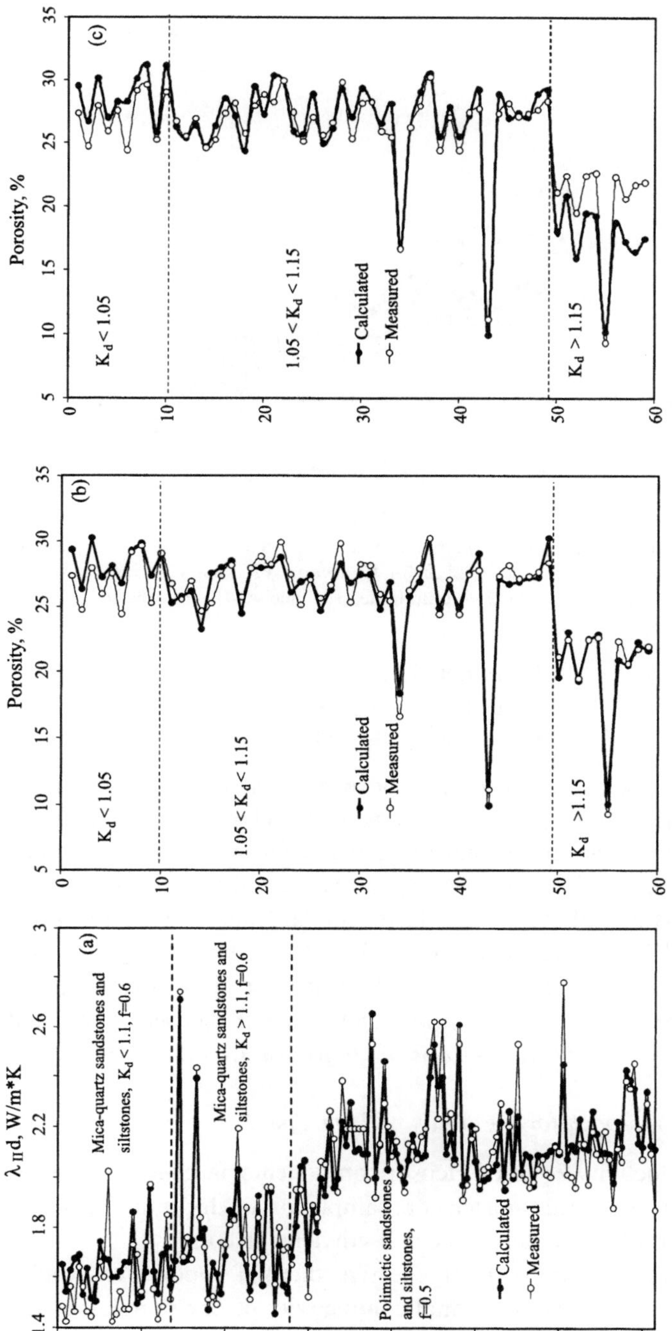

Figure 8
Results of measurements and calculations of thermal conductivity (a) and porosity (b, c).

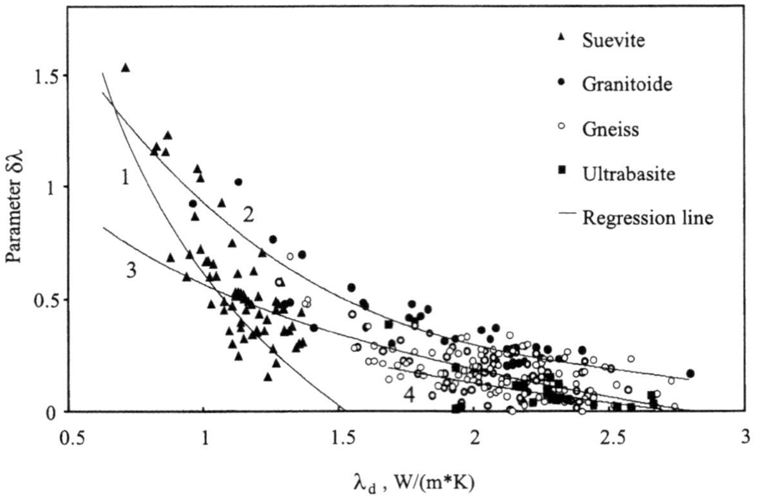

Figure 9
Correlation between the relative change in thermal conductivity after water-saturation (parameter $\delta\lambda_{||}$) and thermal conductivity values for dry cores ($\lambda_{||d}$) for Collection VI. Line 1 corresponds to a regression equation for suevite, 2 — granitoide, 3 — gneiss, 4 — ultrabasites.

$$\begin{cases} \lambda_{\text{eff}}(\alpha, \Psi, \lambda_M, \lambda_{\text{air}}) = \lambda_d, \\ \lambda_{\text{eff}}(\alpha, \Psi, \lambda_M, \lambda_{\text{water}}) = \lambda_s, \\ (\lambda_s - \lambda_d)/\lambda_d = fun(\lambda_d), \end{cases} \tag{8}$$

where α – is a crack aspect ratio, Ψ – porosity, *fun* is a regression function that depends upon structure of pore space. If samples belong to the same petrophysical type, they have the same functional relationship *fun* with similar parameters. It is important to be very careful in choosing samples for the analysis: all samples must belong to the same lithotype and have similar inner structure. For the well "Nördlingen 1973" we selected the rocks based on the lithological description of cores. The set of three nonlinear equations (8) was solved numerically using nonlinear optimisation techniques. Results of inversion for matrix thermal conductivity, aspect ratio, and porosity for different petrophysical groups are plotted in Figure 10.

(e) Three Fluid Approach for the Study of Pore Space

Thermal conductivity $\lambda_{||}$ and $\lambda_{\perp}$ tensor components parallel and perpendicular to stratification were measured for each core sample for air-dry state (λ_d) and after fluid-saturation (λ_s) with fresh water, brine and ethylene glycol.

Using equations (2) and (3) for effective thermal conductivity and results of experimental measurements of thermal conductivity of dry samples λ_d and samples saturated with water λ_s and with ethylene glycol λ_g, we can write the expression for effective thermal conductivity:

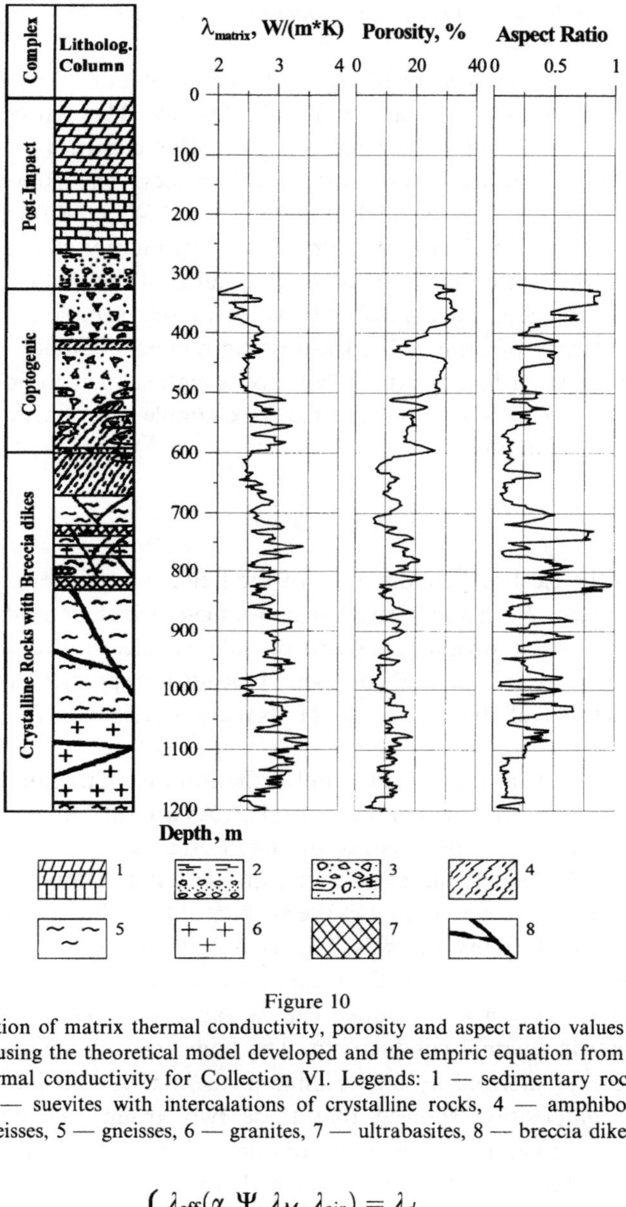

Figure 10
Results of calculation of matrix thermal conductivity, porosity and aspect ratio values from the inverse problem solution using the theoretical model developed and the empiric equation from the experimental data on rock thermal conductivity for Collection VI. Legends: 1 — sedimentary rocks, 2 — suevitic conglomerates, 3 — suevites with intercalations of crystalline rocks, 4 — amphibolites, hornblende gneisses, 5 — gneisses, 6 — granites, 7 — ultrabasites, 8 — breccia dikes.

$$\begin{cases} \lambda_{\text{eff}}(\alpha, \Psi, \lambda_M, \lambda_{\text{air}}) = \lambda_d, \\ \lambda_{\text{eff}}(\alpha, \Psi, \lambda_M, \lambda_{\text{water}}) = \lambda_s, \\ \lambda_{\text{eff}}(\alpha, \Psi, \lambda_M, \lambda_{\text{glycol}}) = \lambda_g, \end{cases} \tag{9}$$

From the equations, it is possible to estimate three independent model parameters: rock matrix conductivity λ_M, aspect ratio of cracks α, and porosity Ψ. The set of non-linear equations (9) was solved numerically, using procedures of non-linear optimisation.

5. Correlation between Thermal Conductivity and other Physical Properties

Measurements of electrical resistivity and thermal conductivity of sedimentary rocks were carried out on siltstones from collection IV. Thermal conductivity was measured on dry and water-saturated samples. The electrical resistivity ρ was measured on brine and fresh water-saturated samples. The contrast in thermal conductivity of minerals (3–4 W/(m*K)) and air (0.024 W/(m*K)) exceeds 100–200. Therefore the thermal conductivity of dry rocks is primarily affected by geometry and volume of pore space. Small fluctuations in mineral thermal conductivity have no influence on the thermal conductivity value of dry samples. The contrast in electrical conductivity of brine and rock-forming minerals is much higher and can exceed 10^{10}–10^{14}. According to Archie law, for a high saline solution the electrical conductivity σ ($\sigma = 1/\rho$) of a fully saturated rock sample is proportional to the fluid conductivity (ELLIS, 1998):

$$\sigma = \frac{1}{F}\sigma_w, \tag{10}$$

where σ_w — is the electrical conductivity of the fluid, F — is the formation factor. The formation factor F is a non-dimensional parameter which depends on pore space morphology and which does not depend on mineral electrical conductivity. This similarity in electrical and thermal conductivity leads to a predictable correlation of thermal conductivity of dry rocks and of formation resistivity of brine saturated samples (Fig. 11a).

In fluid (brine or water) saturated samples, the contrast in thermal conductivity of the mineral matrix and the fluid is much smaller. As a result, the measured thermal conductivity is more readily affected by small fluctuations in mineral thermal conductivity. For low saline saturated cores, the electrical conductivity depends not only on the structure of pore space and the salinity of brine but also on the electrical conductivity of rock forming minerals (Fig. 11b). Experimental results allowed us to distinguish two groups of rocks with similar internal structure but different mineral composition. The low porosity samples (in the range 0–15%) are characterized by low mineral matrix thermal conductivity. The high porosity samples (with porosity >15%) are characterized by higher mineral thermal conductivities. Analysis of mineralogical composition showed that these two groups of rock samples are distinguished by a different clay content that decreases with increasing porosity.

The logging data for the "Nördlingen 1973" well (Ries impact structure, Collection VI) including laterolog deep, laterolog shallow, induction (electric resistivity), and sonic (V_p) logs were used to study the correlation between thermal conductivity and other physical properties. The original data of all physical properties were averaged within depth intervals of 5 m. A close correlation was found between the thermal conductivity λ_d of dry rock samples and other physical properties for all rocks, except for ultramafic rocks (depth intervals of 727–744 m,

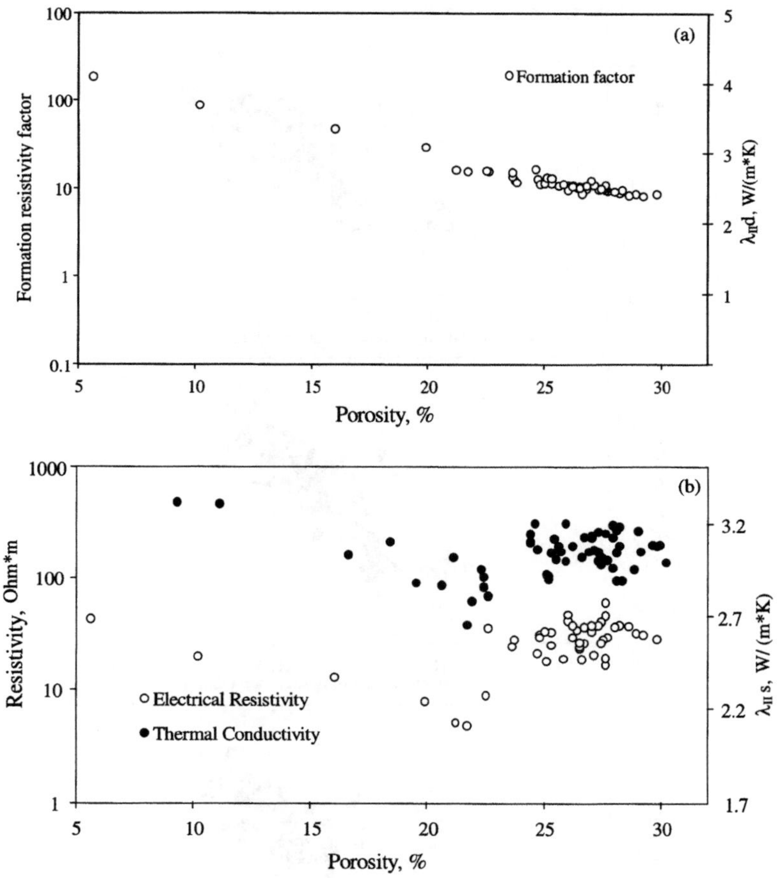

Figure 11

Experimental data on the dependence of thermal conductivity and electrical resistivity on porosity for Collection IV. a — solid circles — $\lambda_{||d}$, empty boxes — electrical resistivity of brine-saturated samples; b — solid circles — $\lambda_{||s}$, empty boxes — electrical resistivity of fresh water-saturated samples.

797–802 m, 814–831 m) and for high-temperature suevites (331–530 m). The physical properties included electric resistivity (induction log, shallow and deep laterologs, with $r = 0.82$, 0.89, and 0.86, Fig. 12) and sonic velocity V_p (sonic log, $r = 0.89$, Fig. 13). Figure 13a illustrates the good agreement in zoning for the vertical distribution of thermal conductivity of dry rock samples λ_d and sonic velocity V_p, except some short depth intervals with ultramafic rocks. A statistically significant correlation between the thermal conductivity λ_s for water-saturated rock samples and sonic velocity V_p was found (Fig. 13b), although this correlation is not as good as for λ_d of dry rock samples. No significant correlation was found between λ_d and results of gamma-ray and spontaneous polarisation logs.

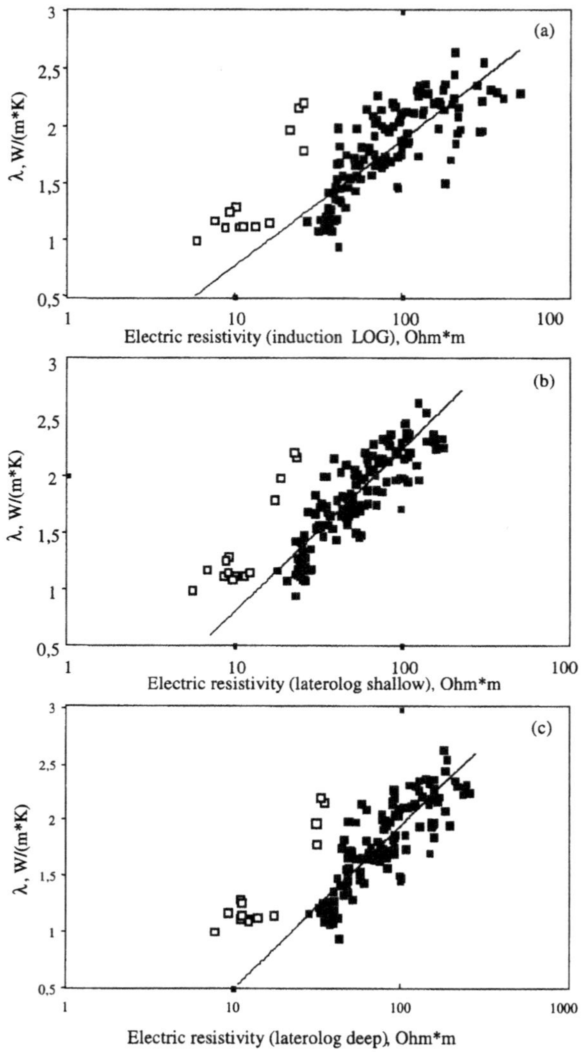

Figure 12
Field of correlation between thermal conductivity λ_d of dry rocks and electric resistivity determined from the induction log (a), the laterolog shallow (b), and the laterolog deep (c) (Collection VI). Empty squares — ultrabasites and high temperature suevites, solid squares — all other rocks. The regression line corresponds to all rocks excluding ultrabasites and high temperature suevites.

6. Conclusions

The optical scanning instrument for the measurement of thermal conductivity of rocks provides in principle a new technique for petrophysical investigations, in

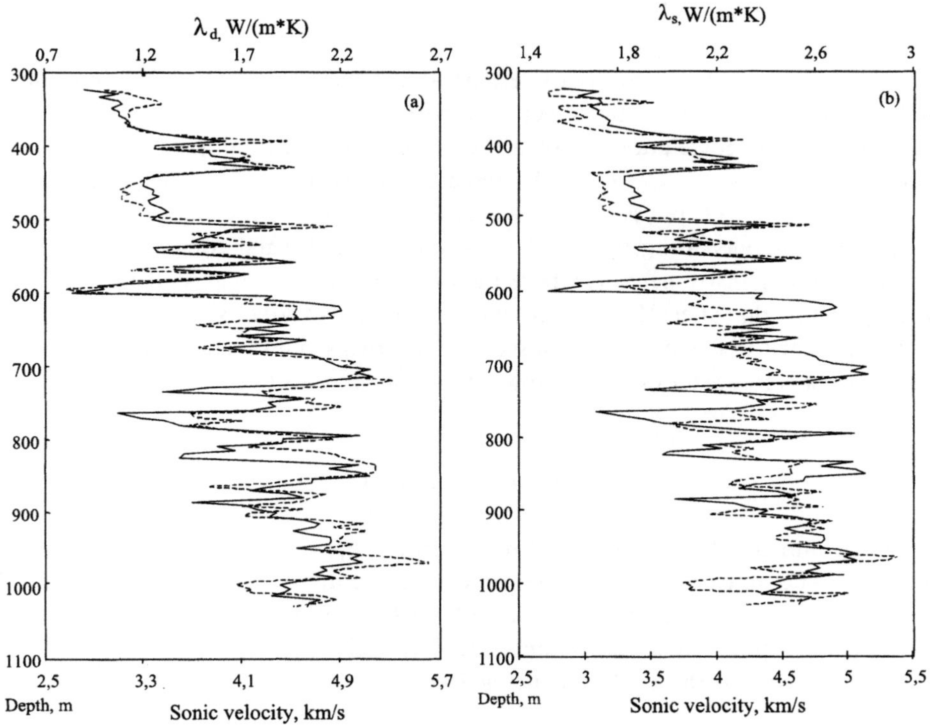

Figure 13

Logs of thermal conductivity λ_d and sonic velocity V_p (Collection VI). Solid line — sonic velocity, dashed line — thermal conductivity. a — thermal conductivity measured on dry cores, b — thermal conductivity measured on water-saturated cores.

particular, for high-precision measurements of the rock thermal conductivity tensor components, taking into account the rock inhomogeneity. The significant effect of thermal anisotropy of sedimentary rocks on the petrophysical characteristics was confirmed and the advantage in using the thermal conductivity component parallel to bedding for petrophysical investigations was established. A substantial inhomogeneity within a rock sample requires the measurement of all physical properties on the same sample in order to provide reliable experimental data.

Conditions and constraints for using the Lichtenecker-Asaad's theoretical model for the estimation of porosity and thermal conductivity of sedimentary rocks were determined. An investigation of the pore space structure using thermal conductivity measurements on cores sequentially saturated with three fluids was performed. It was shown that the new technique provides an opportunity to stabilize the inversion procedures and to proceed to the study of individual samples. The combination of detailed experimental data with theoretical modelling of the effective thermal conductivity gives an estimation of the matrix thermal conductivity, crack aspect ratio, and porosity on the basis of thermal conductivity measurements on dry and fluid-saturated rock samples.

New links between permeability, electrical and thermal conductivity were found and previously known relations between porosity and thermal conductivity were clarified. A close correlation between the permeability and the parameter $\delta\lambda_{||}$ was found. It provides a background for research aimed at the estimation of thermal conductivity from geophysical logs, actual and inferred permeability, porosity and electric conductivity using the rapid measurement of thermal conductivity with an optical scanning instrument. The pore fluid type and pore structure of rocks are the leading factors that affect the sedimentary rock thermal conductivity. The clay content also affects the thermal properties of sedimentary rocks and substantially complicates the theoretical forecast of thermal conductivity of water-saturated samples.

Acknowledgements

We would like to acknowledge the generous support of Schlumberger Oilfield Services. The research of the Ries impact structure was supported by DFG. Constructive criticism and valuable suggestions by Prof. H.-J. Kümpel, Drs. J. Nove and S. Heiramp enhanced the manuscript. The assistance of Dr. Jennifer Tait and Dr. Adrian Muxworthy is also acknowledged.

References

ASAAD, Y., *A Study of the Thermal Conductivity of Fluid-bearing Porous Rock.* (Ph.D. Dissertation, Univ. of Calif., Berkeley, 1955).

BRIGAUD, F., *Conductivité thermique et champ de temperature dans les bassins sédimentaires à partir des données de puits* (Documents et Travaux, Centre Géologique et Géophysique de Montpellier. 23, 1989).

ELLIS, D. V., *Well Logging for Earth Scientists* (Elsevier, 1998).

HORAI, K. (1991), *Thermal Conductivity of Hawaiian Basalt: A New Interpretation of Robertson and Peck's Data*, J. Geoph. Res. *96*, 4125–4132.

LICHTENEKER, K. and ROTHER, K. (1931), *Die Herleitung des logarithmischen Mischungsgesetzes aus allgemeinen Prinzipien des stationären Strömung*, Phys. Zeit. *32*, 255–260.

MCKENNA, TH. E., SHARP, J. M. and LYNCH, F. L. (1996), *Thermal conductivity of Wilcox and Frio Sandstones in South Texas (Gulf of Mexico Basin)*, AAPG Bulletin *8*, 1203–1215.

POHL, J., STÖFFLER, D., GALL, H. and ERNSTSON, K., *The Ries impact crater.* In *Impact and Explosion Cratering* (eds. Roddy, D.J., Pepin, R.O., and Merrill, R.B.) (Pergamon Press, New York, 1977) pp. 343–404.

POPOV, YU. A. (1983), *Theoretical Models of the Method of Determination of the Thermal Properties of Rocks on the Basis of Movable Sources (Part I)*, Geologiya i Razvedka (Geology and Prospecting) *9*, 97–103 (in Russian).

POPOV, YU. A., *Optical scanning technology for nondestructive contactless measurements of thermal conductivity and diffusivity of solid matters.* In *Experimental Heat Transfer, Fluid Mechanics and Thermodynamics* (Proc. of the 4th World Conf. Brussels, Belgium, 1997, 1), pp. 109–117.

POPOV, YU. A. and MANDEL, A. M. (1998), *Geothermal Study of Anisotropic Rock Masses*, Izvestiya, Physics of the Solid Earth. *34* (11), 903–915.

POPOV, YU. A., PIMENOV, V. P., PEVZNER, L. A., ROMUSHKEVICH, R. A. and POPOV, E. YU. (1998), *Geothermal Characteristics of the Vorotilovo Deep Borehole Drilled into the Puchezh-Katunk Impact Structure*, Tectonophysics. *291*, 205–223.

POPOV, YU. A., PRIBNOW, D., SASS, J., WILLIAMS, C. and BURKHARDT, H. (1999), *Characterization of Rock Thermal Conductivity by High-Resolution Optical Scanning*, Geothermics. *28*, 253–276.

PRIBNOW, D. and SASS, J. (1995), *Determination of Thermal Conductivity from Deep Boreholes*, J. Geophys. Res. *100*, 9981–9994.

ROBERTSON, E. C., *Thermal Conductivity of Rocks* (Open File Report 79-356, United States Department of the Interior, US Geological Survey, Reston, Virginia, 1979).

SCHOEN, J. H., *Physical Properties of Rocks: Fundamentals and Principles of Petrophysics* (*Handbook of Geophysical Exploration. Section I, Seismic Exploration*: V.18, Redwood Books, Trowbridge, 1996).

SOMERTON, W. H. (1958), *Some Thermal Characteristics of Porous Rocks*, J. Petr. Techn. *10* (5), 61–64.

SOMERTON, W. H., *Thermal Properties and Temperature-Related Behavior of Rock/Fluid Systems* (Amsterdam, Elsevier, 1992).

TERTYCHNYI, V., POPOV, YU. A. and KOROBKOV, D. (2003), *Influence of Internal Structure on Thermal Conductivity of Porous Rocks*, Pure Appl. Geophys. (this issue).

WILLIAMS, C. F., GALANIS, P. S., GRUBB, A. D. and MOSES, Th. H., *The Thermal Regime of Santa Maria Province (California)*. In *U.S. Geological Survey Bulletin – Evolution of Sedimentary Basins/onshore Oil and Gas Investigations* (ed. Keller M.A.) (Santa Maria Province, USA, 1995).

ZIMMERMAN, R. (1989), *Thermal Conductivity of Fluid-Saturated Rocks*, J. Petr. Sci. Eng. *3*, 219–227.

(Received January 3, 2001, revised July 27, 2001, accepted August 1, 2001)

Notes to Authors

PAGEOPH welcomes original contributions in English (and occasionally in French and German) on all aspects of geophysics. All manuscripts should be submitted to the Regular Issues Editor-in-Chief, in triplicate, formatted with double spacing and wide margins. For further details see the following paragraphs.

Format of Manuscripts

Length and Page Charges: A paper should not exceed 16 printed pages including tables and figures. For articles exceeding 16 printed pages the authors will be charged sFr. 80.00 for each additional page. No page charges, except those for color prints, are required for contributors to special issues.

Title Page: This should include the the complete title, full names and addresses of all authors. In addition, corresponding authors should provide their fax number and e-mail address if they are available.

Abbreviated Title: It is necessary to indicate an abbreviated title, which will be used as a running head (no more than 50 characters including spaces).

Abstract: The abstract should be in English, and in the language of the text, if different. It has to be of no more than 10 sentences and should be concise and self-contained.

Keywords: Up to 6 keywords should be listed, suitable for incorporation into information-retrieval systems.

Text: The text must include a citation for each item listed under References; the approximate position of each figure should be indicated in the text. The metric system should be used throughout the text, figures, and tables.

Tables: Tables are to be presented on separate pages, with a brief title for each.

Figures: Figure captions and legends are to be typed on a separate page or pages as the last element in the manuscript. Make sure that line thickness and lettering allow an adequate size reduction. Heliographic or photocopies are not suitable for reproduction. Highquality, glossy, photographic prints must be submitted. Color prints are permitted but authors will be charged for them.

References: They are to be listed in alphabetical order in the following style:
Journal article: Haurwitz, B., and Cowley, A.D. (1973), The Diurnal and Semidurnal Barometric Oscillations, Global Distribution and Annual Variation, Pure Appl. Geophys. 102, 193-222.

Whole book: Bath, M., Introduction to Seismology (Birkhäuser, Basel 1973).

Article in a book: Haurwitz, B., and Cowley, A.D., Barometric oscillations, In Introduction to Seismology (ed. Bath. M.) (Birkhäuser, Basel 1973) pp. 193-222.

Submission of Manuscripts

Manuscripts must be submitted in triplicate, formatted with double line spacing and wide margins. Copies of the figures should be attached at the end of the manuscript. High-quality, prints may be submitted later. All manuscript pages, including references, tables, and captions, should be numbered consecutively, starting with the title page as page one.

All manuscripts should be submitted to Regular Issues Editor-in-Chief
Brian Mitchell
Department of Earth & Atmospheric Sciences
Saint Louis University
3507 Laclede Avenue
St. Louis, MO 63103, USA
e-mail: mitchbj@eas.slu.edu

The final version of a manuscript accepted for publication in PAGEOPH should be sent as both a hard copy and on diskette to the Editor-in-Chief. Delivering manuscripts in electronic form may substantially facilitate the publication process provided certain points are taken into consideration:

- Texts should be delivered in Ms Word or Wordperfect
- Mathematics in Tex, LaTex and any Tex formats
- Figures in TIFF (high resolution) or EPS

The electronic and printed version must be absolutely identical. All pictorial and graphic illustrations should be delivered as hard copy originals. Do not fail to include a hard copy for ready vicwing. Back-up copies of the diskettes must be kept. Diskettes must be adequately protectedfor transport.

Galley Proofs

Unless indicated otherwise, galley proofs will be sent to the first-named author directly from Birkhäuser Verlag AG and should be returned with the least possible delay. Textual alterations made in the galley proof stage will be charged to the author. One copy of the corrected proof is to be returned immediately.
The editorial office assumes no responsibility for delayed proofs, errors in the original manuscript, or major alterations in proofs for any reason.

Reprints

The authors will receive 50 reprints of each article without charge. Additional reprints may be ordered in lots of 50 when the final corrected page proofs are returned. Orders submitted thereafter are subject to considerably higher rates.

GPSR Compliance

The European Union's (EU) General Product Safety Regulation (GPSR) is a set of rules that requires consumer products to be safe and our obligations to ensure this.

If you have any concerns about our products, you can contact us on ProductSafety@springernature.com

In case Publisher is established outside the EU, the EU authorized representative is:

Springer Nature Customer Service Center GmbH
Europaplatz 3
69115 Heidelberg, Germany

Batch number: 10726322

Printed by Printforce, the Netherlands